Fiber Optics
Standard
Dictionary

Fiber Optics Standard Dictionary

Second Edition

Martin H. Weik, D.Sc.
Consultant in Fiber Optics Research,
Systems, and Standards
for the United States Navy
Dynamic Systems, Inc.
Reston, Virginia

VNR VAN NOSTRAND REINHOLD
——————————— New York

Printed in the United States of America

Van Nostrand Reinhold
115 Fifth Avenue
New York, New York 10003

Van Nostrand Reinhold International Company Limited
11 New Fetter Lane
London EC4P 4EE, England

Van Nostrand Reinhold
480 La Trobe Street
Melbourne, Victoria 3000, Australia

Nelson Canada
1120 Birchmount Road
Scarborough, Ontario M1K 5G4, Canada

16 15 14 13 12 11 10 9 8 7 6 5 4 3 2 1

Library of Congress Cataloging-in-Publication Data

Weik, Martin H.
    Fiber optics standard dictionary / Martin H. Weik.—2nd ed.
        p.   cm.
    Rev. ed. of: Fiber optics and lightwave communications standard
dictionary.
        ISBN 0-442-23387-6
    1. Optical communications—Dictionaries. 2. Fiber optics-
-Dictionaries. I. Weik, Martin H. Fiber optics and lightwave
communications dictionary. II. Title.
    TK5102.W45   1989   89-30814
    621.36'92'0321—dc19   CIP

To my wife, Helen,
for the love and encouragement she has given me during
the preparation of this and the first edition.

And God said, Let there be light:
and there was light. And God saw the light,
that it was good: and God divided the light from the darkness.
Genesis 1:3-4

Preface

The first edition of this dictionary was written during the years preceding 1980. No fiber optics glossary had been published by any recognized standards body. No other dictionaries in fiber optics had been published. A significant list of fiber optics terms and definitions, NBS Handbook 140, *Optical Waveguide Communications Glossary,* was issued in 1982 by the National Bureau of Standards, now the National Institute of Standards and Technology. Since then several publications by standards bodies contained fiber optics terms and definitions. In 1984 the Institute of Electrical and Electronic Engineers published IEEE Standard 812-1984, *Definitions of Terms Relating to Fiber Optics.* In 1986 the National Communication System published Federal Standard FED-STD-1037A, *Glossary of Telecommunication Terms,* containing about 100 fiber optics terms and definitions. In 1988 the Electronic Industries Association issued EIA-440A, *Fiber Optic Terminology.* All of these works were based on NBS Handbook 140 compiled 10 years earlier.

Currently the International Electrotechnical Commission is preparing IEC Draft 731, *Optical Communications, Terms and Definitions.* Work in fiber optics terminology is being contemplated in the International Organization for Standardization and the International Telecommunications Union. None of these works constitutes a comprehensive coverage of the field of fiber optics. Each was prepared by professional people representing specific interest groups. Each work was aimed at specific audiences: research activities, development activities, manufacturers, scientists, engineers, and so on. Their content is devoted primarily to fundamental scientific and technical principles and theory rather than state-of-the-art and advanced technology. Also, for a definition to be approved by committee and survive "public" review, quite often a great deal of explanatory material, examples, illustrations, and cross-references were not appended to what was called "the defining phrase." The author of this dictionary chaired over a dozen vocabulary committees in computers, communication, and information processing at international, national, federal, military, and technical society levels. He is currently serving as a consultant in fiber optics and is engaged in the preparation of military handbooks, specifications, and standards that include fiber optics glossaries such as MIL-STD 2196(SH), *Glossary Fiber Optics,* for the U.S. Navy.

Every effort was made to ensure that this dictionary be consistent with published and draft international, national, federal, industrial, and technical society standards. This comprehensive dictionary goes far beyond the combined scopes and content of the published and pending vocabulary standards. The combined standards embrace several hundred terms and definitions. This dictionary embraces several thousand. Obsolete material in the first edition has been omitted from this edition. Thousands of examples, explanations, diagrams, illustrations, and tables have been included. The latest fiber optics scientific and technical literature have been scoured for new terms and concepts. Fiber optic component and system manufacturer's technical characteristics and specifications were screened for common terms and definitions. Though not strictly fiber optics terms, some terms were included to support the definitions of fiber optics terms. These efforts have yielded hundreds of new fiber optics terms for which definitions were written based on context and the author's knowledge of the subject. Textbooks also provided source material for this dictionary. Lastly, much material originated from the spoken and written words of researchers, designers, developers, operators, manufacturers, and educators with whom the author has associated during the 10 years he has worked in the fiber optics field. The author developed terminology for the U.S. Military Communications-Electronics Board, and the National Communications System. He supported the Fiber Optic Sensors Program at the U.S. Naval Research Laboratory, the U.S. Navy Undersea Surveillance Program in fiber optics at the Space and Naval Warfare Systems Command, and the Fiber Optics Program of the Naval Sea Systems Command.

A final word on organization of this dictionary. Italicised words used in definitions are used in the sense defined in this dictionary. Each word in a multiple-word entry is cross-referenced to the definition of the complete entry. For example, under *modal power distribution* there is a "See" reference to *nonequilibrium modal power distribution.* "Also see" is used to refer the reader to closely related terms and definitions. Synonyms are referenced to the preferred term, where the definition will be found. Only the preferred terms are used in definitions. Preference is based on the recommendations of standards bodies contained in their published works.

The writing and publishing of this comprehensive dictionary took several years, during which time the author encountered many new terms and definitions. These appear in the appendix so that the reader will have the latest word in fiber optics. The author is convinced, and hopes the reader will agree, that this dictionary will serve as the best ready-reference available today for learning, understanding, and retaining concepts in fiber optics science and technology.

MARTIN H. WEIK, D.SC.
Arlington, Virginia

Introduction

Fiber optic technology has advanced faster and further during the 1980s than even the most optimistic workers in the field dared to predict. Hundreds of thousands of kilometers of fiber optic cable containing millions of kilometers of operational optical fibers have been installed in over a hundred countries. Fiber optic cables pulled in underground conduits, buried in ploughed trenches, strung on overhead poles, and installed in building shafts and plenums are carrying analog and digital audio, video, and data signals in long-haul trunks, metropolitan-area networks, local-area networks, and local loops. Each cable is capable of carrying tens of thousands of messages simultaneously, using time-division, frequency-division, wavelength-division, and space-division multiplexing schemes. Current trends show that optical fiber is being installed in local loops entering homes and offices. Optical fiber is competing heavily with twisted-pair wire, coaxial cable, microwave, and satellite communication systems. Studies have shown that the cost of replacing electronic systems with fiber optic systems can be recovered in a relatively short time. One such study showed that a single optical fiber to a home or office can provide sufficient transmission capacity for 50 analog broadcast-quality video channels, 4 high-definition television (HDTV) channels, 4 switched digital video channels, 25 digital audio channels, and many other miscellaneous voice, data, and service channels simultaneously without interference or crosstalk.

Experts predict that the majority of communication traffic will be fiber optic by the year 2010. Over $200 billion in new broadband systems will be installed by then. Bandwidth, a severely limited commodity at premium cost in electronic systems will effectively be "free" in fiber optic systems once a system is installed. For example, by the year 2003, experts predict integrated-services digital networks (ISDNs) will provide many gigahertz of bandwidth and many gigabits per second of data signaling rate per dollar of investment. Integrated optical circuits (IOCs) will sell for $5 each. Optical fiber will be down to 5 cents a meter, cheaper than wire and with almost unlimited traffic capacity. By the year 2000, experts predict over 600 billion voice circuit-kilometers will be on fiber, 160 billion on satellite, 16 on microwave, and 3 billion on coaxial cable, worldwide.

Why such a rapid conversion and installation rate? The primary force is economic. Fiber presents a lower cost per message-kilometer and a lower cost per bit-kilometer/second. Overall costs, including acquisition, installation, and operational, for fiber systems will be far below those of their electronic counterparts. Besides the advantage of lower cost, fiber systems have many other advantages over electronic systems. Fiber systems show almost unlimited traffic capacity; long repeaterless links; no interference or crosstalk between channels; no return circuits and ground loops; improved safety; lighter weight; reduced space; less power; longer life; high resistance to fire damage; less hazard to personnel, explosives, combustibles, and other equipment; easier installation; fewer strategic materials; and reduced documentation requirements. These advantages of fiber systems create a tremendous pressure to move rapidly toward "fiberization." The conversion and installation pace depends on many factors, including availability of capital, off-the-shelf components, knowledgeable personnel, and documentation, including standards. Confidence in the reliability and dependability of fiber optic systems has to be proven, just as in the case of electronic computers, before a particular group or organization will use fiber optic systems to meet its requirements.

Optical fibers have many other uses besides communication and data transfer. Some of these applications include sensing systems, medical instrumentation, illumination devices, display systems, and inspection devices. Components for these applications are already on the market, and the list is growing. As in all new and expanding technologies, starting in earlier days with power machinery, telegraph, and telephone, and later radio and television, the percentage of the population engaged in activities related to fiber optics will continue to increase for decades to come.

This comprehensive dictionary provides sufficient information to thoroughly familiarize anyone with the science and technology of fiber optics. At the same time it provides proper terminology and usage for students, teachers, scientists, engineers, writers, sales persons, designers, developers, manufacturers, installers, repairers, operators, and others in related fields, such as law and medicine. It is written with technical precision in theory and principles for the experienced professional and with explanations, examples, illustrations, and cross-references in laymen's language for the newcomer to the field of fiber optics. Most importantly it is consistent with international, national, federal, military, industrial, and technical society fiber optics vocabulary standards.

Acknowledgments

The author extends his appreciation to all members of the staffs of the Defense Communications Agency (DCA), the National Communications System (NCS), the Military Communications-Electronics Board (MCEB), the Naval Research Laboratory (NRL), the Space and Naval Warfare Systems Command (SPAWARSYSCOM), the Naval Sea Systems Command (NAVSEASYSCOM), and Dynamic Systems, Incorporated (DSI), with whom he was associated during the many years of preparation of the first and second editions of this dictionary.

Many technical discussions with Dr. George Hetland, Jr., and Mr. Jack Donovan, NRL; Capt. Kirk E. Evans, SPAWARSYSCOM; and Mr. Lonnie D. Benson and Mr. Leslie Tripp, DSI resulted in valuable contributions to this dictionary. The editor gratefully appreciates the sharing of expertise and the many constructive comments in the area of fiber optics made by Mr. James H. Davis, NAVSEASYSCOM, during his review of the manuscript.

Appreciation is also extended to the members of the American National Standard Institute's Technical Committee X3K5, Vocabulary for Computing, Communications, and Information Processing, particularly for passing on expertise during the 22 years the author served as chairman.

Appreciation is especially extended to my wife, Helen Harrison Weik, RNC, Ph.D., for her assistance in organizing material and her contributions in those areas of communication systems related to the human factors in the man-machine interface.

The author gratefully acknowledges the technical and editorial assistance given by Alberta Gordon, Managing Editor of Van Nostrand Reinhold.

MARTIN H. WEIK, D.SC.

Fiber Optics
Standard
Dictionary

A

aberration. See *chromatic aberration.*

absolute luminance threshold. The lowest limit of *luminance* necessary for visual perception to occur in a person with normal or average vision.

absolute luminosity curve. The plot of *spectral luminous efficiency* versus *optical wavelength.*

absorptance. See *spectral absorptance.*

absorption. In the *transmission* of *signals,* such as electrical, *electromagnetic optical,* and acoustic signals, the conversion of the transmitted energy into another form of energy, such as heat. Signal *attenuation* is not only a consequence of absorption, but also of other phenomena, such as *reflection, refraction, scattering, diffusion,* and spatial spreading. In the transmission of *electromagnetic waves,* such as *light-waves,* absorption includes the transference of some or all of the energy contained in the wave to the substance or medium in which it is *propagating* or upon which it is *incident.* Absorbed energy from a transmitted or incident lightwave is usually converted into heat with a resultant attenuation of the *power* or energy in the wave. In *optical fibers,* intrinsic absorption is caused by parts of the *ultraviolet* and *infrared* absorption bands. Extrinsic absorption is caused by impurities, such as hydroxyl, transition metal, and chlorine ions; silicon, sodium, boron, calcium, and germanium oxides; trapped water molecules; and defects caused by thermal and nuclear radiation exposure. Synonymous with *material absorption.* See *band-edge absorption; hydroxyl ion absorption; overtone absorption; selective absorption.*

absorption coefficient. The coefficient in the exponent of the *absorption* equation that expresses *Bouger's law,* namely the *b* in the equation:

$$F = F_0 e^{-bx}$$

where F is the *electromagnetic (light) field strength* at the point x, and F_0 is the initial value of field strength at $x = 0$.

absorption index. The ratio of the *electromagnetic radiation absorption constant* to the *refractive index,* given by the relation:

$$K' = K\lambda/4\pi n$$

where K is the *absorption coefficient,* λ is the *wavelength* in vacuum, and n is the refractive index of the absorptive material.

absorption loss. When an *electromagnetic wave propagates* in a *propagation medium,* the loss of wave energy caused by intrinsic *absorption,* that is, by material absorption, and by impurities consisting primarily of metal and hydroxyl ions in the medium. Absorption losses may also be caused indirectly when *light* scattered by atomic defects is also absorbed.

acceptance angle. In *fiber optics,* half the vertex angle of that cone within which *optical power* may be *coupled* into *bound modes* of an *optical waveguide.* For an *optical fiber,* it is the maximum angle, measured from the longitudinal axis or centerline of the fiber to an *incident ray,* within which the ray will be accepted for *transmission* along the fiber, that is, total *(internal) reflection* of the incident ray will occur for long distances within the *fiber core.* If the acceptance angle is exceeded, optical power in the incident ray will be coupled into *leaky modes* or rays or lost by *scattering, diffusion,* or *absorption* in the *cladding.* For a cladded fiber in air, the sine of the acceptance angle is given by the square root of the difference of the squares of the *refractive indices* of the fiber core and the cladding, that is, by the relation:

$$\sin A = (n_1{}^2 - n_2{}^2)^{1/2}$$

where A is the acceptance angle and n_1 and n_2 are the refractive indices of the core and cladding, respectively. If the refractive index is a function of distance from the center of the core, then the acceptance angle at a given distance from the center is given by the relation:

$$\sin A_r = (n_r{}^2 - n_2{}^2)^{1/2}$$

where A_r is the acceptance angle at a point on the entrance face of the fiber at a distance r from the center, and n_2 is the minimum refractive index of the cladding. Sin A and sin A_r are the *numerical apertures (NA).* Unless otherwise stated, acceptance angles and numerical apertures for optical fibers are those for the center of the end face of the fiber, that is, where the refractive index, and hence the NA, is the highest. Power may be coupled into leaky modes at angles exceeding the acceptance angle, that is, at internal incidence angles less than the *critical angle.* See *maximum acceptance angle.*

acceptance-angle plotter. A device capable of varying the *incidence angle* of a narrow *light beam* that is incident upon a surface, such as the end face of an *optical fiber.* The device measures the *intensity* of the *transmitted light,* namely the light *coupled* into the fiber for each angular position of the *light source* relative to the face of the fiber, that is, the incidence angle. As the incidence angle approaches half the *acceptance cone* apex angle, the *optical power* measured by a *photodetector* approaches zero. This condition defines the limit of the *acceptance angle.*

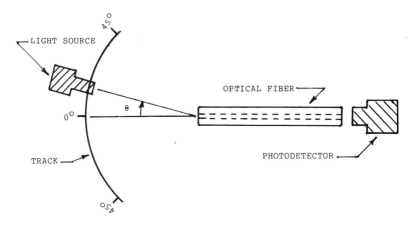

A–1. An **acceptance-angle plotter.**

acceptance cone. In *fiber optics,* that cone within which *optical power* may be *coupled* into the *bound modes* of an *optical waveguide.* The acceptance cone is derived by rotating the acceptance angle, that is, the maximum angle within which light will be coupled into a bound mode, about the *fiber axis.* The acceptance cone for a round *optical fiber* is a solid angle whose included apex angle is twice the acceptance angle. Rays of light that are within the acceptance cone can be coupled into the end of an optical fiber and be *totally internally reflected* as it *propagates* along the *core.* Typically, an acceptance cone apex angle is 40°. For noncircular waveguides, the acceptance cone transverse cross section is not circular, but is similar to the cross section of the fiber. (See Fig. A–2, p. 4)

acceptance pattern. For an *optical fiber* or *fiber bundle,* a plot of *transmitted optical power* versus the *launch angle.* The total power coupled into the fiber is a function of the launch angle, the *transmission coefficient* at the fiber face, the illumination area, the *light source wavelength,* and other *launch conditions.*

access coupler. A device placed between the ends of two *waveguides* to allow *signals* to pass from one waveguide to the other.

acoustic sensor. See *optical fiber acoustic sensor.*

acoustic transducer. See *optoacoustic transducer.*

acoustooptic effect. A variation of the *refractive index* of a material caused by acoustic waves. The changes are also produced in *diffraction gratings* or *phase* pat-

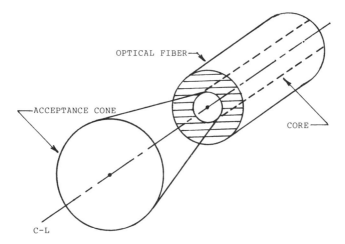

A–2. The **acceptance cone** of an *optical fiber.* The *acceptance angle* is one-half the acceptance cone apex angle. All *light rays* within the cone will be accepted by the fiber. The amount of light that can be *coupled* into and *transmitted* by the fiber is a function of the *launch conditions,* the *fiber numerical aperture,* and the *wavelength.*

terns produced in a *propagation medium* in which a *lightwave* is propagating when the medium is subjected to a sound wave, due to *photoelastic* changes that occur in the material composing the propagation medium. The acoustic waves may be created by a force developed by an impinging sound wave, for example by the piezoelectric effect, or by magnetostriction. The effect can be used to *modulate* a *light* beam in a material because of the changes that occur in light velocities, *reflection* and *transmission coefficients, acceptance angles, critical angles,* and transmission *modes* resulting from changes in the refractive index caused by the acoustic wave.

acoustooptics. The branch of science and technology devoted to the interactions between sound waves and *light* in a solid medium. Sound waves can be made to *modulate,* deflect, and focus *lightwaves* by causing a variation in the *refractive index* of the medium. Also see *electrooptics; magnetooptics; optooptics; photonics.*

acquisition time. The elapsed time between the instant of application of the leading edge of an input *signal* and stabilization of the corresponding output signal of a device.

action. In *quantum* mechanics, the product of the total energy in a stream of *photons* and the time during which the flow occurs, expressed by the relation:

$$A = h \sum_{i = 1}^{m} f_i n_i t_i$$

where f_i is the ith frequency; n_i is the number of photons of the ith frequency; t_i is the time duration of the ith frequency summed over all the frequencies, photons, and time durations of each in a given light beam or beam pulse; and *h* is *Planck's constant.*

activated chemical-vapor-deposition process. See plasma-activated chemical-vapor-deposition (PACVD) process.

active connector. See optical fiber active connector.

active device. A device that contains a source of energy or that requires a source of energy other than that contained in input *signals,* the output of which is a function of present and past input signals that *modulate* the output of the energy source. Examples of *fiber optic* active devices include operational amplifiers, repeaters, oscillators, phototransistors, *lasers, optical masers,* photomultipliers, and *photodetectors.*

active laser medium. The material in a *laser,* such as a crystal, gas, glass, liquid, or semiconductor, that emits *optical radiation.* Radiation from a laser is usually *coherent,* that is, has a high *coherence degree,* and results from stimulated electronic, atomic, or molecular energy transitions from higher to lower energy levels. The action is maintained by causing *population inversion.*

active material. See *optically active material.*

active optical device. A device capable of performing one or more operations on *lightwaves* with *wavelengths* in or *near* the *visible region* of the *electromagnetic spectrum* through the use of input energy, such as electrical or acoustic energy, in addition to that contained in the waves being operated upon. Examples of active optical devices include *fiber optic transmitters, receivers, repeaters, switches, active multiplexers,* and *active demultiplexers.*

active optical fiber. An *optical fiber* designed to be used as the gain (*lasing*) medium in a *fiber laser,* that is, an optical fiber amplifier.

active optics. The development and use of *optical* components whose characteristics are controlled during their operational use in order to modify the characteristics of *lightwaves propagating* within them. Controlled lightwave characteristics include *wavefront* direction, *polarization, modal power distribution, electromagnetic field strength,* or the *path* they take. Also see *fixed optics.*

actuation method. The way in which a motive force must be applied to a switch to place it into its various states.

actuator. A device that provides the motive force that must be applied to a switch to place it into its various states.

adjusting. See *self-adjusting.*

advanced television system. A television system in which an improvement in performance has been made to an existing television system to create the advanced system. The improvement may or may not result in a system that is compatible with the original system or with other present systems. Some features and technical characteristics of the original system on which the advanced system is based may be included in the advanced system. Also see *enhanced-definition television (EDTV) system; high-definition television (HDTV) system; improved-definition television (IDTV) system.*

aerial insert. In a buried cable run, a raising of the cable followed by an overhead run usually on poles, followed by a return to the ground, in places where it is not possible or practical to bury a cable, such as might be encountered in crossing a deep ditch, canal, river, or subway line.

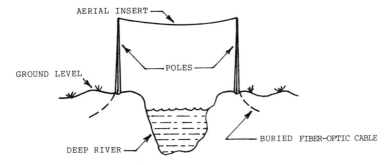

A-3 An **aerial insert** in a buried *fiber optic cable.*

A-4. A figure 8 aerial messengered *fiber optic cable* with rugged tight *buffering.* The cable can be used for fully-aerial systems or for **aerial inserts.** (Courtesy Optical Cable Corporation).

aerial optical cable. An *optical cable* designed for use in overhead suspension devices, such as towers or poles.

aligned bundle. A bundle of *optical fibers* in which the relative spatial coordinates of each fiber are the same at the two ends of the bundle. Synonymous with *coherent bundle.*

alignment sensor. See *fiber axial-alignment sensor.*

all-glass fiber. An *optical fiber* whose *core* and *cladding* consist entirely of glass.

all-plastic fiber. An *optical fiber* whose *core* and *cladding* consist entirely of plastic. Most fibers are of glass core and glass cladding.

alternative test method (ATM). In *fiber optics,* a test method in which a given characteristic of a specified class of fiber optic devices, such as *optical fibers, fiber optic cables, connectors, photodetectors,* and *light sources,* is measured in a manner consistent with the definition of this characteristic; it gives reproducible results that are relatable to the reference test method and is relatable to practical use. Synonymous with *practical test method.* Also see *reference test method.*

ambient light susceptibility. The *optical power* that enters a device from ambient illumination *incident* upon the device. It may be measured in absolute power, such as microwatts, or in *dB* relative to the incident ambient optical power. For example, in a *fiber optic connector* or *rotary joint,* it is the ambient optical power that leaks into the *optical path* in the component.

amplification by stimulated emission of radiation. See *light amplification by stimulated emission of radiation (LASER).*

amplifier. See *fiber amplifier.*

amplitude. See *pulse amplitude.*

amplitude distortion. *Distortion* of a *signal* in a system, subsystem, or device when the output amplitude is not a linear function of the input amplitude under specified operating conditions.

amplitude modulation. See *pulse-amplitude modulation (PAM).*

analog data. *Data* represented by a physical quantity that is considered to be continuously variable and whose magnitude is made directly proportional to the data or to a suitable function of the data. Also see *digital data.*

analog signal. A nominally continuous *signal* that varies in a direct correlation with the instantaneous *value* of a physical variable. For example, the *optical* output *signal* of a *light source* whose *intensity* is a function of a continuous acoustic or electrical input signal, or the continuous *photocurrent* output signal of a *photodetector* whose photocurrent is a function of a continuous optical input signal. Also see *digital signal.*

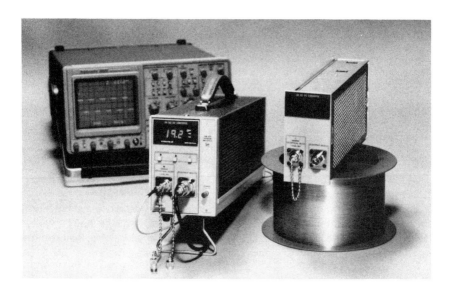

A–5. The OR500 Series Receivers and OT500 Series Transmitters, a set of electrical-optical and optical-electrical converters for use in analyzing *optical* **analog signals** in time and *frequency* domains. The *receivers* cover the 0.700-0.1550-µ *(micron) wavelength* range with a frequency response out to 1.5 GHz on *multimode* or *single-mode optical fibers*. *Transmitters* operate at 0.825, 0.850, or 11.3 µ on multimode fibers. (Courtesy Tektronix).

analyzer. See *lightwave-spectrum analyzer*.

angle. See *acceptance angle; borescope articulation angle; borescope axial viewing angle; Brewster angle; deviation angle; exit angle; critical angle; incidence angle; launch angle; maximum acceptance angle; radiation angle; refraction angle*.

angle of incidence. See *incidence angle*.

angle plotter. See *acceptance-angle plotter*.

angstrom (Å). A unit of *wavelength*. 1 Å $= 10^{-10}$ m (meter). The angstrom is not an SI (International System) unit.

angular misalignment loss. An *optical power loss* caused at a joint, such as at a *splice* or *connector,* because the axes of the *optical fibers* are not parallel, that is, there is an angular deviation of the optical axes of the two fibers being joined. The loss can be from source to fiber, fiber to fiber, or fiber to detector. It is considered extrinsic to the fiber.

anisotropic. Pertaining to material whose properties, such as *electric permittivity, magnetic permeability,* and *electrical conductivity,* are not the same in every direction in which they are measured, that is, not all their spatial derivatives are 0 in every direction. Therefore, an *electromagnetic wave propagating* in the material will be affected in different ways depending on the direction of propagation and direction, or type, of *polarization* of the wave. Also see *isotropic; birefringent medium.*

anisotropic propagation medium. A *propagation medium* in which properties and characteristics of the medium are not the same in reference to a specific phenomenon, such as a medium whose *electromagnetic* properties, like the *refractive index,* at each point are different for different directions of propagation or for different *polarizations* of a wave propagating in the medium.

antireflection coating. One or more thin *dielectric* or metallic films, such as *index-matching material* films, applied to an *optical surface* to reduce the *reflectance* and thus increase the *transmittance.* Ideally, the value of the *refractive index* of a single-layered film is the square root of the product of the refractive indices on either side of the film. The ideal thickness of an antireflection coating is one-quarter of the *wavelength* of the *incident light.* A precise thickness of coating will cancel the reflection from the *interface* surface by *interference.* Also see *index-matching material.*

APD. *Avalanche photodiode.*

aperture. The portion of a plane surface near a directional source of *radiation,* such as an antenna or an orifice, normal to the direction of maximum *irradiation,* that is, radiation intensity, through which the major part of the radiation passes. In the movement or passage of any entity, such as a fluid, an *electromagnetic wave,* a sound wave, or time, an aperture is an opening, or window, in the path or length of time through which or during which the entity is constrained to flow or occur. Apertures can be controlled so as to perform *coupling,* guidance, and control functions. The size of the aperture is a function of its physical dimension, the direction of flow, the boundary conditions, and parameters of the moving entity, such as the viscosity of fluids or the *wavelength* of *light.* The dimensions of an aperture may be expressed in terms of length, width, area, planar angle, solid angle, and time units, as well as in terms of dimensionless or normalized numbers. The numerical values may be relative, that is, normalized, or they may be absolute. Apertures usually occur at *interfaces* or transitional points, such as at a radiating antenna or at the entrance to an *optical fiber.* Apertures usually introduce losses and restrict passage. For example, the aperture of a lens depends on the size of the uncovered portion of the lens, the *refractive index* of the glass, and the *color* (wavelength) of the *incident light.* The aperture presented by a device may also depend on the *incidence angle* of the entering light. The aperture of an electromagnetic wave antenna is limited by the wavelength and the dimensions of the feeder, antenna array, or dish. For a phased-array antenna, the aperture is limited by the dimensions of the array, the *launch angle,* and the wavelength. In an *optical fiber,* the aperture is the *acceptance cone angle* as determined by the numerical aperture. For a sound wave, the aperture may be determined by the size of the hole in an inelastic material and the wavelength. For fluids, the aperture might be the size of a hole in the side of a container or constriction in a pipe, in which case the size of the aperture may depend on the dimensions of the hole or constriction, turbulence, and viscosity. For granular substances, the hole must be large enough to sustain flow or else the effective aperture is zero. See *launch numerical aperture (LNA); numerical aperture; output aperture.*

application layer. In *open-systems architecture,* the layer that is directly accessible to, visible to, and usually explicitly defined by, users. It provides all the functions and services needed to execute their programs, processes, controls, and data exchanges. It is considered the highest layer. Also see *Open-Systems Interconnection (OSI) Reference Model (RM).*

application parameter. In the design of an *optical station/regenerator section,* a performance parameter specified by the manufacturer or the user, e.g., application type (aerial, buried, underwater (salt, fresh, or brackish), or underground), temperature range, cabled fiber reel length, *nominal central wavelength,* central wavelength range, type of *splice, splice loss,* cable designation, maximum *cable cutoff wavelength,* maximum *cable attenuation rates* and expected increases, *dispersion* parameters, interconnection parameters *(cladding diameter, cladding ovality, core/cladding concentricity errors, mode field diameter),* and *global fiber parameters* (standard and extended).

architecture. See *open-systems architecture.*

area. See *cladding tolerance area; coherence area; coherent area; core area; core tolerance area.*

area network. See *local-area network (LAN).*

articulation angle. See *borescope articulation angle.*

assembly. See *optical harness assembly.*

ATM. *Alternative test method.*

atmosphere laser. A *laser* in which the *active laser medium* is a gas. See *longitudinally excited atmosphere laser; transverse atmosphere laser.* (See Fig. A–6)

atomic laser. A *laser* in which the *active laser medium* is an element in atomic form, such as sodium or boron. Also see *molecular laser.*

attenuate. To decrease the *power* of a *signal, light* beam, or *lightwave propagating* in an *optical device,* the decrease usually being the reduction that occurs between the input and output of the device, such as over the entire length of an *optical fiber* or between the input and output of a *fiber optic connector.* The device *attenuates*

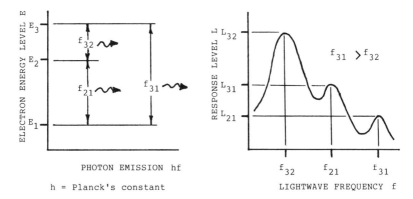

A-6. In an **atmosphere laser** electrons returning to lower energy levels after excitation *emit photons* at energy levels (shown at left) produce the *spectral response* (at right) when the energies are summed over their average number per unit time for each level. For example, photon with energy hf_{32} is emitted when electrons drop from energy level E_3 to E_2.

the signal by *absorption, reflection, scattering, deflection, dispersion,* or *diffusion,* rather than by *geometric spreading.* Synonymous with *pad.*

attenuation. 1. In an *optical* component, the decrease in *power* of a *signal, light* beam, or *lightwave* measured from input to output, such as measured over the entire length of an *optical fiber* or between the input and output of a *fiber optic connector.* The *attenuation* can be in absolute power units, a fraction of a reference value, or in *decibels.* Attenuation in a fiber occurs as a result of *absorption, reflection, scattering, deflection, dispersion,* or *diffusion,* rather than as a result of *geometric spreading.* Attenuation for an optical fiber is usually expressed as an *attenuation rate,* that is, in the units of decibels/kilometer (dB/km). **2.** The loss of *power* that occurs between the input and output of a device, usually expressed in *dB.* See *differential mode attenuation (DMA); induced attenuation; light attenuation; macrobend attenuation; optical dispersion attenuation; path attenuation; transient attenuation.* (See Figs. A–7 and A–8, p. 12)

attenuation constant. See *attenuation term.*

attenuation-limited operation. The condition that prevails in a device or system when the magnitude of the *optical power* at a point in the device or system limits its performance or the performance of the device or system to which it is connected. For example, the condition that prevails in an *optical link* when the *power level* at the receiving end is the predominant mechanism that limits link performance. When an optical link is *attenuation*-limited, the *optical power level* at the *receiver* is below the *sensitivity threshold* of the *photodetector* because the *transmitter power* is not high enough or the attenuation between transmitter and receiver is too high or both. The attenuation is the sum of all the component *insertion losses* and the *attenuation rate* of the fiber times its length.

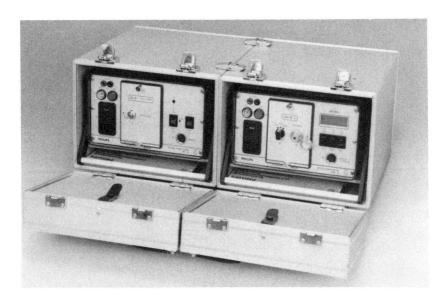

A-7. The OPM/OPS-10 Precision Attenuation Kit in their transport boxes. The kit is used to measure *optical* **attenuation**. (Courtesy Philips Telecom Equipment Company, a division of North American Philips Corporation).

A-8. The OPM/OPS-10 Precision **Attenuation** Kit out of their transport boxes. (Courtesy Philips Telecom Equipment Company, a division of North American Philips Corporation).

attenuation rate. In an *electromagnetic wave* that is *propagating* in a *waveguide,* such as *lightwave* in an *optical fiber,* the *attenuation* in *signal power* that occurs per unit of distance along its length. In dB/km, the attenuation rate is given by the relation:

$$\alpha = [10 \log_{10}(P_i/P_0)]/L$$

where L is given length of the waveguide in kilometers, P_i is the input power for the length L, and P_0 is the output power for the length L. In this form, the attenuation rate will be a positive value because P_i is the larger of the two powers. Also see *attenuation.*

attenuation-rate characteristic. See *fiber global attenuation-rate characteristic; wavelength-dependent attenuation-rate characteristic.*

attenuation term. In an *electromagnetic wave* that is *propagating* in a *waveguide,* such as a *lightwave* in an *optical fiber,* in the expression for the exponential variation characteristic of guided waves, the *a* in the relation:

$$A = A_0 e^{-pz} = A_0 e^{-jb-az}$$

where a represents the attenuation, or pulse-amplitude diminution per unit of propagation distance, z, of the wave and A_0 is the pulse amplitude at $z = 0$. The reduction in amplitude is in the form of an exponential decay. The attenuation term is the real part of the *propagation constant, p.* In a given guide the *phase term, b,* is initially assumed to be independent of the attenuation term, *a,* which is then found separately, assuming b does not change with *losses,* for an optical fiber. The attenuation term, *a,* is supplied by the manufacturer because it can be measured experimentally. Synonymous with *attenuation coefficient; attenuation constant.*

attenuator. A device capable of reducing the *power level* of a *signal,* usually without also creating any other *distortion* of the *signal.* See *continuously variable optical attenuator; fiber optic attenuator; fixed attenuator; optical attenuator; pad; stepwise-variable optical attenuator.* (See Fig. A–9, p. 14)

audio. In communication systems, pertaining to a capability or characteristic to *detect, transmit,* or process *signals* having *frequencies* that lie within the range that can be heard by the human ear. Audio frequencies range from 30 to 15,000 *Hz.*

avalanche multiplication. In semiconductors, the sudden or rapid increase in the number or density of hole-electron pairs that is caused when the semiconductor is subjected to high, that is, near-breakdown, electric fields. This increase in charge carriers causes a further increase in their numbers and density. Incident *photons* of sufficient energy can further multiply the carriers. These effects are used in *avalanche photodetectors.*

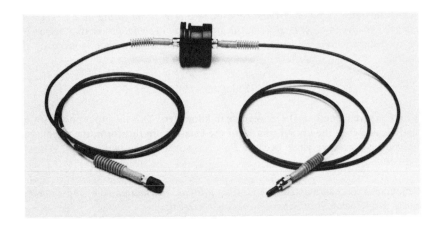

A–9. The A438 *continuously-variable optical* **attenuator** to *induce loss* to simulate long *fiber optic cable* runs and other applications. (Courtesy fotec incorporated).

avalanche photodetector. A *photodetector* that uses the *avalanche multiplication* effect to increase the output *photocurrent*.

avalanche photodiode (APD). A *photodiode* that operates with a reverse-bias voltage that causes the primary *photocurrent* to undergo amplification by cumulative multiplication of charge carriers. As the reverse-bias voltage increases toward breakdown, hole-electron pairs are created by absorbed *photons*. An avalanche effect occurs when hole-electron pairs acquire sufficient energy to create additional pairs when they collide with ions. Silicon and germanium APDs are inherently electrically noisy. Gallium indium arsenide phosphide on indium phosphide substrates and gallium antimonide APDs are being made.

average diameter. In a round *optical fiber,* the mean value of a measurable diameter, such as the *core, cladding,* or *reference surface diameter.*

AVPO. *Axial vapor-phase oxidation.*

AVPO process. See *axial vapor-phase-oxidation (AVPO) process.*

axial-alignment sensor. See *fiber axial-alignment sensor.*

axial-deposition process. See *vapor-phase axial-deposition (VAD) process.*

axial-displacement sensor. See *fiber axial-displacement sensor.*

axial interference microscopy. See *slab interferometry.*

axial propagation constant. See *propagation constant.*

axial ratio. In an *electromagnetic wave* having *elliptical polarization,* the ratio of the major axis to the minor axis of the ellipse described by the tip of the *electric field vector.*

axial ray. A *light ray* that *propagates* along and is coincident with the *optical axis* of a *waveguide.* See *paraxial ray.*

axial slab interferometry. See *slab interferometry.*

axial vapor-phase-oxidation (AVPO) process. A *vapor-phase-oxidation (VPO) process* for making *graded-index (GI) optical fibers* in which the glass *preform* is grown radially, rather than longitudinally as in other processes. The *refractive-index profile* is controlled in a spatial domain, rather than in a time domain. The chemical gasses are burned in an oxyhydrogen flame, as in the *outside vapor-phase-oxidation (OVPO) process,* to produce a stream of soot particles to create the *graded refractive-index* profile.

axial viewing angle. See *borescope axial viewing angle.*

axis. See *optic axis; optical axis.*

B

backscatter. **1.** The deflection, by *reflection* or *refraction,* of a *ray* of an *electromagnetic wave* or *signal* in such a manner that a component of the ray is deflected opposite to the direction of *propagation* of the *incident wave* or signal. The term "scatter" can be applied to reflection or refraction by relatively uniform *propagation media,* but is usually taken to mean propagation in which the *wavefront* and direction are modified in a somewhat disorderly fashion. **2.** The component of an *electromagnetic wave* or *signal* that is deflected by *reflection* or *refraction* opposite to the direction of *propagation* of the *incident wave* or signal. **3.** To *scatter* or randomly *reflect electromagnetic waves* in such a manner that a component is directed back toward the *source* of the waves when resolved along a line from the source to the point of scatter. **4.** The components of *electromagnetic waves* that are directed back toward their *source* when resolved along a line from the source to the point of scatter. Backscatter occurs in *optical fibers* because of the *reflecting* surfaces of particles or occlusions in the *propagation medium,* resulting in *attenuation.* Backscatter of lightwaves and radio waves also occurs in the atmosphere and the ionosphere. Also see *forward-scatter.*

backscattering. *Lightwave* and radio *wave propagation* in which the directions of the *incident* and *scattered* waves, resolved along a reference direction, usually horizontal, are opposite. A *signal* received by backscattering is often referred to as *"backscatter."*

backscattering technique. See *optical time-domain reflectometry.*

backshell. In a *fiber optic connector,* the portion of the shell (housing) immediately to the rear of the fastening mechanism, which is considered the front. The backshell contains the *optical fibers* and attaches to the bend limiter.

bait tube. The basic structure upon which a *fiber optic preform* is built.

band. **1.** In communication, the *frequency spectrum* between two defined frequency limits. **2.** In an atom, the difference between two electron energy levels. See *frequency band; narrow-band; stop-band; wideband.*

band-edge absorption. Absorption of energy in *lightwaves* that occurs in the *visible region* and extends from the *ultraviolet* to the *infrared* regions. In glass, band-edge absorption is usually caused by oxides of silicon, sodium, boron, calcium, germanium, and other elements, and by the hydoxyl ion.

band-gap energy. The difference between any pair of allowable energy levels of the electrons of an atom. An electron that absorbs a *photon* absorbs the energy equivalent to one or more band gaps. When an electron loses energy, a photon is emitted with energy equivalent to the energy of one or more band gaps.

bandwidth. 1. The difference between the limiting *frequencies* within which performance of a device, in respect to some characteristic, falls. The particular performance characteristic, such as the gain of an amplifier, the *responsivity* of a device, or the *bit-error ratio (BER)*, and the limiting frequencies must be specified. **2.** The difference between limiting *frequencies* of a continuous frequency *band*. **3.** A range of *frequencies*, in a given *frequency spectrum*, usually specifying the number of *hertz* of the band or of the upper and lower limiting frequencies. **4.** The range of *frequencies* that a device is capable of generating, handling, passing, or allowing, usually the range of frequencies in which the *responsivity* is not reduced greater than 3 dB from the maximum response. **5.** In *fiber optics,* the value numerically equal to the lowest *modulation frequency* at which the magnitude of the *baseband transfer function* of an *optical fiber* decreases to a specified fraction, generally to one-half of the zero-frequency value. In *multimode fibers,* bandwidth is limited mainly by *multimode distortion* and *material dispersion,* which cause *intersymbol interference* at the receiving end of an optical fiber and thus places a limit on the *bit-error ratio (BER)*. In *single-mode fibers,* it is limited mainly by material and waveguide dispersion. In *fiber optics,* the concept of bandwidth, as applied to electronic systems, is not entirely useful in describing the *data signaling rate* capability or the transmission capacity of an optical fiber, because the optical frequencies involved range over a spectrum about a decade wide near 10^{15} Hz, or 1 PHz. The *bandwidth-distance factor,* that is, the *bit-rate-length product (BRLP),* is a better measure of the signal-carrying capacity of an optical fiber than the bandwidth. However, the modulation scheme must also be specified to make the value of the BRLP meaningful. See *fiber bandwidth; Nyquist bandwidth.* (See Fig. B–1, p. 18)

bandwidth compression. A reduction of the *bandwidth* needed to *transmit* a given amount of information in a given time or a reduction of the time needed to transmit a given amount of information in a given bandwidth. The reduction must be accomplished by a means that does not reduce the information content of a *signal* and that usually does not increase the bit-error ratio (BER).

bandwidth-distance factor. In *fiber optics,* a figure of merit used to express the *signal*-carrying capacity of an *optical fiber*. The figure indicates that *bandwidth* and distance may be reciprocally related, though the tradeoff is not necessarily linear, especially for short lengths, and thus the operating values, namely bandwidth and distance, should both be specified. The bandwidth-distance factor is usually ex-

B-1. The OF192 Fiber Optic Bandwidth Test Set (BWTS), consisting of the OF192T Transmitter and the OF192R Receiver. The BWTS is capable of making **bandwidth** and *loss* measurements on *graded-index multimode optical fibers* up to 1.450 GHz at 0.850 μ *(microns)* and/or 1.3 μ automatically. (Courtesy Tektronix).

pressed as the product of the bandwidth and the distance between end-points, normally expressed in megahertz-kilometers. The associated *bit-error ratio (BER)* should also be given along with the bandwidth-distance factor. Synonymous with *bandwidth-length product*. Also see *bit-rate-length product (BRLP)*.

bandwidth-length product. See *bandwidth-distance factor*.

bandwidth-limited operation. See *bit-rate-length-product-limited operation*.

bandwidth rule. See *Carson bandwidth rule (CBR)*.

barrier layer. In the manufacture of *optical fibers*, a layer that prevents hydroxyl ion diffusion into the *fiber core*.

baseband. 1. In *fiber optics*, the *spectral band* occupied by a *signal* that neither requires *carrier modulation* nor describes the signal state prior to modulation or following *demodulation*. It is usually characterized by being much lower in the *frequency spectrum* than the resultant signal after it is used to modulate a carrier or subcarrier. 2. The *band* of *frequencies* associated with or comprising an original *signal* from the *source* that generated it. In the process of *modulation,* the baseband is occupied by the aggregate of the transmitted signals used to modulate a *carrier.* In *demodulation,* it is the recovered aggregate of the transmitted signals.

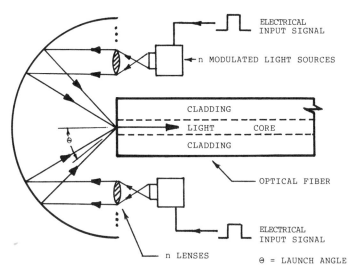

B-2. *Multiplexing* n **baseband** *optical signals* into a single *optical fiber* by means of a *fiber optic multiplexer.*

baseband response function. See *transfer function.*

baud. 1. A unit of *modulation* rate. One baud corresponds to a rate of one unit interval per second, where the modulation rate is expressed as the reciprocal of the duration in seconds of the shortest unit interval. **2.** A unit of signaling speed equal to the number of discrete *signal* conditions, variations, or events, per second.

beam. See *Gaussian beam; laser beam; light beam.*

beam diameter. The distance between two diametrically opposed points in a plane normal to the beam axis at which the *irradiance,* that is, the *optical power density,* is a specified fraction of the beam peak irradiance that is, the *peak spectral power.* The specified fraction of the beam peak irradiance is typically stated as being 1/2, 1/e, 1/e², or 1/10 of the peak irradiance at the point where the beam diameter is measured. The beam profile of the *emission* from a *multimode optical fiber* is assumed to be Gaussian, and the beam width (diameter) is given by the relation:

$$D = [2d \tan (\sin^{-1} NA]/1.73 \approx 2d (NA)/1.73$$

where d is the distance from the end of the fiber, and NA is the *numerical aperture* measured at the 5%-of-peak irradiance points. The beam diameter for a *single-mode optical fiber* is given by the relation:

$$D = 2d\lambda/\pi d_0$$

where d_0 is the *mode field (core) diameter,* and λ is the *wavelength.* The numerical aperture is defined at the 5%-of-peak irradiance. To calculate the axial beam irradiance of a *Gaussian beam,* the beam cross-sectional area at the $1/e$ points is used. The 1.73 accounts for the difference. Synonymous with *beam width.*

beam divergence. **1.** In an *electromagnetic* beam, such as a *light* beam, the increase in diameter with increase in distance along the beam axis from the appropriate *aperture.* **2.** For beams that are circular or nearly circular, the angle subtended by the *far-field* beam diameter. Generally, for noncircular beams, only the maximum and minimum divergences corresponding to the major and minor diameters of the far-field *irradiance* need be specified. **3.** The *far-field* angle subtended by two diametrically opposed points in a plane perpendicular to the beam axis, at which points the *irradiance,* that is, the power density, is a specified fraction of the beam peak irradiance. The divergence, usually expressed in milliradians, is measured at specified points along the beam and radially from the beam central axis, such as where the energy density is $1/2$ or $1/e$ of the maximum value at that point along the beam. The divergence must be specified as the half-angle or as the full angle. Generally, only the maximum and minimum divergences, corresponding to the major and minor diameters of the far-field irradiance, need be specified. The *emission* from a *multimode optical fiber* is assumed to be *Gaussian,* in which case the beam divergence is given by the relation:

$$\phi = [\tan (\sin^{-1} NA)]/1.73 = NA/1.73$$

where NA is the *numerical aperture,* measured at the 5%-of-peak irradiance points. The corresponding beam divergence for a *single-mode fiber* is given by the relation:

$$\phi = 2\lambda/\pi d_0$$

where d_0 is the *mode field (core) diameter* and λ is the *wavelength.*

beam splitter. A device that divides a beam, such as a *light* beam, into two or more beams whose total power is less than or equal to the original beam. For example, a beam splitter might consist of a flat parallel transparent plate coated on one side with a *dielectric* or metallic material that *reflects* a portion and *transmits* the remaining portion of a light beam. Also see *partial mirror.*

beam width. See *beam diameter.*

bend. See *E-bend; H-bend; microbend.*

bending. See *macrobending; microbending.*

bending loss. In an *optical fiber,* the *radiation emitted* at bends due to the escape of discrete *modes* of the transmitted light. The bends may be *microbends,* which occur along the *core-cladding interface,* causing *incidence angles* to occur that are greater than the *critical angle* for *total reflection;* or they may be *macrobends* with radii of curvature less than the *critical radius,* in which case *evanescent waves* can no longer remain coupled to *bound modes* and hence *radiate* laterally away from the fiber. The outside edges of the evanescent waves will radiate because their *wavefronts* cannot exceed the velocity of light as they sweep around the outside of the bend, that is, they cannot maintain a constant *phase* relationship with the wave inside the fiber. Thus, they become uncoupled or unbound and hence radiate away.

bend limiter. In *fiber optic cable* systems, a device, such as a rod, bracket, fixture, or tube, that reduces the tendency of a fiber optic cable to bend when bending forces are applied. The purpose of the bend limiter is to assist in preventing the cable from bending at a radius of curvature less than the *critical radius* or the *minimum bend radius,* whichever is greater.

bend sensor. See *microbend sensor.*

BER. See *bit error ratio.*

bidirectional optical cable. A *fiber optic cable* that can handle *signals* simultaneously in both directions and that has the necessary components to operate successfully and avoid *crosstalk* between the oppositely directed *signals.* Components that are used in, or in conjunction with, bidirectional cables include *beam splitters,* entrance and exit *ports, mixing rods,* couplers, and *interference filters.*

bidirectional transmission. In an *optical waveguide, signal transmission* in both directions.

birefringence. 1. That property of a material that causes a *light* beam passing through the material to be split into two beams with orthogonal *polarizations.* **2.** The *splitting* of a *light* beam into two divergent components upon passage into and through a *doubly refracting propagation medium,* with the two components propagating at different velocities in the medium. The medium is characterized by having two different *refractive indices* in the same direction, which causes the existence of the different velocities of propagation for the different orthogonal polarizations.

birefringence sensor. A *fiber optic sensor* in which *light* entering an *optical fiber* of *birefringement* material, that is, material demonstrating two *refractive indices,* is split into two beams, each propagating in a different path of different *optical path length.* When the beams are recombined, *interference* patterns are produced. External stimuli to be sensed are used to alter the optical path lengths to produce reinforcement or cancellation of the *lightwaves* at a *photodetector.* Thus, the output *signal* of the photodetector is a function of the type and degree of applied stimuli, such as force, pressure, *electric field,* and temperature. Examples include the *Pockels-effect* sensor and the *Kerr-effect* sensor.

birefringent medium. A material that exhibits different *refractive indices* for orthogonal linear *polarizations* of *incident light.* The *phase velocity* of a *wave* in a birefringent medium depends on the polarization of the wave. *Optical fibers* may be made of *birefringent* material. Thus, if light in a *propagation medium* can be decomposed into two perpendicular directions of polarization that propagate at different velocities, the medium is said to be birefringent. Also see *anisotropic; isotropic.*

bit-error rate (BER). See *bit-error ratio (BER).*

bit-error ratio (BER). The number of erroneous bits divided by the total number of bits in a *propagating signal* over some stipulated span of time. For example, the number of erroneous bits received divided by the total number of bits *transmitted.* The BER is usually expressed as a coefficient and a power of 10. For example, 2.5 bits that are erroneously detected by a *photodetector* out of 100,000 bits properly transmitted by a signal source would correspond to a system or link BER of 2.5 × 10^{-5}. Synonymous with *bit-error rate (BER).*

bit-rate-length product (BRLP). For an *optical fiber* or *cable,* the product of the bit rate, that is, the *data signaling rate,* the fiber or cable can handle for tolerable *dispersion* or for a given *bit-error ratio (BER).* The BRLP is usually stated in units of megabit-kilometers/second. A typical value for *graded-index fibers* with a *numerical aperture* (NA) of 0.2 is 1000 Mb-km/s, higher values to be expected in the future. The value of the product is a good indicator of fiber performance in terms of *transmission* capability. Another useful measure of optical fiber performance or transmission capacity is the bit rate at which the *full-wave-half-power point* occurs at the receiving end of a given length of the fiber.

bit-rate-length-product-limited operation. The condition that prevails in a device or system when *distortion* of the *signal* caused by *dispersion,* that is, *waveguide* and *material dispersion,* limits its performance or the performance of the device or system to which it is connected. For example, the condition that prevails in an *optical link* when *intersymbol interference* at the receiving end is the predominant mechanism that limits link performance. Synonymous with *bandwidth-limited operation.*

blank. See *optical blank; optical fiber blank.*

bobbin-wound sensor. A *distributed-fiber sensor* in which the *optical fiber* is wound on a bobbin in such a manner as to be differentially, sequentially, or selectively stimulated by the parameter to be measured, such as a spatially shaded winding or a segmented in-line winding of many coils on a single bobbin, the bobbin serving only to support the distributed fiber but playing no other role, such as stretching or squeezing the fiber.

bolometer. A device used to measure the *radiant* energy from an *emitter* by measuring the change in resistance of a temperature-sensitive element exposed to the radiation. The bolometer is usually used to measure temperatures high above ambient, such as that of a furnance or other body heated to incandescence.

Boltzmann's constant. A physical constant defining the ratio of the universal gas constant to Avogadro's number. Usually represented by the symbol **k**, it is equal to 1.38042×10^{-16} erg/K. For example, in a *photoemissive photodetector,* the *dark current* is given by the relation:

$$I_d = AT^2 e^{qw/\mathbf{k}T}$$

where A is the surface area constant, T is the absolute temperature, q is the electron charge, w is the *work function* of the photoemissive surface material, and **k** is Boltzmann's constant; the *signal-to-noise ratio* for a photodetector is given by the relation:

$$\text{SNR} = (N_p/B)\,(1 - e^{-hf/\mathbf{k}T})$$

where N_p/B is the number of incident *photons* per unit of *bandwidth,* h is *Planck's constant,* f is the *frequency* of the incident photons, **k** is Boltzmann's constant, and T is the absolute temperature; and for the noise generated by thermal agitation of electrons in a conductor, the *noise* power, P, in watts, is given by the relation:

$$P = \mathbf{k}T(\Delta f)$$

where **k** is Boltzmann's constant, T is the conductor temperature in kelvins, and Δf is the bandwidth in *hertz.*

borescope. See *fiber optic borescope; flexible borescope; rigid borescope.*

borescope articulation angle. In a *flexible borescope,* the angle between the axis of the articulating tip of the borescope at its maximum deflection and the axis of the tip at the undeflected position, i.e., when the tip is collinear with the straight-line working length. A typical articulation angle is 90°, though borescopes with larger articulation angles are available.

borescope axial viewing angle. The angle by which the axis of the *borescope view field* deviates from the nominal straightforward axis of the working section of a borescope.

borescope magnification. The ratio of the apparent size of an object seen through a borescope to that of the size of the same object viewed directly by the unaided eye, with the object-borescope distance equal to the object-eye distance.

borescope overall length. The length of a borescope, including the *borescope working length,* which is the length of *fiber optic cable* to the tip, the *tip length,* and the eyepiece assembly length, combined.

borescope target resolution. The *resolving power* of a borescope, usually given in line pairs per millimeter of a backlighted, high-contrast resolution target (object) placed a selected nominal operating distance from the objective end (distal tip).

borescope tip length. The distance between the center *objective* window and the objective end (distal tip) of a *flexible* or *rigid borescope*. This distance determines how close to the bottom of a cavity an image can be observed when the borescope is completely inserted. The tip length should be compatible with the diameter of the probe.

borescope view field. The area or solid angle that can be viewed through a borescope. The solid angle is determined by the *aperture*. The area is determined by the aperture and the focal length.

borescope working diameter. The diameter of the working section, i.e., the diameter of the *fiber optic cable* and tip, of a *flexible borescope* or the diameter of the probe of a *rigid borescope*. The diameter must be sufficiently smaller than the smallest opening it must pass through so that it reaches the object to be viewed without undue friction or binding.

borescope working length. In a *fiber optic borescope,* the distance between the distal end of the hand-held body portion and the *optical axis* of the tip. The *borescope tip length* is not included.

Bouger's law. In the *transmission* of *electromagnetic radiation* through a *propagation medium,* the *attenuation* of the *irradiance,* that is, the *electromagnetic field strength* or *optical power density,* as an exponential function of the product of a constant coefficient, which is dependent upon the material and its thickness, given by the relationship:

$$I = I_0 e^{-ax}$$

where I is the *irradiance* at distance x, I_0 is the irradiance at $x = 0$, and a is a material constant coefficient that depends upon the *scattering* and *absorptive* properties of the propagation medium. If only absorption takes place, a is the *spectral absorption coefficient* and is a function of wavelength. If only scattering takes place, it is the *scattering coefficient.* If both absorption and scattering occur, it is the *extinction coefficient,* being then the sum of the absorption and scattering coefficients. The irradiance, or optical power density, is usually measured in watts per square meter.

bounce time. See *switch operating bounce time; switch release bounce time.*

bound mode. In an *optical waveguide,* a *mode* whose field decays monotonically in the transverse direction everywhere external to the *core* and does not lose power to *radiation.* For an *optical fiber* in which the *refractive index* decreases with distance from the *optical axis* and where there is no central *refractive-index dip,* a bound mode is governed by the relation:

$$n_a k < \beta < n_0 k$$

where β is the imaginary part, that is, the *phase term,* of the *axial propagation constant;* n_a is the refractive index at $r = a$, the *core radius;* n_0 is the refractive index at $r = 0$, the core axis; k is the angular *free space wave number,* $2\pi/\lambda$, and λ is the *wavelength.* Bound modes correspond to *guided rays* in *geometric optics.* When more than one mode is propagating in a fiber, the *optical power* in bound modes is largely confined to the core. Synonymous with *guided mode; trapped mode.* Also see *cladding mode; guided wave; leaky ray; unbound mode.*

bound ray. See *guided ray.*

box. See *optical fiber distribution box; optical fiber interconnection box; optical mixing box.*

branch. In *network* topology, a direct path, such as a *bus,* joining two adjacent *nodes* of a network.

branching device. See *fiber optic branching device; optical branching device.*

Brewster angle. The angle, measured with respect to the normal, at which an *electromagnetic wave incident* upon an *interface* surface between two *dielectric* media of different *refractive indices,* is totally *transmitted* from one *propagation medium* into another. The reflectance will be zero only for *light* that has its *electric field vector* in the plane defined by the direction of *propagation,* that is, the direction of the *Poynting vector,* and the normal to the surface. This implies that the *magnetic field component* of the incident wave must be parallel to the interface surface. The Brewster angle is given by the relation:

$$\tan B = (\epsilon_1 \epsilon_2)^{1/2} = n_1/n_2$$

where B is the Brewster angle, ϵ_1 is the *electric permittivity* of the incident-wave propagation medium and ϵ_2 is the electric permittivity of the transmitted-wave propagation medium, and n_1 and n_2 are their corresponding refractive indices. The Brewster angle is a convenient angle for transmitting all the energy in an *optical fiber* to a *photodetector* or from a *source* to a fiber. There is no Brewster angle when the electric field component is parallel to the interface, except when the electric permittivities are equal, in which case there is no interface. For entry into a denser medium, such as from air into an optical fiber, B is less than $45°$; and from a denser medium into a less dense medium, such as fiber to air, B is greater than $45°$. The Brewster angle is measured with respect to the normal-to-the-interface surface.

Brewster's law. When an *electromagnetic wave* is *incident* upon a surface and the angle between the *refracted* and *reflected ray* is $90°$, maximum *polarization* occurs in both rays. The reflected ray has its maximum polarization in a direction normal to the incidence plane and the refracted ray has its maximum polarization in the incidence plane.

brightness. An attribute of the physiological sensation of *light,* that is, a visual perception or appearance of a higher or lower level of light. In photometrics, brightness was used as a synonym for luminance. In radiometrics, brightness was incorrectly used as a synonym for radiance. See *radiance.*

brightness conservation. See *radiance conservation.*

brightness theorem. See *radiance conservation.*

Brillouin diagram. A diagram showing the allowable and unallowable *frequencies* or energy levels of *electromagnetic waves* that can pass through materials that have certain periodic microstructures. When an *electric field* is applied to the material, electrons experience periodic attractive and repulsive forces as they approach and depart from atomic nuclei in their path as they move through the material. The resulting electron vibration results in a field that interacts or resonates with the *propagating* electromagnetic wave, causing energy from the wave to be absorbed by the electrons at the resonant frequencies. The net result is that electromagnetic waves of the resonant frequencies are absorbed while others are passed. Crystals and certain glasses have these periodic microstructures, and therefore Brillouin diagrams can be drawn for them.

Brillouin scattering. The *scattering* of *lightwaves* in a *propagation medium* caused by thermally driven density fluctuations that cause frequency shifts of several *gigahertz* at room temperature.

BRLP. *Bit-rate-length product.*

broadcast satellite. See *direct-broadcast satellite (DBS).*

broadening. See *pulse broadening.*

budget. See *error budget; loss budget; optical power budget; power budget.*

budget constraint. See *loss-budget constraint.*

buffer. **1.** Material placed on or around an *optical fiber* to protect it from mechanical damage. Buffers can be bonded to the *cladding.* **2.** In a *fiber optic cable,* material used to cushion and protect the *optical fibers,* particularly from mechanical damage, such as *microbends* and *macrobends,* caused during manufacture, spooling, subsequent handling, and when pressure is applied during use. **3.** A data-storage device used to compensate for differences in the rate of flow of *data,* or the time of occurrence of events, when transferring data from one device to another. **4.** An *isolating* circuit used to prevent a driven circuit from influencing the driving circuit. **5.** To allocate and schedule the use of buffers. See *fiber buffer.*

buffered fiber. An *optical fiber* that has a *coating* over the *cladding* for protection, increased visibility, and ease of handling.

bulk optical glass. Large quantities of glass pure enough and sufficiently free of contaminants to be used for optical elements, such as lenses, prisms, and mirrors, compared with raw glass used for making containers, window panes, and building blocks. Bulk optical glass is used as the starting glass for making *optical fiber preforms.* The glass is further purified, particularly dehydrated. *Dopants* are added to achieve desired values of *refractive indices.*

bundle. A group of conductors, such as wires, *optical fibers,* or coaxial cables, associated together and usually in a single sheath. Optical fibers are nonconductors of electric currents, but, when properly designed, will *transmit optical signals.* See *aligned bundle; unaligned bundle; optical fiber bundle; ray bundle.* Also see *cable.*

bundle of rays. See *ray bundle.*

bundpack. In *fiber optics,* an *optical fiber* wound in a spherical or conical shape after the *fiber-drawing* process in such a manner that it can be payed out, that is, unwound, from either the inside or the outside of the winding. The conical shape permits easy removal of the collapsible winding mandrel after winding. The bundpack can be wrapped in a tight plastic skin for protection during storage, shipping, and handling. There are no spools or reels involved during payout, though one may be used for outside payout if convenient to do so. A light bonding cement is used to control payout, so that payout is steady and without bunching. A twist is placed in the fiber when winding so that payout is without strain, and hockels, that is, kinks, do not occur.

Burrus diode. See *surface-emitting LED.*

Burrus LED. See *surface-emitting LED.*

bus. One or more conductors, such as wires, *coaxial cables,* or *optical waveguides,* that serve as a common connection for a related group of devices. See *optical data bus.*

butt coupling. In *fiber optics,* the *coupling* of one *optical fiber,* or other optical element, to another by placing the end face of one against the end face of the other so that an *electromagnetic wave* can be *transmitted* with a minimum loss of power and a maximum *transmission coefficient* at the *interface.* The interface is usually transverse to the direction of *propagation.* The *Brewster angle* can be used for maximum coupling and total transmission, that is, zero *reflection,* when *lightwaves* pass from one *propagation medium* to another, each medium having a different *refractive index.* However, when optical fibers are butt-coupled, the end faces are perpendicular to the *optical axis* because they are easy to make and the joint can be rotated and shifted for maximum coupling, with *Fresnel reflection* normally being about 4% of the *incident irradiance,* that is, the incident *optical power.*

C

c. The symbol used for the speed of an *electromagnetic wave,* such as a *lightwave,* in a vacuum, that is, 3×10^8 m/s (meters per second). The speed in a material *propagation medium* is given by the relation:

$$v = c/n$$

where n is the *refractive index* of the medium relative to that of a vacuum.

cable. In *fiber optics,* an assembly of one or more *optical fibers,* strength members, buffers, and perhaps other components all within an enveloping sheath or *jacket,* constructed so as to permit access to the fibers singly or in groups. See *aerial optical cable; bidirectional optical cable; central-strength-member optical cable; fiber optic cable; filled cable; grooved cable; hybrid cable; loose-tube cable; multifiber cable; optical station cable; optoelectrical cable; peripheral-strength-member optical cable; ribbon cable; Tempest-proofed fiber optic cable; tactical fiber optic cable; tight-jacket cable.* Also see *bundle.* (See Fig. C–1)

cable assembly. See *fiber optic cable assembly.*

cable component. See *optical fiber cable component (OFCC).*

cable core. See *fiber optic cable core.*

cable cutoff wavelength. For a cabled *optical fiber,* the *wavelength* region above which the fiber supports the *propagation* of only one *mode* and below which multiple modes are supported. Operation of the fiber below the *cutoff wavelength* usually results in modal noise, *modal distortion* (increased *pulse broadening*), and unsatisfactory operation of *connectors, splices,* and *wavelength-division multiplexing (WDM) couplers.* The operating wavelength range, as determined by the *transmitter nominal central wavelength* and the transmitter *spectral width,* must be greater than the maximum cutoff wavelength to ensure that the cable is operating entirely in the fiber *single-mode* region. Usually, the highest value of cabled fiber cutoff wavelength occurs in the shortest cable length. A criterion that will ensure that a system is free from high-cutoff wavelength problems is given by the relation:

$$\lambda_{cc\,max} > \lambda_{t\,min}$$

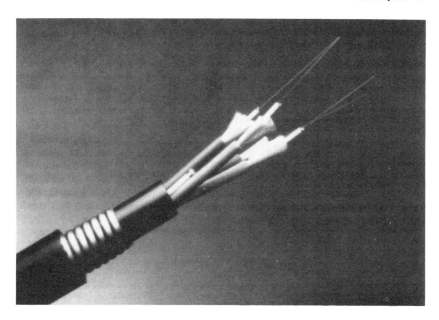

C-1. An armored *fiber optic* **cable** for added protection and strength in harsh environments. (Courtesy Optical Cable Corporation).

where $\lambda_{cc\,max}$ is the maximum cutoff wavelength, and $\lambda_{t\,min}$ is the minimum value of the transmitter central wavelength range, which is also caused by the worst-case variations introduced by manufacturing, temperature, aging, and any other significant factors determined when the cable is operated under *standard* or *extended operating conditions*. Also see *fiber cutoff wavelength*.

cable facility. See *optical cable facility*.

cable facility loss. See *optical cable facility loss; statistical optical cable facility loss*.

cable feed-through. See *fiber optic cable feed-through*.

cable interconnect feature. See *optical cable interconnect feature*.

cable pigtail. See *optical cable pigtail*.

cable splice. In *fiber optics,* a connection between two cables, consisting of one or more *optical fiber splices* within a *cable splice closure*. It provides *optical* continuity between cables, protection against environmental conditions, and mechanical strength to the cable joint. The cable splice is used primarily to complete a cable span or to repair a cable.

cable splice closure. The portion of a *fiber optic cable splice* that covers the fiber splice housings, seals against the outer *jackets* of the joined cables, provides protection against the environment, and provides mechanical strength to the joint.

cabling process. In the manufacture of *fiber optic cables,* starting with one or more reels of *optical fiber* or *fiber ribbon,* usually with appropriate *coatings* and *buffers* already applied, the assembly of a fiber optic cable by wrapping additional buffers, *strength members, jackets,* and perhaps other materials, such as *overarmor* and *harnesses.* The process includes putting coatings on the jacket, adding markings to indicate the type of cable, and winding the finished cable on spools or other forms for internal or external payout. Care must be taken to ensure that the cabling process does not introduce *microbends* in the optical fibers and does not bend the fibers with a radius of curvature less than the *minimum bend radius* or the *critical radius,* whichever is larger.

carrier. A wave, *pulse* train, or other *signal* suitable for *modulation* by an information-bearing signal in a *modulator* for the purpose of representing information and capable of being transported or *transmitted* by a communication system.

Carson bandwidth rule (CBR). A rule defining the approximate *bandwidth* requirements of communication system components for a *carrier signal* that is *frequency-modulated* by a continuous, that is, a broad spectrum of *frequencies,* rather than a single frequency. The rule is expressed by the relation:

$$CBR = 2(\Delta f + f_m)$$

where Δf is the *carrier* peak deviation frequency, and f_m is the highest modulating frequency. *Bandwidth* requirements for modulating carriers and equipment capabilities impose limits on the extent of *multiplexing* that can be accommodated. The Carson bandwidth rule is often applied to *transmitters,* antennas, *light sources, receivers, photodetectors,* and other communication system components.

cavity mirror sensor. See *optical cavity mirror sensor.*

cavity. See *optical cavity; resonant cavity.*

CBR. *Carson bandwidth rule.*

cell. See *Kerr cell; Pockels cell.*

cement. See *optical cement.*

center. See *cladding center; core center; network control center; processing center; reference surface center; scattering center; switching center.*

central-strength-member optical cable. A *cable* containing *optical fibers* that are wrapped around a high-tensile-strength material, such as stranded steel, Kevlar, and nylon, with a crush-resistant *jacket* wrapped around the outside. Also see *peripheral-strength-member optical cable.*

central wavelength. See *nominal central wavelength.*

central-wavelength measurement. For the *transmitter terminal/regenerator facility* of an *optical station/regenerator section,* the measurement of the *wavelength* that identifies where the effective *optical power* of the transmitter resides using either the peak mode or the power-weighted method.

central wavelength range. See *transmitter central wavelength range.*

change in transmittance. See *induced attenuation.*

channel. 1. A connection between initiating and terminating *nodes* of a circuit. **2.** A single path provided by a *transmission medium* either by physical separation, e.g., multipair cable, or by electrical separation, e.g., *frequency* or *time-division multiplexing.* **3.** A single unidirectional or bidirectional path for *transmitting,* receiving, or both electrical or *electromagnetic signals,* usually distinct from other parallel paths. **4.** Used in conjunction with a predetermined letter, number, or codeword to reference a specific radio or video *frequency.* **5.** A path along which *signals* can be sent, e.g., *data channel* or output channel. **6.** The portion of a storage medium that is accessible to a given reading or writing station, e.g., a track or band. **7.** In information theory, that part of a communication system that connects the message source to the message sink.

characteristic. See *fiber global dispersion characteristic; fiber global attenuation-rate characteristic; global dispersion characteristic; wavelength-dependent attenuation-rate characteristic.*

characteristic impedance. 1. In a *uniform electromagnetic plane wave propagating* in *free space,* or in *dielectric propagation media,* the ratio between the *electric* and *magnetic field strengths,* given by the relation:

$$Z = E/H$$

in which E and H are orthogonal and are in a direction perpendicular to the direction of propagation, namely perpendicular to the *Poynting vector,* which is the vector obtained from the cross (vector) product of the *electric* and *magnetic field vectors,* with the direction of a right-hand screw obtained when rotating the electric vector into the magnetic vector over the smaller angle. **2.** From *Maxwell's equations,* the impedance of a linear, homogeneous, isotropic, dielectric, and electric-charge-free propagation medium, given by the relation:

$$Z = (\mu/\epsilon)^{1/2}$$

where μ is the magnetic permeability and ϵ is the electric permittivity. For free space, $\mu = 4\pi \times 10^{-7}$ H/m and $\epsilon = (1/36\pi) \times 10^{-9}$ F/m, from which 120π, or 377 ohms is obtained. For dielectric media, the permittivity is the dielectric constant, which, for glass, ranges from about 2 to 7. However, the *refractive index* is the square root of the dielectric constant. Thus, the characteristic impedance of dielectric materials is $377/n$ ohms, where n is the refractive index.

chemical-vapor-deposition (CVD) process. A method of making *optical fibers* in which *silica* and other glass-forming oxides and *dopants* are deposited at high temperatures on the inner wall of a fused silica tube, called a *bait tube,* which is then collapsed into a short, thick, *preform* from which a long thin fiber is pulled at a high softening temperature using a *fiber-pulling machine.* See *modified chemical-vapor-deposition (CVDP) process; plasma-activated chemical-vapor-deposition (PACVD) process.*

chemical-vapor-phase-oxidation (CVPO) process. A method of making *low-loss* (less than 10 dB/km), high *bit-rate-length product* (greater than 300 Mb-km/s), *multimode, graded-index (GI), optical fiber* involving either the *inside vapor-phase-oxidation (IVPO) process,* the *outside vapor-phase-oxidation (OVPO) process,* the *modified chemical-vapor-deposition (MCVD) process,* the *plasma-activated chemical-vapor-deposition (PCVD) process,* the *axial vapor-phase-oxidation (AVPO) process,* or a combination or variation of these, by soot deposition on a glass substrate followed by oxidation and *pulling* of the fiber.

chief ray. The central *ray* of a *ray bundle.*

chirp. See *laser chirp.*

chirping. 1. A sudden shift from one set of *spectral lines,* that is, in the *wavelengths* emitted by a *source,* to another set of lines. Chirping is often observed when sources are *pulsed,* that is, *modulated* by pulses. **2.** Rapid fluctuations in *frequency* or *wavelength* of an *electromagnetic wave,* usually caused by some form of *modulation* at the *source* or of the wave after *emission.*

chopper. See *optical chopper.*

chromatic aberration. Image imperfections caused by *light* of different *wavelengths* following different paths through an *optical* system due to the *dispersion* caused by the optical elements of the system.

chromatic dispersion. *Dispersion* or *distortion* of a *pulse* in an *optical waveguide* due to differences in wave velocity caused by variations in the *refractive indices* in different portions of the guide. The *propagation* velocity varies inversely with the refractive index. The propagation time from the beginning to a point in the guide varies directly as the *optical path length,* the rate of change of refractive index with respect to *wavelength,* and the *spectral line width* of the *source.* Differences in delay, that is, differences in propagation time, cause the distortion.

chromatic dispersion coefficient. The derivative with respect to *wavelength* of the normalized *group delay* of a *waveguide*, such as an *optical fiber*, given by the relation:

$$D(\lambda) = d\tau(\lambda)/d\lambda$$

where $\pi(\lambda)$ is the normalized group delay, and λ is the wavelength at which the normalized group delay is taken.

chromatic dispersion slope. The derivative with respect to *wavelength* of the *chromatic dispersion coefficient* of a *waveguide*, such as an *optical fiber*, given by the relation:

$$S(\lambda) = dD(\lambda)/d\lambda$$

where $D(\lambda)$ is the chromatic dispersion coefficient, and λ is the wavelength at which the slope is taken.

chromatic distortion. Synonymous with *intramodal distortion*.

chromaticity. The *frequency*, or *wavelength*, composition of *electromagnetic waves* in the *visible region* of the *spectrum*. Chromaticity is normally characterized by the dominant frequencies or wavelengths, which describe the quality of *color*. It affects *propagation* in *filters*, *attenuation* in *optical fibers*, and the sensitivity of photofilm.

chromatic radiation. See *polychromatic radiation*.

chromatic resolving power. 1. The ability of an instrument to separate two *electromagnetic waves* of different *wavelengths*. It is equal to the ratio of the shorter wavelength divided by the difference between the wavelengths. **2.** A measure of the ability of an *optical* component to separate or differentiate two or more points on the object that are close together relative to the distance between the optical component and the object points.

circuit. See *integrated optical circuit (IOC)*.

circular dielectric waveguide. A *waveguide* consisting of a long circular cylinder of concentric *dielectric materials*, capable of sustaining and guiding an *electromagnetic wave* in one or more *propagation modes*. For example, most *optical fibers* are circular *dielectric waveguides* that are capable of *transmitting*, or transporting, *lightwaves*. Some *polarization-preserving (PP) fibers* are somewhat elliptical or rectangular in cross section, especially the *core*.

circularity. See *cladding noncircularity; core noncircularity*.

circular polarization. *Polarization* of the *electric* and *magnetic field vectors* in a *uniform plane-polarized electromagnetic wave*, such as some *lightwaves* or radio waves, in which the two arbitrary, sinusoidally varying rectangular components of the electric field vector are equal to each other in amplitude but are 90° out of *phase*. This causes the electric field vector to rotate, with the direction of rotation depending on which component leads or lags the other, that is, the tip, or head, of the electric field vector rotates in a circle. The magnetic field vector is at the center of this circle and perpendicular to the plane of the circle. See *left-hand circular polarization; right-hand circular polarization.*

clad. See *cladding.*

cladding. A layer of a lower *refractive-index* material, in intimate contact with a *core* of higher refractive-index material, used to achieve *reflection*. The cladding confines *lightwaves* to the core, provides some protection to the core, and also *transmits evanescent waves* that are usually *bound* to waves in the core. These evanescent waves will *propagate* in the cladding and even extend beyond. They will not *radiate* away if the cladding is sufficiently thick. They will radiate away if the fiber is bent too sharply, that is, bent sharper than a certain radius called the *critical radius*. Synonymous with *clad*. See *depressed cladding; extramural cladding; fiber cladding; homogeneous cladding; matched cladding.*

cladding center. In a round *optical fiber,* the center of the circle that best fits the outer limit of the cladding. The cladding center may not be the same as *cladding* and *reference surface* centers.

cladding concentricity. See *core-cladding concentricity.*

cladding diameter. In a round *optical fiber,* the diameter of the circle that best fits the outer limit of the *cladding*. The center of this circle is the *cladding center*.

cladding guided mode. See *cladding mode.*

cladding mode. In an *optical waveguide,* a *propagation mode* supported by the *cladding,* that is, a mode in which the *electromagnetic field* is confined to the cladding and the *core* because there is a lower *refractive-index propagation medium* surrounding the outermost cladding. These modes are in addition to the modes supported only by the core. Cladding modes may be *attenuated* by using lossy media to absorb the energy propagating in the mode, thus preventing reconversion of energy to core-guided modes and thereby reducing *dispersion*. Synonymous with *cladding guided mode; cladding ray.* Also see *bound mode; leaky mode; leaky ray; unbound mode.*

cladding-mode stripper. A device in which *cladding modes* are converted to *radiation modes*. A material is usually applied to *cladding* to allow *electromagnetic* energy *propagating* in the cladding to leave the cladding by having its *refractive index* be equal to or greater than that of the cladding.

cladding noncircularity. **1.** In the cross section of a round *optical fiber,* the difference between the diameters of the smallest circle that can circumscribe the *core area* and the largest circle that can be inscribed within the *cladding,* both circles concentric with the cladding center, divided by the *cladding diameter.* **2.** In a round *optical fiber,* the percentage of deviation of the *cladding* cross section from a circle. Synonymous with *cladding ovality.* Also see *optical fiber concentricity error.*

cladding ovality. See *cladding noncircularity.*

cladding power distribution. See *core-cladding power distribution.*

cladding ray. See *cladding mode.*

cladding tolerance area. In the cross section of a round *optical fiber,* the region between the smallest circle that can circumscribe the *core area* and the largest circle that can be inscribed within the *cladding,* both circles concentric with the *cladding center.*

clad fiber. See *quadruply clad fiber.*

clad silica fiber. See *hard-clad silica (HCS) fiber.*

clockwise-polarized wave. An *elliptically* or *circularly polarized electromagnetic wave* in which the direction of rotation of the *electric vector* is clockwise as seen by an observer looking in the direction of *propagation* of the *wave.* Also see *counterclockwise-polarized wave.*

closed waveguide. A *waveguide* that has electrically conducting walls, thus permitting an infinite but discrete set of *propagation modes* of which relatively few are practical. Each discrete mode has its own *propagation constant.* The *electric* and *magnetic fields* at any point can be described in terms of these modes. There is no *radiation* from the guide. Examples of a closed waveguide include a metallic rectangular-cross-section pipe not unlike an ordinary rectangular rain downspout on a house and an *optical fiber* with a *reflective coating* to obtain *internal reflection* and used as a *light pipe* for interior illumination. Also see *open waveguide.*

closure. See *cable splice closure.*

coated-fiber sensor. A *fiber optic sensor* in which a special *coating* is used to apply stress to an *optical fiber* resulting in fiber strain. For example, a *magnetoscrictive* coating may be used to elongate the fiber proportionally to an applied longitudinal *magnetic field.* The variations in length can be measured using *interferometric* techniques. If the magnetic field is made proportional to an electric current, the electric current can be measured. If the coating is thermally sensitive, the coefficient of thermal expansion can be used to measure temperature, or the thermal expansion effect of an electric current in the coating can be used to measure the electric current. A narrow *spectral-line-width light source;* a length of optical fiber, serving as a sensing leg, with a bonded coating sensitive to the parameter being measured; a shielded length of optical fiber, serving as a reference leg, that is not exposed to the parameter being measured to serve as a reference; a *fiber optic splitter* to split the same wave into both legs; a *coupler* to combine the *waves* from both legs; and a *photodetector* responsive to the *light intensity* produced by the two interfering waves, all connected together constitute a coated-fiber sensor.

coated optics. The use of *optical* elements or components whose optical *refractive* and *reflective* surfaces have been *coated* with one or more layers of *dielectric* or metallic material for reducing or increasing reflection from the surfaces, either totally or for selected *wavelengths,* and for protecting the surfaces from abrasion and corrosion. Some antireflection coating materials that can serve as optical coatings are magnesium fluoride, silicon monoxide, titanium oxide, and zinc sulfide.

coating. See *antireflection coating; optical protective coating; optical fiber coating; primary coating; secondary coating.*

coefficient. See *absorption coefficient; chromatic dispersion coefficient; dispersion coefficient; extinction coefficient; global-dispersion coefficient; scattering coefficient; spectral absorption coefficient; reflection coefficient; transmission coefficient; worse-case dispersion coefficient.*

coherence. 1. The phenomenon pertaining to a correlation between the *phase* points of two *waves* at the same instant or at the same point in space. **2.** The phenomenon pertaining to a correlation between the phase points of one wave at two different instants of time at the same point in space or at two points in space at the same instant. In practice, coherence is a matter of degree. If coherence were perfect, the phase relationships within a wave, and the phase relationships among waves, would be always and everywhere precisely predictable. See *spatial coherence; coherence degree; partial coherence; time coherence.*

coherence area. In *optical fiber communications,* the area in a plane perpendicular to the direction of *propagation* over which *light* may be considered highly *coherent.* It is usually considered to be the area over which the *coherence degree* exceeds 0.88.

coherence degree. A measure of the *coherence* of a *light source,* equal to the visibility, V, given by the relation:

$$V = (I_{max} - I_{min})/(I_{max} + I_{min})$$

where V is the visibility of the fringes of a two-beam interference test, I_{max} is the *irradiance,* that is, the *optical power density,* at a maximum of the *interference* pattern, and I_{min} is the irradiance at a minimum of the interference pattern. Light may be considered to be coherent if the coherence degree exceeds 0.88 and incoherent if significantly less than 0.88. Synonymous with *degree of coherence.*

coherence length. The *propagation* distance over which a *light* beam may be considered *coherent*. The coherence length in a *propagation medium* is given by the relation:

$$L_c = \lambda_0^2/n\Delta\lambda$$

where λ_0 is the *central wavelength* of the *source, n* is the *refractive index* of the medium, and $\Delta\lambda$ is the *spectral line width* of the source.

coherence time. The time over which a *propagating electromagnetic wave* may be considered *coherent*. It is equal to the *coherence length* divided by the *phase* velocity of *light* in a *propagation medium,* approximately given by the relation:

$$T_c = \lambda_0^2/c\Delta\lambda$$

where λ_0 is the *central wavelength,* $\Delta\lambda$ is the *spectral line width,* and c is the velocity of light in vacuum. In long-distance transmission systems, the coherence time may be degraded by other propagation factors.

coherent. 1. Pertaining to a fixed *phase* relationship between points on an *electromagnetic wave*. A truly coherent wave would be perfectly coherent at all points in space. In practice, however, the region of high *coherence* may extend only a finite distance from a *source*. **2.** Pertaining to an *electromagnetic wave* in which the *electric* and *magnetic field vectors* are uniquely and specifically definable always and everywhere within specified tolerances. An electromagnetic wave of single *frequency,* that is a purely *monochromatic wave,* would be coherent at all points in free space and in a homogeneous *isotropic propagation medium*. Thus, the electric and magnetic field values could be precisely predicted at every point in space and time using the solutions to *Maxwell's equations*. In actual practice monochronism is not perfect. There is a *spectral width*. The region of high coherence may extend only a short distance from the source. The area on the surface of a practical *wavefront* over which the wave may be considered highly coherent is the *coherent area* or is a coherence patch. The distance in the direction of *propagation* in which the wave is highly coherent is the *coherence length,* in which case the wave is phase- or length-coherent. This coherence length divided by the velocity of the light in the propagation medium is the *coherence time*. Thus, phase and time coherence are related. Phase coherence may also be called time coherence because both are taken in the direction of propagation, rather than transverse to the direction of propagation as in area coherence. Area coherence is taken at an instant of time. Time, or temporal, and phase coherence of a wave occur over the length of time that corresponds to its coherent length.

coherent area. The area in a plane perpendicular to the direction of *propagation* over which *radiation* may be considered to be *coherent,* that is, the locus of all points in a plane at which the *coherence degree* is greater than 0.88.

coherent bundle. See *aligned bundle.*

coherent detection. In *fiber optics,* optical detection in which a low-power *phase-modulated lightwave* is *coherently* mixed with an *optical signal* to produce an amplitude-modulated signal that can be directly detected downstream by conventional means.

coherent light. *Light* in which all parameters are predictable and correlated at any point in time or space, particularly over an area in a plane perpendicular to the direction of *propagation* or over time at a point in space. See *time-coherent light.*

coherent radiation. At a given point, *electromagnetic radiation* that has a *coherence degree* greater than 0.88.

collimation. **1.** The process by which divergent or convergent *light,* that is, a beam of *electromagnetic radiation,* is converted into a beam with the minimum divergence or convergence possible for that system. Ideally, a bundle of parallel rays is a collimated radiation. **2.** The process of making *electromagnetic rays,* such as *light rays,* parallel. For example, the use of a lens system to make divergent or convergent rays parallel. Light rays from a very distant, not necessarily point source are practically parallel.

collimator. An *optical* device that renders diverging or converging light rays parallel. The degree of *collimation,* that is, the extent to which they are parallel, should be stated in terms of an angle of convergence or divergence.

color. **1.** The sensation produced by *light* of a given *wavelength,* or group of wavelengths, in the *visible region* of the *electromagnetic spectrum.* **2.** A given *wavelength,* or set of wavelengths, in the *visible spectrum.*

color-division multiplexing. See *wavelength-division multiplexing.*

combiner. See *optical combiner.*

communication network. See *long-haul communication network; short-haul communication network.*

communications. See *optical fiber communications.*

compensated optical fiber. An *optical fiber* whose *refractive-index profile* has been adjusted so that *light rays propagating* in the higher-*refractive-index* portion of the *core,* that is, near the core center, thus propagating a shorter distance at lower speeds due to the high-index material, and the rays propagating in the outer lower index material, thus undergoing *total (internal) reflection,* or bending back and forth, and thus traversing a longer path but propagating faster due to the lower index, arrive at the end of the fiber at the same time, along with the *skew rays* that travel a helical path, thus reducing *modal dispersion* to nearly zero. *Higher order modes* have higher eigenvalues in the *wave equation* solutions, higher *frequencies,* and shorter *wavelengths* than *lower order modes.* Also see *overcompensated fiber; undercompensated fiber.*

component. See *magnetic field component; optical fiber cable component (OFCC); optical path component.*

compression. See *bandwidth compression; optical fiber pulse compression.*

compression sensor. See *fiber longitudinal-compression sensor; fiber transverse-compression sensor.*

computer. See *optical computer.*

concatenation. The linking of a series of devices or components, such as *optical waveguides,* such that the output of the first is the input to the second, the output of the second is the input to the third, and so on.

concentrator. See *optical fiber concentrator.*

concentric-circle near-field template. See *four-concentric-circle near-field template.*

concentric-circle refractive-index template. See *four-concentric-circle refractive-index template.*

concentricity. See *core reference-surface concentricity; core-cladding concentricity.*

concentricity error. See *optical fiber concentricity error.*

condition. See *extended operating condition; launch condition; standard operating condition.*

conduction. See *optical conduction.*

conductivity. See *electrical conductivity; photoconductivity.*

conductor. See *optical conductor.*

cone. See *acceptance cone.*

configuration scattering. Scattering of *electromagnetic radiation* caused by variations in the configuration of a *propagation medium,* such as *scattering* caused by variations in geometry and *refractive-index profile* of an *optical fiber.* Synonymous with *fiber scattering.*

confinement factor. For a given *guided mode* of an *electromagnetic wave propagating* in a *waveguide,* the ratio between the power within the guiding layer, such as the *core* of an *optical fiber* or inner layer of a *planar waveguide,* and the total guided power.

connection. See *fiber optic cross-connection; fiber optic interconnection.*

connector. See *expanded-beam connector; fiber optic connector; fixed connector; free connector; heavy-duty connector; light-duty connector; optical waveguide connector; optical fiber active connector; receiver optical connector; transmitter optical connector; wet-mateable connector.* See *hybrid connector.* (See Figs. C–2–C–7)

connector set. See *optical fiber connector set.*

connector variation. See *optical connector variation.*

conservation law. See *radiance conservation law.*

conservation of radiance law. Synonymous with *radiance conservation law.*

constant. See *Boltzmann's constant; Planck's constant; propagation constant; transverse propagation constant; Verdet's constant.*

constitutive relations. A set of three equations pertaining to the properties of a *propagation medium* in which *electric* and *magnetic fields,* electric currents, and

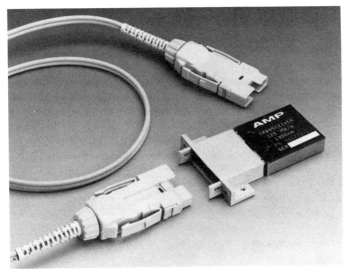

C–2. The AMP Fixed Shroud Duplex **Connector** for protecting *optical fiber* ferrules from mishandling damage. *Insertion loss* is less than 1 *dB,* with a typical forward *loss* of less than 0.6 dB. Also shown is the 125-Mbps *Transceiver* directly connected for space saving. (Courtesy AMP Incorporated).

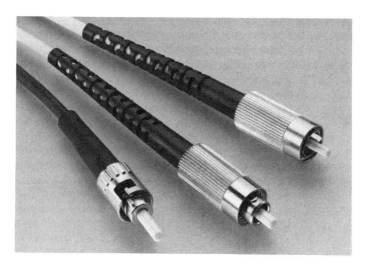

C-3. Ceramic-ferrule *single-mode* **connectors** having *insertion losses* averaging less than 0.3 *dB* and *return loss* averaging −36 dB. The OPTIMATE™ 2.5 mm threaded, 2.5 mm bayonet, and 2.0 mm threaded styles are compatible, respectively, with most FC/PC, ST/PC, and D4/PC types. (Courtesy AMP Incorporated).

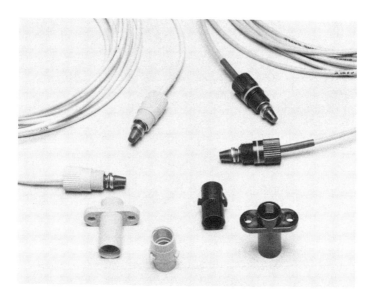

C-4. Biconic Field-Mountable **Connector** Assemblies for *single-mode* and *multimode fiber optic cable* connections. (Courtesy 3M Fiber Optic Products, EOTec Corporation).

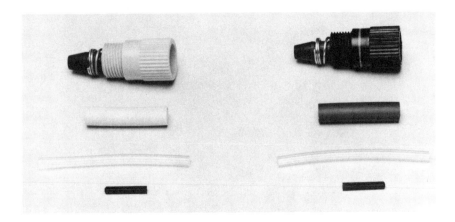

C-5. Biconic Field-Mountable **Connectors** for *single-mode* and *multimode fiber optic cable* connections. (Courtesy 3M Fiber Optic Products, EOTec Corporation).

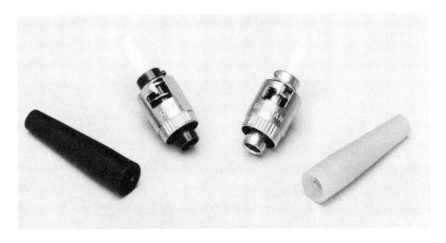

C-6. Dorran™ ST-compatible field-mountable *fiber optic* **connectors** for *single-mode* and *multimode* applications. (Courtesy 3M Fiber Optic Products, EOTec Corporation).

electromagnetic waves exist and *propagate.* The relations are given as $D = \epsilon E,$ $B = \mu H,$ and $J = \sigma E,$ where D is the electric flux density or electric displacement vector, E is the *electric field strength,* B is the magnetic flux density, H is the *magnetic field strength,* J is the electric current density, ϵ is the *electric permittivity,* μ is the *magnetic permeability,* and σ is the *electrical conductivity.* The relations are used in conjunction with *Maxwell's equations* for electromagnetic wave propagation. In *dielectric* materials, such as *optical fibers,* $\sigma = 0$ and $\mu = 1.$

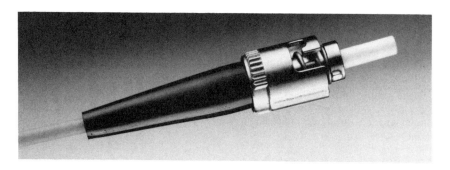

C-7. This fiber optic ceramic-tip (FOCT) **connector** is designed for single-*channel* fiber optic interconnection for *cable*-to-cable or cable-to-equipment applications. It is small, rugged, and precision-made for *networking,* telecommunications, and other *data,* voice, or *video*commercial and military applications. (Courtesy ITT Cannon, Military/Aerospace Division, Fiber Optic Products Group).

constraint. See *loss-budget constraint.*

contact. See *terminus.*

continuously variable optical attenuator. A device that *attenuates* the *irradiance,* that is, the *power density,* of *lightwaves* over a continuous range, rather than in discrete steps. The attenuator may be inserted in an *optical link* to control the irradiance of light at the receiving *photodetector.* Also see *stepwise-variable optical attenuator.*

contrast. See *refractive-index contrast.*

control center. See *network control center.*

control link. A dedicated *data link* in which the *transmitted* information is used to control the operation of a device, such as a ship or aircraft propulsion system or subsystem. Usually the control link is used to *transmit data* from the device being controlled to a control point, such as a control panel, or to transmit data, usually in the form of a control *signal,* to the device where the signal is converted to an electric current or mechanical actuating device of some type to control the operation of that device and hence the system.

convergent light. *Light* consisting of a bundle of *rays* each of which is *propagating* in such a direction as to be approaching every other ray, that is, they are not parallel to each other and are propagating toward a line, or are heading toward a point of intersection or focus. The *wavefront* of convergent light is somewhat spherical, that is, convex on the *incidence* side. If a *collimated* beam enters a convex lens, the emerging light is convergent. Also see *divergent light.*

conversion. See *mode conversion.*

core. The center region of an *optical waveguide* through which *light* is *transmitted.* In a *dielectric waveguide,* such as an *optical fiber,* the *refractive index* of the *core* must be higher than that of the *cladding.* Most of the *optical power* is confined to the core. Unlike the magnetic material at the center of a relay or coil winding, the core of an optical fiber is a *dielectric,* that is, it is an electrically nonconducting, nonmagnetic, transparent material such as glass. See *fiber core; fiber optic cable core.*

core area. 1. That part of the cross-sectional area of a *dielectric waveguide,* such as an *optical fiber,* within which the *refractive index* is everywhere greater than that of the adjacent, that is, innermost, homogeneous *cladding* by a given fraction of the difference between the maximum refractive index of the core and the refractive index of the homogeneous cladding. Any *refractive-index dip* is excluded. Thus, the core area is the smallest cross-sectional area within which the refractive index is given by the relation:

$$n_3 = n_2 + m(n_1 - n_2)$$

where n_1 is the maximum refractive index of the core, n_2 is the refractive index of the homogeneous cladding adjacent to the core, and m is a fraction, usually not greater than 0.05. **2.** In an *optical fiber,* the cross-sectional area enclosed by the line that connects all points nearest to the *optical axis* on the periphery of the core where the refractive index of the core exceeds that of the homogeneous cladding by k times the difference between the maximum refractive index in the core and the refractive index of the homogeneous cladding, where k is a specified positive or negative constant $|k| < 1$.

core center. In a round *optical fiber,* the center of the circle that best fits the outer limit of the *core area.* The core center may not be the same as *cladding* and *reference surface centers.*

core-cladding concentricity. 1. In a *multimode* round *optical fiber,* the distance between the *core* and *cladding centers,* divided by the *core diameter.* **2.** In a *single-mode optical fiber,* the distance between the *core* and *cladding centers.*

core-cladding power distribution. For a given *mode* of an *electromagnetic wave* propagating in an *optical fiber,* the fraction of the total power in the mode that is within the *core.* The remaining fraction is in the *cladding.* Thus, the fraction of power in the core is given by the relation:

$$P_{core}/P_T = 1 - P_{clad}/P_T$$

where P_{clad} is the power propagating in the cladding, and P_T is the total power in the mode. The power distribution for a given mode can change with distance along the guide, especially for the *higher order modes* wherein the cladding power can be lost to lateral *radiation.*

core diameter. In a round *optical fiber,* the diameter of the circle that best fits the outer limit of the *core area.* The center of this circle is the *core center.*

core eccentricity. See *optical fiber concentricity error.*

core mode filter. A device, such as one turn of an *optical fiber* wrapped around a mandrel, that removes *high-order propagation modes* from the *core* of the optical fiber.

core noncircularity. 1. In a round *optical fiber,* the difference between the diameters of the smallest circle that can circumscribe the *core area* and the largest circle that can be inscribed within the core area, divided by the *core diameter.* **2.** In a round *optical fiber,* the percentage of deviation of the *core* cross section from a circle. Synonymous with *core ovality.* Also see *optical fiber concentricity error.*

core ovality. See *core noncircularity.*

core reference-surface concentricity. 1. In a *multimode* round *optical fiber,* the distance between the *core* and *reference surface centers,* divided by the *core diameter.* **2.** In a *single-mode* round *optical fiber,* the distance between the *core* and *reference surface centers.*

core tolerance area. In the cross section of a round *optical fiber,* the region between the smallest circle that can circumscribe the *core area* and the largest circle that can be inscribed within the core area, both circles concentric with the *core center.*

cosine emission law. See *Lambert's cosine law.*

cosine law. See *Lambert's cosine law.*

counterclockwise-polarized wave. An *elliptically* or *circularly polarized electromagnetic wave* in which the direction of rotation of the *electric vector* is counterclockwise as seen by an observer looking in the direction of *propagation* of the wave. Also see *clockwise-polarized wave.*

coupled mode. In *fiber optics,* a *mode* that shares energy *propagating* in an *electromagnetic wave* with one or more other modes in such a manner that the coupled modes propagate together. The propagating energy is distributed among the coupled modes. The energy distribution among the modes changes with distance along an *optical fiber,* especially before the *equilibrium length,* that is, the *equilibrium modal-power-distribution length* is reached.

coupled power. See *source-to-fiber coupled power.*

coupler. **1.** In *fiber optics,* a device that enables the transfer of *optical* energy from one *optical waveguide* to another. **2.** A *passive optical device* that distributes or combines *optical power* among two or more *ports,* including protective housing and *pigtails.* See *directional coupler; fiber optic coupler; fiber optic multiport coupler; nonuniformly distributive coupler; optical waveguide coupler; optoelectronic directional coupler; reflective star coupler; star coupler; tee coupler; wye coupler; uniformly distributive coupler.* **3.** In electronics, the means by which energy is transferred from one conductor, including a fortuitous conductor, to another. Types of coupling include: (a) capacitive coupling—the linking of one conductor with another by means of capacitance (also known as electrostatic coupling); (b) inductive coupling—the linking of one conductor with another by means of magnetic inductance; (c) conductive coupling—hard-wire connection of one conductor to another.

C-8. A packaged tree **coupler.** (Courtesy Aster Corporation).

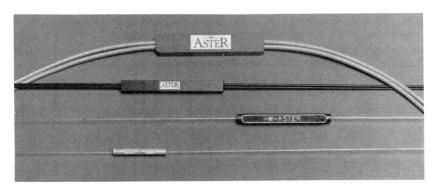

C-9. A few of the various ways *fiber optic* **couplers** are packaged. (Courtesy Aster Corporation).

coupler excess loss. *Fiber optic coupler loss* given by the relation:

$$P_{ex} = \sum_i^1 P_j - \sum_i^1 P_{ij(id)}$$

where P_{ex} is the coupler excess loss, i is the number of input *ports,* P_j is the output power at port j, and $P_{ij(id)}$ are the *transmittances* that would be obtained from an ideal *coupler* in which all the input power is coupled to the nonisolated output ports.

coupler loss. The *insertion loss,* that is, the power loss, that occurs when power is transferred from a given input *port* to a given output port with all other ports properly *terminated.*

coupler transmittance. For a *fiber optic coupler,* the *optional transmittance* measured between an input *port* and an output port of the *coupler* by launching a known *optical power* level into each input port i, one at a time, and measuring the power at each of the j output ports for each of the i input ports. For example, in a 2 × 4 coupler there are six ports. For each of the two input ports, the power at each of the four output ports would be measured to determine the eight coupler transmittances. Also see *coupler transmittance matrix.*

coupler transmittance matrix. For a *fiber optic coupler,* a matrix $L \times L$ dimensions, where L is the total number of *ports,* no distinction being made between input and output ports. For example, in a 2 × 4 coupler L would be 6, there would be six ports, numbered 1 to 6. Each of the *transmittances,* in *dB,* in the matrix is represented by the relation:

$$P_{ij} = 10 \log_{10} (P_i/P_j)$$

where P_{ij} is the transmittance in dB from the input port i to the output port j, P_i is the *optical power* launched into port i, and P_j is the output power at port j. The subscripts of the elements of the transmittance matrix correspond to the numbers labeled on the coupler. The format of the transmittance matrix is defined as:

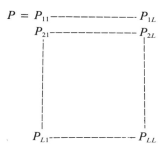

coupling. 1. In *fiber optics,* the transferring of *optical power* or energy from one device to another, for example, the transferring of an *optical signal* from one *optical fiber* to another. **2.** In *fiber optics,* the part of a *fiber optic connector* that aligns the *optical termini.* See *butt coupling; free-space coupling; mode coupling; proximity coupling.*

coupling coefficient. See *coupling efficiency.*

coupling efficiency. In *fiber optics,* the efficiency with which *optical power* is transferred between two components. Coupling efficiency is usually expressed as the percentage of power output from the *transmitting* component that is actually received at the receiving component. If the coupling efficiency is expressed in *dB,* it corresponds to the coupling loss. The maximum light coupling efficiency from a *Lambertian radiator,* that is, a *light source,* such as an *LED* and a *graded-index optical fiber,* is given by the relation:

$$\Gamma = A_c(\text{NA})^2/2A_s n_0^2$$

where A_c is the *core area,* A_s is the source area, n_0 is the *refractive index* of the *propagation medium* between the *source* and fiber, and NA is the *numerical aperture.* Synonymous with *coupling coefficient.* See *source coupling efficiency.*

coupling loss. The power *loss* that occurs when power is transferred from one circuit or *optical waveguide* to another. Coupling loss may be expressed in absolute units, such as watts, or as a ratio, that is, a fraction or percentage of the power in the originating circuit. If expressed as 10 times the logarithm to base 10 of the ratio, that is, in *dB* or *dBm,* it is called *coupling efficiency.* See *intrinsic coupling loss; extrinsic coupling loss.*

covering. See *protective covering.*

crack. See *microcrack.*

criterion. See *Nyquist criterion.*

critical angle. In *optics,* the least *incidence angle* between one *medium* and another at which *total reflection* takes place, that is, total (internal) reflection, wherein the incidence medium is considered as the inside. The critical angle varies with the *refractive indices* of the two media according to the relationship:

$$\sin A_1 = n_2/n_1$$

where A_1 is the critical angle; n_2 the refractive index of the less dense medium, that is, the lower refractive-index material; and n_1 the refractive index of the denser medium, that is, n_1 is greater than n_2, and the incident wave is in the denser medium. Thus, when an *electromagnetic wave* is incident upon an *interface* surface between two *dielectric media,* the critical angle is the angle at which total reflection of the incident ray first occurs as the incidence angle is increased from zero, and beyond which total internal reflection continues to occur. Also, the incident ray must be in the medium with the greater refractive index. For angles less than the critical angle, there will be a *refracted ray* that is bent away from the normal in the less dense medium. If the incidence angle is equal to the critical angle, the *reflection coefficient* becomes unity, all the incident power is reflected, the *transmission coefficient* becomes zero, and none of the incident power is transmitted into the less dense medium. In an *optical fiber,* the wave will be confined to the fiber for all incidence angles greater than the critical angle. Some *bound modes* will remain and *propagate* in the *cladding.* However, the bulk of the *optical power* will be confined to the *core.* In an optical fiber, the incidence angles are measured with respect to the normal to the core-cladding interface surface.

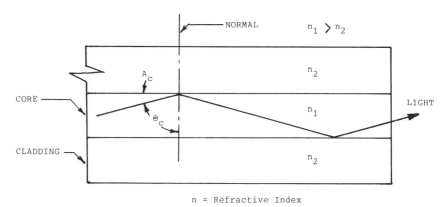

n = Refractive Index

C-10. For a *step-index optical fiber* with *refractive indices* of $n_1 = 1.500$ and $n_2 = 1.485$, the complement, A_c, of the **critical angle,** $\theta_c = \sin^{-1} 1.485/1.500$, is about 84°. *Total (internal) reflection* will occur for all *light rays* making an angle A_c less than 6° with the *core-cladding interface* surface.

critical-angle sensor. A *fiber optic sensor* in which the amount of *light coupled* into an *optical fiber* is a function of the angle at which the light enters the fiber's end face. The light intensity that reaches a *photodetector* at the end of the fiber will depend on the entry angle, permitting the angle between the direction of the *source* beam and the fiber axis to be measured.

critical radius. In an *optical fiber,* the radius of curvature of the fiber at which the field of an *electromagnetic wave propagating* in the fiber detaches itself from the fiber and *radiates* out into space, because the velocity of the portion of the *wavefront* on the outside part of the curve must increase beyond the speed of light in the given *propagation medium* in order to maintain the proper *phase* relationship with the wave inside the fiber. Thus, it no longer can remain *bound* to the inside wave. For example, after the velocity of an *evanescent* wavefront element farthest from the central axis of the fiber begins to exceed the velocity of light as it sweeps around the bend and it can no longer keep up with the part of the wave inside the fiber, the energy begins to radiate out from the fiber, resulting in *attenuation* of the total energy in the propagating wave. Also, if the radius is less than the *critical radius,* the *incidence angles* at the bend become less than the *critical angle,* causing part of the energy in the wave to enter the *cladding.* For radii of curvature, that is, *macrobends* below the critical radius, nearly all of the energy *radiates* from the fiber.

Below the critical radius, *mode conversion* also occurs, because of the many *reflections* at the bend. Also see *critical-radius sensor.*

critical-radius sensor. A *fiber optic sensor* in which an *optical fiber* is bent in such a manner that increased energy in a *lightwave propagating* within it is *radiated* out of the fiber as the bend radius is decreased. The amount of *light* radiated away at the bend increases sharply from 0 to 100% as the radius of the bend is decreased to and passes through the critical value. The loss is due to *mode-conversion loss* in the transition from the straight fiber to the curved fiber and to separation of the *evanescent wave* into space when the outer edge of the wave cannot keep up with the internal wave, that is, the outer edge has reached the speed of light as it swings around the bend. A *photodetector* is used to detect the variation in light intensity at the end of the fiber, which is a function of the bend radius. Thus, the displacement that is causing the bend can be measured after the device is calibrated. Also see *critical radius.*

cross-connection. See *fiber optic cross-connection.*

crosstalk. 1. In *fiber optics,* the undesired *optical power* that is transferred from, that is, leaks from, one *optical waveguide* to another. It is often a measure of the optical power picked up by an *optical fiber* from an adjacent energized fiber. The crosstalk in *dB* is then given by the relation:

$$CT = 10 \log_{10}(P_d/P_e)$$

where P_e is the *optical-power output* at the receiving end of the energized fiber, and P_d is the optical-fiber power output of the unenergized disturbed fiber at the corresponding end. Because P_d is always less than P_e, the crosstalk is always negative. The negative sign is commonly ignored; for example, the crosstalk may be given as simply 90 dB. **2.** The phenomenon in which a *signal transmitted* on one circuit or *channel* of a *transmission system* creates an undesired effect in another circuit or channel. See *fiber crosstalk.*

crystal. See *multirefracting crystal.*

current. See *dark current; noise current; photocurrent; threshold current.*

curvature loss. In *fiber optics,* the *optical power loss* caused by *macrobends* in an *optical fiber.* Power will *radiate* away from the fiber if the radius of the bend is less than the *critical radius.* Synonymous with *macrobend loss.*

curve. See *spectral-loss curve.*

cut. See *runner-cut.*

cutback technique. A technique for measuring certain *optical fiber transmission* characteristics, such as the *attenuation rate, insertion loss, bit-rate-length product, bandwidth,* and *dispersion,* by performing two transmission measurements. One measurement is made at the output of a length of the fiber, and the other is made at the output of a shorter length of the same fiber without changing the *launching conditions,* that is, without changing the entrance-end conditions; or one measurement is made without an inserted component and another measurement is made after a component whose insertion loss is to be measured, is inserted, that is, *spliced* in. Access to the shorter length is obtained simply by cutting off a length of the exit end of the fiber. The two transmission measurements and the measured lengths can be used to calculate the transmission characteristic, such as the attenuation rate or the insertion loss. Several such cutback measurements can be made on the same fiber for the same or different types of measurements, all measurements being made with the same entrance-end launch conditions.

cutoff. See *low-frequency cutoff.*

cutoff frequency. **1.** In an *optical fiber,* the *frequency* below which a specified *propagation mode* fails to exist. As the frequency is decreased, the *wavelength* increases until the wavelength is too long for the dimensions of the fiber to support the wave. **2.** The *frequency* above or below which the output electrical current in a circuit, such as a line or a filter, is reduced to a specified level. **3.** The frequency below which a radio wave fails to penetrate a layer of the ionosphere at the *incidence angle* required for *transmission* between specified points by *reflection* from the layer.

cutoff mode. The *highest order mode* that will *propagate* in a given *waveguide* at a given *frequency.*

cutoff wavelength. In a *waveguide,* the *wavelength* corresponding to the *cutoff frequency,* that is, the *free-space* wavelength above which a given *bound mode* cannot exist in the guide. Thus, it is the wavelength greater than which a particular *mode* ceases to be a bound mode, that is, the mode ceases to *propagate* within the guide. In a *single-mode* waveguide, concern also is with the cutoff wavelength of the second-order mode, in order to ensure single-mode operation. For a short, uncabled, single-mode *optical fiber* with a specified large radius of curvature, that is, *macrobend,* the cutoff wavelength is the wavelength at which the presence of a second-order mode introduces a significant *attenuation* increase when compared with a fiber whose *differential mode attenuation* is not changing at that wavelength. Because the cutoff wavelength of a fiber is dependent upon length, bend, and cabling, the cabled *fiber cutoff wavelength* may be a more useful value for cutoff wavelength from a systems point of view. In general, the cable cutoff wavelength is less than the fiber cutoff wavelength. See *cable cutoff wavelength.*

cutting tool. See *fiber-cutting tool.*

CVD. *Chemical-vapor deposition.*

CVD process. See *chemical-vapor-deposition (CVD) process.*

CVPO. *Chemical-vapor-phase oxidation.*

CVPO process. See *chemical-vapor-phase-oxidation (CVPO) process.*

D

D*. See *specific detectivity.*

damage. See *optical fiber radiation damage.*

dark current. The external current that, under specified biasing conditions, flows in a *photodetector* when there is no *incident radiation.* The dark current usually increases with increased temperature for most photodetectors. For example, in a *photoemissive* photodetector, the dark current is given by the relation:

$$I_d = AT^2 e^{qw/kT}$$

where A is the surface area constant, T is the absolute temperature, q is the electron charge, w is the work function of the photoemissive surface material, and k is *Boltzmann's constant.*

data. **1.** Representation of facts, concepts, or instructions in a formalized manner suitable for communication, interpretation, or processing by humans, or by automatic means. **2.** Any representations, such as characters or analog quantities, to which meaning is or might be assigned. See *analog data; digital data.*

data bus. See *optical data bus.*

data-interface standard. See *fiber distributed-data-interface (FDDI) standard.*

data link. **1.** A communication *link* suitable for *transmitting data.* **2.** A *link* capable of *transmitting analog* or *digital data,* or both, from one point to another, such as a connection by wire, *optical fiber,* radio, microwave, or other *propagation medium,* that allows information to be transferred between two locations. See *dedicated data link; fiber optic data link.*

data-link layer. In *open-systems architecture,* the layer that provides the functions, procedures, and protocol needed to establish, maintain, and release *data-link* connections between user end-instruments and *switching centers* and between two switching centers of a *network.* It is a conceptual level of data processing or control logic existing in the hierarchical structure of the station that is responsible for maintaining control of the data link. Data-link-layer functions provide an interface between the station high-level logic and the data link, including bit injection at the *transmitter* and bit extraction at the receiver; address and control field interpretation; command-response generation, transmission and interpretation; and frame check sequence computation and interpretation. Also see *Open-Systems Interconnection (OSI) Reference Model (RM)*.

data network. See *integrated services digital (or data) network (ISDN)*.

data processing. See *optical data processing*.

data rate. See *data signaling rate*.

data signaling rate (DSR). The aggregate *signaling rate* in the *transmission* path of a *data* transmission system, usually expressed in bits/second. The data signaling rate is given by the relation:

$$DSR = \sum_{i=1}^{m} (1/T_i)\log_2 n_i$$

where m is the number of parallel *channels*, T_i is the minimum time interval for the ith channel expressed in seconds, and n_i is the number of significant conditions of the modulation signal in the ith channel. Each significant condition occurs at a *significant instant*. Thus, for a single channel, the DSR reduces to $(1/T)\log_2 n$. With a two-condition channel, $n=2$, the DSR is $1/T$. For a parallel transmission with equal minimum intervals and equal number of significant conditions on each channel, the DSR is $(m/T)\log_2 n$. Synonymous with *data transmission rate*. Also see *pulse-repetition rate (PRR); range designation of data signaling rates (DSRs)*.

data-transfer network (DTN). A *network* that transports *data* in any form among a group of interconnected stations or users. The stations or users may be *switching centers,* switchboards, devices being controlled, or other systems, subsystems, or user end-instruments, such as telephones, computers, video terminals, and control panels. See *fiber optic data-transfer network*.

data transmission rate. See *data signaling rate (DSR)*.

dB. *Decibel.*

dBm. Abbreviation for dBmW, that is, *dB* referred to 1 milliwatt. The unit is used in communication work as a measure of absolute power values. Zero dBm equals 1 milliwatt.

DBS. *Direct-broadcast satellite.*

DC. *Double crucible.*

DC process. See *double-crucible (DC) process.*

DDS. *Doped-deposited silica.*

dead zone. The region in an *optical waveguide,* such as an *optical fiber,* in which a measurement cannot be made with a given *optical time-domain reflectometer (OTDR).* In an optical fiber, the distance is usually from immediately beyond the *light source,* or the entrance face of the fiber to which the OTDR is connected for the fiber portion of the dead zone, to a point, some distance into the fiber, where the measurement is 3 dB *(decibels)* down from measurements made a much greater distance into the fiber where measurements are their normal values. Dead zones are of the order of 1 m *(meter).*

decibel (dB). A gain or an *attenuation* measured as 10 times the logarithm to the base 10 of a power ratio or 20 times the logarithm to the base 10 of the voltage or *electric field strength* ratio. Attenuation in dB in *optical fibers* is measured as an optical power ratio between two points and therefore is expressed as dB/km.

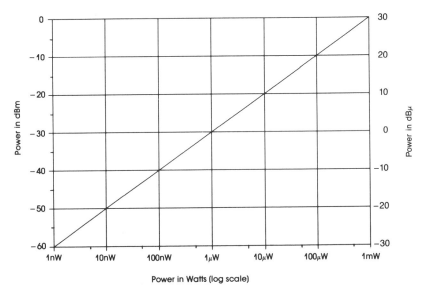

D–1. This power level conversion chart permits quick conversion from dB **(decibel)** referred to milliwatts and microwatts, to and from nanowatt, microwatt, and milliwatt *power levels.* The chart is useful for preparing *optical power budgets* in *fiber optic systems.* (Courtesy fotec incorporated.)

decollimation. In *optics,* that effect wherein a beam of parallel *light rays* is caused to diverge or converge from parallelism. Any of a large number of factors may cause this effect, such as *refractive-index* inhomogeneities, occlusions, *scattering,* deflection, *diffraction, reflection,* and *refraction.*

dedicated data link. A *data link* used for one and only one purpose. For example, a data link from a *sensor* to a display panel or a data link between a control panel and a motor control actuator. No other connections are made to the dedicated data link.

defect. See *vacancy defect*.

degree. See *coherence degree*.

degree of coherence. See *coherence degree*.

delay. See *group delay; group-delay time; multimode group delay; propagation delay*.

delay distortion. The *distortion* of a *wave* form or *signal* made up of two or more different *frequencies,* caused by the difference in arrival time of each frequency at the output of a *transmission system*. In an *optical fiber,* the *spectral width* gives rise to delay distortion because the different *wavelengths* in an *optical pulse* arrive at the end of a fiber at different times. See *waveguide delay distortion*.

delay line. See *optical fiber delay line*.

delta-beta switch. A *fiber optic switch* in which a two-branch Y-splitter lithium niobate *waveguide* is made so that one path is made of titanium lithium niobate and the other of just lithium niobate. An applied voltage can be manipulated to cause one path to shift the *wavelength* of light passing through relative to the other path. The switch thus acts as a fast *wavelength-division multiplexer.* Electrons line up preferentially along the length of the titanium-doped channel, which changes its *refractive index from n_1 to n_2,* the nonlinear refractive index. This shifts the wavelength of *light* down that channel. If a specific voltage is applied to one waveguide, the *polarization* state of light in that waveguide is shifted. The shifted *mode* is coupled instantly with the wave going down the second waveguide, thus shutting off the first channel. Because this switch operates by change of beta, the nonlinear coefficient of the waveguide, it is referred to as a delta-beta switch. In another variety of lithium niobate *electrooptic* behavior, the device can be made into a picosecond-speed on-off switch. Also, by replacing the Y-*splitters* on a common *Mach-Zehnder* intensity modulator with 3-dB couplers, a 2-by-2, high-speed, balanced-bridge switch can be mass produced to perform the same function as the delta-beta switch.

demodulate. The opposite of *modulate,* that is, to recover an intelligence-bearing *signal* from a *wave* that has been modulated by the signal. Also see *modulate*.

demodulation. The process wherein a *signal* resulting from previous *modulation* is processed to derive a signal having substantially the characteristics of the original modulating signal.

demultiplexer.   See *fiber optic demultiplexer (active); fiber optic demultiplexer (passive); optical demultiplexer (active); optical demultiplexer (passive).*

density.   See *electromagnetic-energy density; optical-energy density; optical density; packing density; reflectance density; spectral density; transmittance density.*

departure.   See *phase departure.*

departure angle.   See *launch angle.*

dependent attenuation-rate characteristic.   See *wavelength-dependent attenuation-rate characteristic.*

deposition.   See *inside vapor deposition (IVD); outside vapor deposition (OVD).*

deposition process.   See *plasma-activated chemical-vapor-deposition (PACVD) process; modified chemical-vapor-deposition (MCVD) process; vapor-phase axial-deposition (VAD) process; chemical-vapor-deposition (CVD) process; plasma-activated chemical-vapor-deposition (PACVD) process.*

depressed cladding.   *Cladding* in which the region adjacent to the *core* has a *refractive index* less than that of outer regions.

depressed-cladding fiber.   An *optical fiber,* usually a *single-mode fiber,* that has two *claddings,* the outer cladding having a *refractive index* intermediate between that of the *core* and the inner cladding. Also see *doubly clad fiber.*

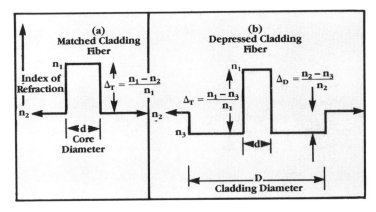

D-2.   *Matched-cladding optical fiber* and **depressed-cladding optical fiber** *refractive index profiles.* Typical *losses* are 0.35 dB/km at 1.31 μ *(microns)* and 0.20 dB/km at 1.55 μ. *Dispersion* is less than 3.2 ps/(nm-km) from 1.285 μ to 1.330 μ. (Courtesy AT&T.)

diagram   57

descriptor. See *receiver-information descriptor; transmitter-information descriptor.*

designation of data signaling rates (DSRs). See *range designation of data signaling rates (DSRs).*

designation of frequency. See *spectrum designation of frequency.*

detection. See *coherent detection; fiber optic illumination detection; homodyne detection.*

detection threshold. See *sensitivity.*

detectivity. The reciprocal of the *noise-equivalent power.* See *specific detectivity.*

detector. 1. A transducer. 2. A *signal*-conversion device that converts power from one form to another, such as from *optical power* to electrical power, from sound power to electrical power, or mechanical power to electrical power, while preserving the information in, or extracting the information from, the input signal. For example, a device that accepts a *modulated carrier* and emits only the modulation signal. See *fiber optic photodetector; internal photoeffect detector; optical detector; optoelectronic detector; phase detector; photodetector (PD); photoelectromagnetic photodetector; photoemissive photodetector; photon detector; photovoltaic photodetector; video optical detector.*

detector-emitter (DETEM). An *optoelectronic* transducer in which the functions of an *optical detector* and an optical *emitter* are combined in a single device or module.

detector signal-to-noise ratio. See *photodetector signal-to-noise ratio.*

detector type. See *optical detector type.*

DETEM. *Detector-emitter.*

detem. An *optoelectronic* device that combines the function of a *photodetector* with that of an *emitter.*

deviation angle. In *optics,* the net angular deflection experienced by a *light ray* after one or more *refractions* or *reflections.* At a single *interface,* that is, at a step-change in *refractive index,* it is the difference between the *incidence angle* and the *refraction angle.* For a lens, prism, or other *optical* element, it is the difference in direction between the *incidence ray* and the emergent ray.

device. See *active device; active optical device; linear device; network-interface device; nonlinear device; optical branching device; passive optical device.*

diagram. See *Brillouin diagram.*

diameter. See *average diameter; beam diameter; borescope working diameter; cladding diameter; core diameter; fiber diameter; mode field diameter.*

diameter tolerance. In a round *optical fiber,* the maximum allowable deviation from the nominal values of the *core, cladding,* or *reference surface diameters.*

dichroic filter. An *optical filter* capable of *transmitting* all *frequencies* in an *electromagnetic wave* above a certain *cutoff frequency* and *reflecting* all lower frequencies, being either a high-pass or low-pass filter, depending on whether the transmitted or reflected wave is used. Thus, the dichroic filter is designed to separate *radiation* into two *spectral bands.*

dichroic mirror. A mirror that *reflects light* selectively according to its *wavelength* and not its *polarization plane.*

dichroism. 1. In *anisotropic propagation media,* the *absorption* of *light rays propagating* in only one particular plane relative to the crystalline axes of the material media. **2.** In *isotropic propagation media, the selective reflection* and *transmission* of *light* as a function of *wavelength* regardless of the direction of the polarization plane. The *color* of such materials, as seen by transmitted light, varies with the thickness of the material examined. Synonymous with *dichromatism; polychromatism.*

dielectric. Pertaining to material composed of atoms whose electrons are so tightly bound to their atomic nuclei that electric currents are negligible even when high *electric field strengths,* that is, near-breakdown voltages, are applied to the material. Thus, in the *constitutive relation* $J = \sigma E,$ where J is the electric current density, σ is the *electrical conductivity,* and E is the electric field strength, $\sigma = 0$ for dielectric materials. Therefore $J = 0$ for these materials.

dielectric constant. See *electric permittivity.*

dielectric waveguide. A *waveguide* consisting of *dielectric* material surrounded by air or other material having a lower *refractive index* than that of the surrounded material. *Electromagnetic waves propagate* through the waveguide similar to the way sound *waves propagate* through a speaking tube. In *fiber optics,* the *cladding* is considered to be a part of the waveguide. In an all-dielectric *fiber optic cable,* strength members, sheaths, *buffers,* and *jackets,* if any, consist only of dielectric materials. Some fiber optic cables may use metals as electrical conductors, strength members, jackets, sheathing, and overarmor. If there are electrical conductors in the cable, it is considered to be an *optoelectrical* cable. See *slab-dielectric waveguide; circular dielectric waveguide.* (See Fig. D-3)

differential mode attenuation (DMA). In an *electromagnetic wave propagating* in a *waveguide,* such as an *optical fiber,* the differences in *attenuation* that occur to

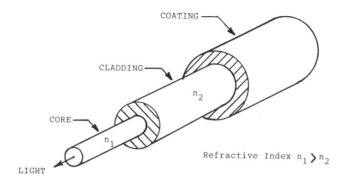

D-3. An *optical fiber,* a typical circular **dielectric waveguide.**

each of the *modes* comprising the wave. For example, *high-order modes* experience a higher attenuation in a given fiber than *low-order modes.*

differential mode delay. See *multimode group delay.*

differential quantum efficiency. In a *quantum device,* the derivative, or slope, of the characteristic graph or equation that defines the countable elementary events at the output as a function of the countable events at the input. The function is the device *transfer function* for the countable events. For example, in a *photodetector,* the slope at a given point in the curve in which the number of electrons generated per unit time for the *photocurrent* is plotted against the number of *incident photons* per unit time.

diffraction. The deviation of a *wavefront* from the *optical path* predicted by *geometric optics* when a wavefront is restricted by an opening, edge, obstruction, or inhomogeneity in a *propagation medium.* Diffraction is usually most noticeable for *apertures* of about a *wavelength.* However, diffraction may still be important for apertures many orders of magnitude larger than the wavelength.

diffraction grating. An array of fine, parallel, equally spaced *reflecting* or *transmitting* ruled lines that mutually enhance the effects of *diffraction* at the edges of each so as to concentrate *light* diffracted from an *incident* beam in a few directions characteristic of the spacing of the lines and the *wavelengths* in the incident beam. Diffraction gratings produce *diffraction patterns* of many orders, each numbered. The order number is given by the relation:

$$N = (d/\lambda)(\sin \theta_i + \sin \theta_d)$$

where N is the order number of the *spectrum,* i is the *incidence angle,* d is the diffraction angle, s is the center-to-center distance between successive rulings, and λ is the wavelength of the incident light. If a large number of narrow, close, equally spaced rulings are made on a *transparent* or reflecting substrate, the grating will be capable of dispersing incident light into its wavelength component spectrum like the dispersion of white light into its constituent *colors* caused by a prism.

diffraction-limited. **1.** In ordinary *optics,* pertaining to a *light* beam in which the divergence is equal to that predicted by diffraction theory. **2.** In focusing *optics,* pertaining to a light beam in which the *resolving power,* that is, the resolution limit, is equal to that predicted by diffraction theory.

diffraction pattern. See *Fraunhofer diffraction pattern; Fresnel diffraction pattern; far-field diffraction pattern; near-field diffraction pattern.*

diffused optical waveguide. An *optical waveguide* consisting of a substrate into which one or more *dopants* have been diffused to a depth of a few *microns,* thus producing a lower *refractive-index* material on the outside, as in an *optical fiber.* The result is an *optical waveguide* with a *graded refractive-index profile.* Diffused optical waveguides can be made using single crystals of zinc selenide or cadmium sulfide. Dopants include cadmium for the zinc selenide and zinc for the cadmium sulfide.

diffused waveguide. See *planar diffused waveguide.*

diffuse relection. *Reflection* of an *incident,* perhaps *coherent,* group of *light rays* from a rough surface, such that different parts of the same group are reflected in different directions, as *Snell's laws* of reflection and *refraction* are microscopically obeyed at each undulation of the rough surface. Diffuse reflection prevents formation of a clear image of the *light source* or of the illuminated object.

digital data. **1.** *Data* represented by discrete values or conditions, as opposed to *analog data.* **2.** A discrete representation of a quantized value of a variable, for example, the representation of a number by numerals or the representation of a letter of the alphabet by a pattern of *pulses.* Also see *analog data.*

digital network. See *integrated-services digital (or data) network (ISDN).*

digital signal. A nominally discontinuous *signal* that changes from one state to another in discrete steps. For example, the *electromagnetic wave pulses, phase shifts,* or *frequency shifts* that represent 0s or 1s; or electrical output pulses from a *photodetector* caused by light input pulses. Also see *analog signal.*

diode. See *laser diode; light-emitting diode (LED); PIN diode; PIN photodiode; restricted edge-emitting diode (REED); stripe laser diode.*

diode laser. See *laser diode.* (See Fig. D–4)

D-4. The PLS10 and PSL20 series *pulsed* **diode lasers** *emit* ultra-short high-power pulses at *pulse repetition rates* up to 30kHz. The *laser* is driven by a LCU10 or LCU20 Laser Control Unit (shown). (Courtesy Opto-Electronics Inc.).

diode photodetector. See *photodiode.*

dip. See *refractive-index dip.*

direct-broadcast satellite (DBS). Pertaining to the broadcasting of television programming via satellite to terrestrial receiving antennas. The programming may be rebroadcast on earth to conventional receiving antennas at homesites, received directly from satellites at homesite's receiving antennas, or distributed locally via co-axial cable (CATV). *Fiber optic cables* could handle the television programming.

directional coupler. A *coupler* in which *optical power* applied to certain input *ports* is transferred to only one or more specified output ports. See *optoelectronic directional coupler.*

direct ray. A *ray* of *electromagnetic radiation* that follows the *path* of least possible *propagation* time between *transmitting* and receiving antennas. The least time path is not always the shortest distance path. For example, in a *graded-index optical fiber,* a *light* ray propagating in the lower *refractive-index* regions propagates faster than light rays propagating in higher index regions, but the higher speed ray might travel a longer distance, particularly if it is a *skew ray* and takes a helical path around the fiber *optical axis.*

discontinuity. A sudden or step-change of a value, such as an interruption, discontinuation, or dropout of an *optical signal.* See *optical impedance discontinuity.*

disk. See *optical disk; optical video disk (OVD).*

dispersion. 1. The phenomenon in which *electromagnetic wave propagation* parameters are dependent upon *frequency* or *wavelength.* For example, an electromagnetic *signal* may be distorted because the various frequency components of the signal have different *propagation* characteristics. The variation of *refractive index* with frequency (or wavelength) also causes dispersion. In *optical fibers,* the total dispersion is the combined result of *material dispersion, waveguide dispersion,* and *profile dispersion.* Dispersion will cause *distortion* of a *transmitted signal* because of the different times at which different wavelengths that comprise a *pulse* arrive at the end of a *waveguide.* **2.** *Pulse broadening,* that is, the spreading of a *signal* in time and space as a result of transversing *transmission* paths of different physical, electrical, or *optical path lengths.* **3.** The process by which *electromagnetic waves* of different *wavelength* are deviated angularly by different amounts because the *refractive index* and other *optical* properties vary with wavelength, as occurs in a prism. For example, if a *lightwave pulse* of given *pulse duration* and given *spectral line width* is launched into an *optical fiber,* the different wavelengths experience different refractive indices and therefore take different *paths* as well as *propagate* at different speeds, causing pulse broadening as it propagates along the fiber because the different wavelengths that make up the spectral line width arrive at the far end at different times *(material dispersion).* At the same time, energy in the different wavelengths distributes among different *modes,* also causing pulse broadening *(modal dispersion).* Dispersion imposes a limit on the *bit-rate-length product* for *digital data* and *bandwidth* for *analog data transmitted* in an optical fiber for a given *bit-error ratio (BER)* at the limit of *intersymbol interference.* **4.** In communication systems, the allocation of circuits between two points over more than one geographical or physical route. See *chromatic dispersion; fiber dispersion; intermodal dispersion; material dispersion; maximum transceiver dispersion; modal dispersion; optical multimode dispersion; profile dispersion; waveguide dispersion; zero dispersion.*

dispersion attenuation. See *optical dispersion attenuation.*

dispersion characteristic. See *fiber global-dispersion characteristic; global-dispersion characteristic.*

dispersion coefficient. For a *lightwave propagating* in an *optical fiber,* a value that expresses the *dispersion,* usually in *picoseconds* per kilometer of fiber for each *nanometer* of *spectral width* of the lightwave launched into the fiber. The unit for the dispersion coefficient usually is ps/nm-km. See *chromatic dispersion coefficient. global-dispersion coefficient; worst-case dispersion coefficient.*

dispersion equation. An equation that indicates the dependence of the *refractive index* of a *propagation medium* on the *wavelength* of the *light transmitted* by the

medium. The adjustment of the index for wavelength permits more accurate calcula-
tion of *reflection* and *refraction angles, phase shifts,* and *propagation paths* that
are dependent upon the *refractive index* and the refractive-index gradient. Often it
is necessary to obtain a value of the rate of change of refractive index with respect
to the wavelength. There are several useful dispersion equations. One of these relations is:

$$n = n_0 + C/(\lambda - \lambda_0)$$

which is attributed to Hartmann. Another relation is:

$$n = A + B/\lambda^2 + C/\lambda^4$$

which is attributed to Cauchy. In these relations, λ is the wavelength, n is the refractive index, and the other symbols are material-dependent and empirically determined.

dispersion-limited length. The sheathed length of *fiber optic cable,* such as might
be in the cable facility of an *optical station/regenerator section,* whose length is
limited by *dispersion.* The dispersion arises from the combined effects of *chromatic
dispersion* and *modal noise.* The dispersion-limited length is given by the relation:

$$\ell_D = D_{TR}/D_{\max}$$

where D_{TR} is the *maximum transceiver dispersion* (ps/nm), and $D_{\max}$ is the absolute
value of the *worst-case chromatic dispersion coefficient* (in ps/(nm-km)) over the
range of *transmitter wavelength* and over the specified variation in cable dispersion.
A regenerator section will not be limited by dispersion as long as the relation:

$$D(\lambda_t) \cdot \ell < D_{TR}$$

holds, where ℓ is the fiber length in km, $D(\lambda_t)$ is the fiber *chromatic dispersion
coefficient* (in ps/(nm-km)), at wavelength λ_t and D_{TR} is the maximum transceiver
dispersion (in ps/nm). Generally for *single-mode* systems operating at bit rates *(data
signaling rates)* under 0.5 Gb/s, the length of the *optical cable facility* of a regenerator section can be expected to be limited by *attenuation (loss)* rather than by dispersion. At higher bit rates the *optical link* length can be expected to be limited by
dispersion. In the design of an optical station/regenerator section, the effect of
dispersion is accounted for in the equation for the *terminal/regenerator system gain,*
G, via the *dispersion power penalty,* $P_{D'}$ in dB. Also see *terminal/regenerator system
gain.*

dispersion parameter. See *material-dispersion parameter; profile-dispersion parameter.*

dispersion power penalty. In an *optical transmitter,* the worst-case increase in receiver input *optical power (dB)* needed to compensate for the total *pulse distortion*
due to *intersymbol interference* and *mode* partition *noise* at the specific bit rate

(data signaling rate), bit-error ratio (BER), and *maximum transceiver dispersion* specified by the manufacturer when operated under *standard* or *extended operating conditions,* as measured under standard *fiber optic test procedures.*

dispersion-shifted fiber. 1. A *single-mode fiber* that has a nominal *zero-dispersion wavelength* near 1.55 μm. Also see *dispersion-unshifted fiber.* **2.** An *optical fiber* whose *refractive-index profile* is designed in such a manner that because of the variation in *refractive index* with various *wavelengths* in the *spectral width* of the *pulse* and because of the various lengths of the *paths* taken by the various *propagation modes,* at certain median *wavelength* and particular refractive-index profile, all modes can be made to have the same end-to-end propagation time, thus reducing *signal pulse distortion* caused by *material dispersion* to an absolute minimum, actually nearly zero. A fiber can be designed such that the *zero-material-dispersion wavelength* is also at the minimum *attenuation* wavelength, that is, at a *spectral window,* a trough in the *attenuation rate* versus wavelength curve.

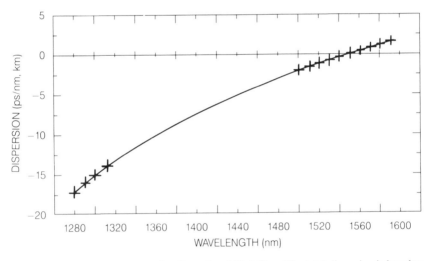

D–5. A typical *dispersion* curve for **dispersion-shifted fiber.** The total dispersion is less than 3.5 ps/(nm – km) over the *wavelength* range 1520–1580 nm (1.52–1.58 μ *(microns).* The fiber is used in *fiber optic systems* that operate in the 1.55 μ *window* and that require very high *data signaling rates (DSRs),* that is, bit rates, over very long distances, such as *long-haul* land and undersea systems. (Courtesy AT&T.)

dispersion slope. See *chromatic dispersion slope; zero-dispersion slope.*

dispersion-unshifted fiber. A *single-mode fiber* that has a nominal *zero-dispersion wavelength* near 1.3 μm. Also see *dispersion-shifted fiber.*

dispersion wavelength. See *zero-dispersion wavelength.*

dispersive medium. In the *propagation* of *electromagnetic waves* in material media, a *medium* in which the *phase velocity* varies with *frequency*. Thus, if an *optical pulse* consisting of more than a single *wavelength* is passed through a dispersive medium, the different wavelengths will arrive at the end at different times. Thus, the phase velocity and the *group velocity* are not the same thing. All materials are dispersive to some extent. Also see *nondispersive medium*.

displacement sensor. See *fiber axial-displacement sensor*.

display device. See *fiber optic display device*.

distance resolution. In an *optical time-domain reflectometer (ODTR),* a rating based on the shortest distance along the length of an *optical waveguide,* such as an *optical fiber,* that the OTDR can distinguish on its display screen, usually a cathode ray tube. It is a measure of how close together two events, or faults, can be distinguished as separate events. Synonymous with *spatial resolution*. Also see *dynamic range; measurement range*.

distortion. Any departure of an output *signal* shape from an input signal shape over a range of *frequencies,* amplitudes, or *phase shifts* during a given time interval. In a signal-*transmission system,* it is the undesirable amount that an output wave shape or *pulse* differs from the input form. It may be expressed as the undesirable difference in amplitude, frequency composition, *phase,* shape, or other attribute between an input signal and its corresponding output signal. See *delay distortion; intermodal distortion; intramodal distortion; multimode distortion; optical distortion; optical pulse distortion; pulse-width distortion; total harmonic distortion; waveguide delay distortion*.

distortion-limited operation. The condition that prevails in a device or system when any form of *distortion* of a *signal* limits its performance or the performance of the device or system to which it is connected. For example, the condition that prevails in an *optical link* when a combination of *attenuation, dispersion,* and *delay distortion* all contribute toward limiting *link* performance.

distributed-data-interface standard. See *fiber distributed-data-interface (FDDI) standard*.

distributed-fiber sensor. A *fiber optic sensor* that uses a spatial distribution of *optical fiber* to sense a parameter, such as a planar distribution, a bobbin wound with spatial shading, a series of windings in line, two or more parallel coils, or an array of wound bobbins. Distributed sensors can sense many parameters, such as a pressure point at a cartesian coordinate, a formed beam direction, a pressure gradient, and an *electric field* gradient using *optical time-domain reflectometry (OTDR), interferometry,* and other *optical* techniques to *modulate irradiance,* that is, the *power density* or *intensity* of *light, incident* upon a *photodetector*.

distributed star-coupled bus. See *star-mesh network.*

distributed thin-film waveguide. See *periodically distributed thin-film waveguide.*

distribution. See *core-cladding power distribution; modal distribution; nonequilibrium modal power distribution; Rayleigh distribution.*

distribution frame. See *fiber optic distribution frame.*

distribution length. See *nonequilibrium modal power distribution length.*

distributive coupler. See *nonuniformly distributive coupler; uniformly distributive coupler.*

divergence. see *beam divergence.*

divergent light. *Light* consisting of a bundle of *rays* each of which is *propagating* in such a direction as to be departing from every other ray, that is, they are not parallel to each other, but are spreading and hence are not *collimated* or convergent, and their *wavefront* is somewhat spherical or concave on the incidence side. If a collimated beam enters a concave lens, the emerging light is divergent. Also see *convergent light.*

diversity. See *polarization diversity.*

DMA. *Differential mode attenuation.*

domain reflectometry. See *optical time-domain reflectometry.*

dopant. A material mixed, fused, amalgamated, crystalized, or otherwise added to another in order to achieve desired characteristics of the resulting material. The germanium tetrachloride or titanium tetrachloride added to pure glass to control the *refractive index* of glass for *optical fibers* or the gallium or arsenic added to germanium or silicon to produce *p-type* or *n-type semiconducting* material for making diodes and transistors are examples of dopants.

doped-deposited-silica (DDS) process. A process for making *optical fibers* in which *dopants* are deposited in inside or outside walls of silica glass tubes or cylinders and fused to produce desired *refractive-index profiles.*

dosimeter. See *fiber optic dosimeter.*

double-crucible (DC) process. In the production of *optical fibers,* a *fiber-drawing process* in which two concentric crucibles are used, one for the *core* glass and one for the *cladding* glass. The cladded fiber is pulled out of the bottom at the convergence or apex of the two crucibles. Diffusion of the *dopant* materials in the glasses produces a *graded refractive-index profile.*

double refraction. That property of a material in which a single *incident light* beam is *refracted* into two *transmitted* beams as though the material had two distinct *refractive indices.*

doubly clad fiber. An *optical fiber,* usually a *single-mode fiber,* that has a *core* covered by an inner low-*refractive-index* material, which, in turn, is covered by a second cladding of a higher refractive-index material, but not higher than that of the core. This is in contrast to an optical fiber having only one lower refractive-index cladding covering the higher (*step* or *graded*) *refractive-index* core. In the doubly clad fiber, inner cladding modes are stripped off because the outer cladding refractive index is higher than the inner cladding refractive index, so that when *rays* are *incident* upon the intercladding interface surface, they are bent toward the normal to the intercladding interface surface and hence escape. Also see *depressed-cladding fiber.*

drawing. See *fiber drawing.*

drawn glass fiber. A glass fiber formed by pulling a strand from a heat-softened *preform* through a die or from a mandrel with a *pulling machine.*

dry glass. Glass from which as much water as possible has been driven out, for example, glass with a water concentration less than 1 ppm. Drying is done primarily to reduce *hydroxyl-ion-absorption loss.* Drying can be accomplished by many days of drying of starting powders in a 205°C vacuum oven, followed by many hours of dry-gas melt bubbling to dry out water to extremely low levels.

DSR. *Data signaling rate.*

D-star. See *specific detectivity.*

DTN. *Data-transfer network.*

duration. See *pulse duration; pulse half-duration; root-mean-square (rms) pulse duration.*

dynamic range. In an *optical time-domain reflectometer (OTDR),* a measure of how far into an *optical waveguide,* such as an *optical fiber,* a measurement can be made. Thus, the dynamic range may be a measure of the *optical power* rating of an OTDR based on the inherent *loss* in *dB*/km of the waveguide, the *pulse width* of the *light source,* the operational *wavelength,* the length of waveguide being tested, the number of times the source *pulse* is launched, and whether the rating is based on a one-way or a round trip of the pulse; or the dynamic range may be the loss through the waveguide that can be measured based on the guide's inherent *attentuation rate,* e.g., if a *multimode optical fiber* has an attenuation rate of 0.8 dB/km and the goal is to measure a 40-km piece of fiber, or it is 40 km to a break, the OTDR has to have a dynamic range of at least 32 dB in order to measure the length of fiber, or the distance to the break. Also see *distance resolution; measurement range.*

drop. See *fiber optic drop.*

E

E-bend. A gradual change in the direction of the *optical axis* of a *waveguide* throughout which the axis remains in a plane parallel to the direction of the *electric field vector (E) transverse polarization*. Synonymous with *E-plane bend*. Also see *H-bend*.

edge absorption. See *band-edge absorption*.

edge-emitting diode. See *restricted edge-emitting diode (REED)*.

edge-emitting LED (ELED). A *light-emitting diode*, with a *spectral* output emanating from between the heterogeneous layers. Thus, the light is *emitted* parallel to the plane of the junction. It has a higher output *radiance*, greater *coupling efficiency* to an *optical fiber* or *integrated optical circuit* than the *surface-emitting LED*, but not as great as the *injection laser diode*. Edge-emitting and surface-emitting LEDs provide several milliwatts of *optical power* output in the 0.8- 1.2-μ *(micron) wavelength* range at drive currents of 100 to 200 mA. Diode lasers at these currents provide 10s of milliwatts.

EDTV system. *Enhanced-definition television system.*

effect. See *acoustooptic effect; electrooptic effect; internal photoelectric effect; Kerr effect; knife-edge effect; magnetooptic effect; photoconductive effect; photoelastic effect; photoelectric effect; photoelectromagnetic effect; photoemissive effect; photovoltaic effect; Pockels effect; speckle effect.*

effective mode volume. In a *waveguide*, the square of the product of the diameter of the *near-field radiation pattern* and the sine of the *radiation angle* of the *far-field radiation patterns*. The diameter of the near-field radiation pattern is the full-width-half-maximum diameter and the radiation angle at half maximum *intensity*. In a *multimode optical fiber*, the effective *mode* volume is proportional to the breadth of the relative distribution of power among the modes *propagating* in the fiber. The effective mode volume is not an actual spatial volume, but rather an optical volume having the units of the product of area and solid angle.

efficiency. See *coupling efficiency; differential quantum efficiency; optical power efficiency; quantum efficiency; radiant efficiency; radiation efficiency; response quantum efficiency; source coupling efficiency; source power efficiency.*

electrical cable. See *optoelectrical cable.*

electrical conductivity. A measure of the ability of a material to conduct an electric current when under the influence of an applied *electric field*. The conductivity is the constant of proportionality in the *constitutive relation* between the electrical current density and the applied *electric field strength* at a point in a material. It is expressed mathematically as $J = \sigma E$, where J is the current density, σ is the *electrical conductivity,* and E is the electric field strength. For example, if J is in amperes per square meter and E is in volts per meter, the conductivity is given as $\sigma = J/E$ amperes/volt-meter, (ohm-meter)$^{-1}$, or mhos/meter.

electrical length. The length of a sinusoidal *wave* expressed in *wavelengths,* radians, or degrees. When expressed in angular units, it is the length in wavelengths multiplied by 2π to give radians, or by 360 to give degrees.

electric field. The effect produced by the existence of an electric charge, such as an electron, ion, or proton, in its surrounding volume of space or material medium. Each of a distribution of charges contributes to the whole field at a point on the basis of superposition. A charge placed in an electric field has a force exerted upon it by the field. The magnitude of the force is the product of the *electric field strength* and the magnitude of the charge. Thus, $\mathbf{E} = \mathbf{F}/q$, where $\mathbf{E}$ is the electric field strength, $\mathbf{F}$ is the force the field exerts on the charge, and q is the charge. The direction of the force is in the same direction as the field for positive charges, and in the opposite direction for negative charges. Moving an electric charge against the field requires work, that is, energy, and increases the potential energy of the charge. Electric field strength is measurable as force per unit charge or as potential difference per unit distance, that is, a voltage gradient, such as volts/meter. It is also expressible as a number of electric lines of flux per unit of cross-sectional area. The force and the electric field are vector quantities. The charge is a scalar quantity.

electric field intensity. See *electric field strength.*

electric field strength. The intensity, that is, the amplitude or magnitude, of an *electric field* at a given point. The term is normally used to refer to the rms value of the electric field, expressed in volts per meter. An instantaneous electric field strength is usually given as a gradient, and therefore is a vector quantity, because it has both magnitude and direction. For example, the units of electric field strength may be volts per meter, kilograms per coulomb, or electric lines of flux per square meter divided by the *electric permittivity* of the medium at the point it is measured. Synonymous with *electric field intensity.* Also see *magnetic field strength.*

electric field vector. In an *electromagnetic wave,* such as a *lightwave propagating* in a *propagation medium,* the vector that represents the instantaneous amplitude and direction of the *electric field strength* at a given point in the medium in which the wave is propagating.

electric permittivity. A parameter of *free space* or a material that serves as the constant of proportionality between the magnitude of the force exerted between two point electric charges of known magnitude separated by a given distance. The magnitude of the force is defined by the relation:

$$F = q_1 q_2 / 4\epsilon \pi d^2$$

where q_1 and q_2 are the point electric charges, F is the force between them, d is their distance of separation, ϵ is the electric permittivity of the medium in which they are embedded, and π is approximately 3.1416. The direction of the force is along the line joining the two point charges, being one of attraction if the charges are of opposite polarity and repulsion if they are of the same polarity. The electric permittivity, *magnetic permeability,* and *electrical conductivity* determine the *refractive index* of a material. However, for *dielectric* materials, that is, electrically nonconducting media, such as the glass and plastic used to make *optical fibers,* the electrical conductivity is zero, and the relative velocities of *electromagnetic waves,* such as *lightwaves, propagating* within them, are given by the relation:

$$v_1 / v_2 = [(\mu_2 \epsilon_2)/(\mu_1 \epsilon_1)]^{1/2}$$

where v is the velocity, and μ and ϵ are the magnetic permeability and electric permittivity, respectively, of media 1 and 2. If propagation medium 1 is free space, the above equation becomes the relation:

$$c/v = [(\mu \epsilon)/\mu_0 \epsilon_0)]^{1/2}$$

where c is the velocity of light in a vacuum, v is the velocity in a material medium, the nonsubscripted values of the constituent constants are for the material medium and the subscripted values for free space. However, c/v is the definition of the refractive index of a medium relative to free space. From this is obtained the relation:

$$n = (\mu_r \epsilon_r)^{1/2}$$

where n is now a dimensionless number and relative to free space. The magnetic permeability and electric permittivity are also relative to free space, or air. As stated above, for dielectric materials, μ_r is approximately 1, being very much greater than 1 only for ferrous substances and slightly greater than 1 for a few metals, such as nickel and cobalt. Hence, for the glass and plastics used to make optical fibers and other *dielectric waveguides,* the refractive index, relative to a vacuum or air is given by the relation:

$$n \approx \epsilon_r^{1/2}$$

where ϵ_r is the electric permittivity of the propagation medium. Unless otherwise stated, refractive indices are usually given relative to a vacuum, or relative to 1.0003 for air near the earth's surface and less as the air gets thinner in the upper atmosphere and outer space. In absolute units for free space, the electric permittivity is $\epsilon = 8.854 \times 10^{-12}$F/m (farads per meter), and the magnetic permeability is $\mu = 4\pi \times 10^{-7}$ H/m (henries per meter). The reciprocal of the square root of the product of these values yields approximately 2.998×10^8 m/s (meters per second), the velocity of light in a vacuum. Because the refractive index for a propagation medium is defined as the ratio between the velocity of light in a vacuum and the velocity of light in the medium, the ratio becomes unity when the propagation medium is a vacuum. See *relative electric permittivity*. Also see *refractive index*.

electroluminescence. The direct conversion of electrical energy into *light,* for example, *photon emission* caused by electron-hole recombination in a pn junction of a *light-emitting diode (LED)*. The conversion process does not produce incandesence in the emitting material. The emitted *optical radiation* is in excess of the radiation caused by thermal emission that would result from the application of electrical energy.

electromagnetic effect. See *photoelectromagnetic effect*.

electromagnetic-energy density. The amount of energy contained per unit volume of *free space* or *propagation medium* at a point at which an *electromagnetic wave* is propagating. The density is scalar and is a function of the *electric* and *magnetic field strengths* at the point. When the propagation velocity is taken into account, the energy flow rate can be considered as the energy per unit volume times velocity, that is, distance per unit time, thus becoming the vector quantity of energy crossing per unit cross-sectional area per second, that is, the *power density,* more properly called the *irradiance,* usually expressed as watts per square meter, or joules/second-meter^2.

electromagnetic field. A field characterized by both *electric* and *magnetic field vectors* that interact with one another rather than exist independently of each other, as in electrostatic, staticmagnetic, and electromagnetostatic fields. The electromagnetic field varies with time at a point and propagates as a wave from its *source,* whereas the others, although they may be made to vary, do not exist very far from their sources and decay rapidly with distance.

electromagnetic photodetector. See *photoelectromagnetic photodetector*.

electromagnetic radiation (EMR). *Radiation* made up of oscillating *electric* and *magnetic fields* that *propagate* with the speed of *light*. It includes *gamma radiation; x-rays; ultraviolet, visible,* and *infrared radiation;* and radar and radio waves. The radiation is propagated with a *phase velocity* given by the relation:

$$v = \lambda f = c/n$$

where λ is the wavelength, f is the frequency, c is the velocity of light in a vacuum (approximately 3×10^8 m/s), and n is the *refractive index* of the *propagation medium*.

electromagnetic spectrum. 1. The *frequencies,* or *wavelengths,* present in given *electromagnetic radiation.* A particular spectrum could include a single frequency or a wide range of frequencies. **2.** The entire range of *electromagnetic wavelengths* that can be physically generated, extending from nearly infinite length to the shortest possible wavelength. It includes the *optical* portion of the *spectrum,* which includes the *visible* and *near-visible* portion of the spectrum, that is, lightwaves. The optical portion extends from the shorter-wavelength far-ultraviolet to the longer-wavelength *far-infrared,* while the visible portion extends from violet to red. Also see *optical spectrum.*

electromagnetic wave. The effect obtained when a time-varying *electric field* and a time-varying *magnetic field* interact, causing electrical and magnetic energy to be *propagated* in a direction that is dependent upon the spatial relationship of the two interacting fields that are interchanging their energies as the wave propagates. The two fields define the *polarization plane* as well as the *wavefront.* The cross-product of the two fields, with the *electric field vector* rotated into the *magnetic field vector,* defines a vector, called the *Poynting vector,* that indicates the direction of propagation and defines a *ray,* which is perpendicular to the wavefront. See *left-hand-polarized electromagnetic wave; plane (electromagnetic) wave; plane-polarized electromagnetic wave; right-hand-polarized electromagnetic wave; trapped electromagnetic wave; uniform plane-polarized electromagnetic wave.*

electron energy. See *emitted-electron energy.*

electronic. See *optoelectronic.*

electronic directional coupler. See *optoelectronic directional coupler.*

electronics. See *optoelectronics.*

electrooptic. Pertaining to the effect that an *electric field* has on a *lightwave,* such as rotating the direction of the *polarization plane* as the lightwave passes through the field, or, if the wave is *propagating* in material media, the effect the electric field has on the *refractive index* of the material. Electrooptic is often erroneously used as a synonym for *optoelectronic.* Also see *optoelectronic.*

electrooptic effect. 1. The change in the *refractive index* of a material when subjected to an *electric field.* The effect can be used to *modulate* a *lightwave propagating* in a material because many lightwave propagation properties, such as propagation velocities, *reflection* and *transmission coefficients* at *interfaces, acceptance angles, critical angles,* and *transmission modes,* are dependent upon the refractive indices of the media in which the lightwave is propagating. **2.** The effect an *electric field* has on the *polarization* of a *lightwave,* such as rotating the *polarization plane.* *Pockels* and *Kerr effects* are *electrooptic* effects that are linear and quadratic, respectively, in relation to the applied electric field. In any case, the applied *electric field strength* adds to the electric field strength of the *electromagnetic wave,* causing rotation as it propagates. Also see *magnetooptic effect; optoelectronic.*

electrooptics. The branch of science and technology devoted to the interaction between *optics* and electronics leading to the transformation of electrical energy into *light,* or vice versa, with the use of an optical device, such as devices whose operation relies on modification of a material's *refractive index* by *electric fields,* particularly the generation and control of *lightwaves* by electronic means and vice versa. For example, in a *Kerr cell,* the *refractive index* change is proportional to the square of the *electric field strength.* The material is usually a liquid. In a *Pockels cell,* a crystal is used whose refractive index change is a linear function of the electric field strength.

electrooptic transmitter. An *optoelectronic transmitter* in which the *electrooptic effect* is used in its operation.

ELED. See *edge-emitting LED (ELED).*

elliptical polarization. In an *electromagnetic wave, polarization* such that the tip, that is, the extremity, of the *electric field vector* describes an ellipse in any fixed plane intersecting the wave and normal to the direction of *propagation.* An elliptically polarized wave may be resolved into two *linearly polarized waves* in *phase* quadrature with their *polarization planes* at right angles to each other.

emission. *Electromagnetic* energy *propagating* from a *source.* The energy thus propagated may be either desired or undesired and may occur at a *frequency,* or *wavelength,* anywhere in the *electromagnetic spectrum.* See *spontaneous emission; spurious emission; stimulated emission.*

emission of radiation. See *light amplification by stimulated emission of radiation (LASER).*

emission wavelength. See *peak emission wavelength.*

emissivity. **1.** The ratio of power *radiated* by a *source* to the power radiated by a blackbody at the same temperature. Power and area units for both bodies must be the same or normalized. **2.** The ratio of the *radiant emittance* of, or radiated flux from, a *source* to the radiant emittance of, or radiated flux from, a blackbody having the same temperature. Emissivity depends on *wavelength* and temperature.

emittance. See *radiant emittance; spectral emittance.*

emitted-electron energy. In a *photoemissive detector,* the remaining energy of an electron that escapes from the emissive material due to the energy imparted to it by an *incident photon,* given by the relation:

$$E_e = hf - qw$$

where E_e is the remaining energy of the electron, h is *Planck's constant,* f is the *photon frequency,* q is the charge of an electron and w is the *work function* of the emissive material. Actually, *hf* is the photon energy and *qw* is the energy required for an electron to escape from the emissive material, that is, to overcome the boundary effects.

emitter. See *detector-emitter (DETEM); optical emitter.*

emitting diode. See *light-emitting diode (LED).*

EMR. See *electromagnetic radiation (EMR).*

end finish. At the end of an *optical fiber,* the condition of the end surface, that is, the surface perpendicular to the *optical axis.*

end-instrument. See *user end-instrument.*

endoscope. See *endoscopic device; fiber optic endoscope.*

endoscopic device (endoscope). A device used to observe specific surfaces, particularly areas that are relatively inaccessible to photographic and television cameras, such as interior parts of functioning equipment and internal organs of the human body. The device usually consists of a lens that focuses an image on a *faceplate* consisting of the originating end faces of an *aligned bundle* of *optical fibers;* the aligned bundle itself that transmits the image to another faceplate at the other end of the bundle, made from the originating end face of the bundle; and perhaps another lens for displaying the image. Arrangements must be made for the object, whose image is to be *transmitted,* to be illuminated. This can be accomplished with additional optical fibers placed in the same *fiber optic cable* with the aligned bundle and energized by a *light source* to illuminate the object.

end-point node. 1. In *network* topology, a *node* connected to one and only one *branch.* 2. A *node* at the end of a *path.*

end-to-end separation sensor. See *fiber end-to-end separation sensor.*

energy. See *band-gap energy; emitted-electron energy; photon energy; radiant energy.*

energy band. A specified range of energy levels that a constituent particle or component of a substance may have. The particles may be electrons, *protons,* ions, neutrons, atoms, molecules, mesons, or others. Some energy bands are allowable and others are unallowable. For example, electrons of a given element at a specific temperature can occupy only certain energy bands, such as the conduction and valence bands. When radiation strikes an electron and imparts enough energy to the electron to cause its energy to rise to a higher energy band, energy is absorbed. If the electron energy level is reduced to that of a lower energy band, energy is emitted in the form of *electromagnetic radiation,* namely a *photon* is emitted. Statistical laws apply to the number of electrons in each of the possible bands, the density of these electrons in the material, and the energies of the emitted photons.

energy density. See *optical energy density; electromagnetic energy density.*

enhanced-definition television (EDTV) system. A television system in which improvements are made that are not compatible with present receivers. Some features and technical characteristics of the original system on which the EDTV system is based may be included in the enhanced system. See *advanced television system; high-definition television (HDTV) system; improved-definition television (IDTV) system.*

E-plane bend. See *E-bend.*

equation. See *dispersion equation; photoelectric equation; Sellmeier equation; wave equation.*

equations. See *Fresnel equations; Helmholtz equations; Maxwell's equations.*

equilibrium coupling length. See *equilibrium length.*

equilibrium distance. See *equilibrium length.*

equilibrium length. The distance within a *multitude waveguide,* measured from the input end, necessary to attain *electromagnetic wave equilibrium modal-power distribution* for a specific excitation condition, that is, a specific set of *launch conditions* at the input end. After the equilibrium length is reached, the fraction of total *radiant power* in each *mode* remains a constant with respect to distance. Usually the equilibrium length is the longest distance obtained from a worst-case, but undefined, excitation condition. Synonymous with *equilibrium coupling length; equilibrium distance; equilibrium modal-power-distribution length.*

equilibrium modal-power distribution. In an *electromagnetic wave propagating* in a *waveguide,* such as a *lightwave* in an *optical fiber,* the distribution of *radiant power* among the various *modes* such that the fraction of total power in each mode is stable and does not redistribute as a function of distance. Equilibrium modal-power distribution will not occur at the beginning of the guide until the *wave* has *propagated* a distance called the *equilibrium length,* because the *launch conditions, wavelength,* and *refractive indices* cannot be controlled with sufficient precision. Synonymous with *equilibrium mode distribution; steady-state condition.* See *nonequilibrium modal-power distribution.* (See Fig. E-1)

equilibrium modal-power distribution length. See *equilibrium length.*

equilibrium mode distribution. See *equilibrium modal-power distribution.*

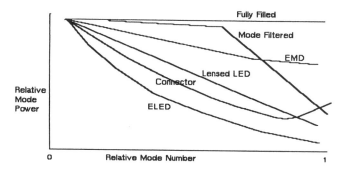

E-1. Relative *modal (power) distribution* in *multimode optical fibers.* A fully filled fiber means that all *modes* carry equal *optical power,* as shown by the line across the top of the graph. A long length of optical fiber loses the *higher order modes* faster, leading to the gently sloping **equilibrium modal (power) distribution** (EDM) curve. Mode *filtering* strips off the higher order modes, but provides only a crude approximation of EMD. The lensed *LED* couples most of its power in *lower order modes.* The *edge-emitting diode (ELED)* couples power even more strongly into the lower order modes. *Connectors* are mode *mixers,* because *misalignment losses* cause some power in lower order modes to be coupled up to the higher order modes. (Courtesy fotec incorporated.)

equilibrium mode simulator. A component or system used to create an approximate *equilibrium modal-power distribution.*

equilibrium radiation pattern. The *radiation pattern* at the output end of an *optical fiber* that is long enough so that *equilibrium modal-power distribution* has been reached, that is, the fiber is longer than the *equilibrium length.*

equipment identification. Information that uniquely characterizes a product and provides a traceable indicator for determining its specifications, features, issue or revision, and manufacturer, e.g., Common Language Equipment Indentification.

equivalent power. See *noise-equivalent power (NEP).*

equivalent step-index (ESI) profile. The *refractive-index profile* of a hypothetical *step-index optical fiber* that has the same *propagation* characteristics as a given *single-mode fiber.* The refractive index of the *cladding* material has the same value everywhere in the cladding.

error. See *optical fiber concentricity error.*

error budget. The allocation of a *bit-error ratio (BER)* requirement to the various segments of a *link,* circuit, or trunk, such as trunk lines, switches, access lines, and terminal devices, in a manner that permits the specified system end-to-end error ratio requirements to be satisfied for message traffic *transmitted* over a real circuit or a postulated reference circuit.

error ratio. See *bit-error ratio (BER)*.

ESI. *Equivalent step-index.*

ESI profile. See *equivalent step-index (ESI) profile*.

ESI refractive-index difference. For an *optical fiber*, the difference between the *refractive index* of the *core* and the refractive index of the *cladding* in its *equivalent step-index (ESI) profile*.

evanescent field. A time-varying *electromagnetic field, propagating* outside a *waveguide* but coupled or *bound* to an *electromagnetic wave* or *mode* propagating inside the waveguide, whose amplitude decreases monotonically without an accompanying *phase shift,* as a function of distance from the axis of the waveguide. In *fiber optics,* this field may provide *coupling* of *signals* from one optical fiber to another by means of proximity coupling or evanescent field coupling.

evanescent-field coupling. *Coupling* between two *waveguides* achieved by allowing *evanescent waves* of one to penetrate and be trapped by the other. Transfer of energy will take place if the guides are placed parallel and close together for a short distance. Evanescent-field coupling is achieved between *optical fibers* by etching away the *cladding* of both fibers and placing their *cores* close together or locally modifying their *refractive indices*.

evanescent wave. In a *transverse electromagnetic wave propagating* in a *waveguide,* a *wave* in the *cladding* or on the outside of the guide. Evanescent waves are *bound,* that is, remain coupled to *modes* in the guide. They will *radiate* away at sharp bends in the guide if the radius of the bend is less than the *critical radius*. They usually have a *frequency* less than the *cutoff frequency,* above which true propagation occurs and below which the waves decay exponentially with radial distance from the guide. Evanescent *wavefronts* of constant *phase* may be perpendicular, or at an angle less than 90°, to the surface of the guide. Evanescent waves will not remain bound to modes inside a guide if the internal modes to which they are bound propagate faster than the evanescent waves propagate in the medium they are in, in which case they will radiate away.

exahertz (EHz). A unit of *frequency* that is equal to 1 million trillion *hertz,* that is, 10^{18} Hz.

excess loss. See *coupler excess loss*.

exchange process. See *ion-exchange process*.

exitance. See *radiant emittance*.

exit angle. When a *light ray* emerges from a surface of a material, the angle between the ray and a normal to the surface at the point of emergence. For an *optical fiber,* the exit angle is the angle between the fiber *optical axis* and the emerging ray. The exit angle of a *light source* is the *launch angle*.

expanded-beam connector. A *fiber optic connector* in which the diameter of the *light* beam and the *launch angle* are increased, so that the *losses* caused by *longitudinal offset, angular misalignment,* and *lateral offset* are reduced to a minimum. In many cases, special lenses and other *passive optical devices* are built into the connector to optimize *coupling.* In some cases, the glass in the fiber itself is used to form the optical elements in the connector. Synonymous with *lensed connector.*

extended operating condition. In the design of an *optical station/regenerator section,* a condition that is more severe, e.g., a higher or lower temperature condition, a higher or lower humidity condition, than a standard operating condition, or any other environmental condition not included in the set of standard conditions, such as high pressures, corrosive atmospheres, severe abrasion, and high tensile stress, that may result in the specified parameters not having the specified performance as given for standard operating conditions. Also see *standard operating condition.*

external photoelectric effect. See *photoemissive effect.*

extinction coefficient. 1. The sum of the *absorption coefficient* and the *scattering coefficient.* The extinction coefficient may be expressed as a fraction or in *decibels.* 2. The ratio of two *optical power levels,* P_1/P_2, of a *digital signal* generated by a *light source,* where P_1 is the optical power level generated when the light source is "on," and P_2 is the power level generated when the light source is "off." Synonymous with *extinction ratio.* Also see *Bouger's law.*

extinction ratio. See *extinction coefficient.*

extramural cladding. A layer of dark or opaque absorbant *coating* placed over the *cladding* of an *optical fiber* to increase *total (internal) reflection,* protect the smooth *reflecting* wall of the cladding, and absorb *scattered* or escaped stray *light rays* that might penetrate the cladding.

extraordinary ray. A *light ray* that has an *anisotropic* velocity in a *doubly refracting* crystal, that is, the crystal is an *anisotropic propagation medium.* An extraordinary ray does not necessarily obey *Snell's law* of *refraction* at the crystal surface. Also see *ordinary ray.*

extrinsic coupling loss. In *fiber optics,* the *coupling loss* that is caused when two *optical fibers* are imperfectly joined, such as by *longitudinal offset, angular misalignment,* and *lateral offset.* Synonymous with *extrinsic joint loss.* Also see *intrinsic coupling loss.*

extrinsic joint loss. See *extrinsic coupling loss.*

F

Fabry-Perot fiber optic sensor. A high-*resolution*, multiple-beam *interferometric sensor* in which two optically flat, parallel *transparent* plates are held a short distance apart, and the adjacent surfaces of the plates, that is, interferometric flats, are made almost totally *reflective* by a thin silver film or a multilayer *dielectric coating*. If one plate is moved relative to the other, *interference patterns* can be made to occur when a *lightwave* is directed at the plates. If the ends of an *optical fiber* are made reflective, moving one end relative to the other will produce an output signal proportional to the relative motion when *monochromatic light,* that is, light with a narrow *spectral linewidth,* is inserted into the fiber.

faceplate. See *fiber faceplate.*

facility. See *optical cable facility; optical station facility.*

facility loss. See *optical facility loss; statistical optical cable facility loss.*

factor. See *bandwidth-distance factor; confinement factor; intrinsic quality factor (IQF).*

fading. See *Rayleigh fading.*

failure. See *mean time to failure (MTTF).*

failures. See *mean time between failures (MTBF).*

fall time. The time it takes for the amplitude of a *pulse* to decrease from a specified value, usually near the peak value, to a specified value, usually near the lowest or zero value. Values of fall time are often specified as the time interval between the 90% and 10% values of the peak value of a pulse. If other than 10 to 90% values are used, they should be specified. Synonymous with *pulse-decay time.* Also see *rise time.*

Faraday effect. See *magnetooptic effect.*

far-field diffraction pattern. The *diffraction pattern* of a *source,* such as a *light-emitting diode (LED), injection laser diode (ILD),* or the output of an *optical waveguide* observed at an infinite distance from the source, that is, in the *far-field region* of the source. A far-field diffraction pattern is considered to exist at distances larger than that given by the relation:

$$d_i = s^2/\lambda$$

where *s* is the characteristic of the source, and λ is the *wavelength.* For example, if the source is a uniformly illuminated circle, then *s* is the radius of the circle. The far-field diffraction pattern is an approximation, for finite distances, for a *diffraction pattern* of a source observed from infinity or in the focal plane of a well-corrected lens, except for scale. The far-field diffraction pattern of a diffracting screen illuminated by a point source may be observed in the image plane of the source. Synonymous with *Fraunhofer diffraction pattern.*

far-field pattern. See *far-field radiation pattern.*

far-field radiation pattern. The *radiation pattern* of a *source* of *electromagnetic radiation* in the far *field region,* for example, the radiation pattern for an *optical fiber* that describes the distribution of *irradiance* as a function of angle in the far-field region of the radiation from the exit face of the fiber. Synonymous with *far-field pattern.*

far-field region. The region far from an *electromagnetic radiation source* or *aperture* where the *radiation pattern* and the *irradiance,* that is, the *incident optical power* per unit area, or the *electric field strength,* is negligibly dependent upon the distance from the source, such as where the field strength varies inversely as the first power of the distance and the *propagation* times from the source to points within the region are not negligible compared with propagation times to points in the *near-field region.* However, because all the points in the far-field region are far from the source, the change in pattern from point to point within the far-field region is negligible.

far-infrared. Pertaining to the region of the *electromagnetic spectrum* that lies between the *longer-wavelength* end of the *middle-infrared* region and the *shorter-wavelength* end of the radio region, that is, from about 30 to 100 μ *(microns).* Thus, the far-infrared region is included in the *optical spectrum.* However, the near-infrared region lies between the longer-wavelength end of the visible spectrum and the shorter-wavelength end of the middle-infrared region, that is, from about 0.8 to 3.0 μ (microns). Thus, the near-infrared is included in the near-visible region, but the middle-infrared and far-infrared regions are not included in the near-visible region. None of the infrared regions—near, middle, or far—are included in the *visible spectrum.*

FDDI. See *fiber distributed-data-interface (FDDI) standard.*

FDHM. *Full-duration half-maximum.*

FDM. *Frequency-division multiplexing.*

feature. See *optical cable interconnect feature.*

feed-through. See *fiber optic cable feed-through.*

femtosecond. A unit of time equal to 10^{-15} seconds.

Fermat's principle. A *light ray* follows the *path* that requires the least time to *propagate* from one point to another, including *reflections* and *refractions* that may occur. The *optical path length* is an extreme path in the terminology of the calculus of variations. Thus, if all rays starting from point *A propagate* via a *propagation medium* to the same point *B,* the optical paths from *A* to *B* are the same optical length, though their geometric lengths may be different. In *optical fibers,* different rays *launched* into the fiber at the same time take different paths. Thus, they arrive at the end at different points and different times, giving rise to *dispersion.*

ferrule. A mechanical device or fixture, generally a rigid tube, used to hold the stripped end of an *optical fiber* or *fiber bundle.* Typically, individual fibers of a bundle are cemented together within a ferrule that has a diameter designed to hold the fibers firmly with a maximum *packing fraction.* Nonrigid materials, such as shrink tubing, may also be used for ferrules. Generally, a ferrule provides a means of positioning fibers within a *connector* by performing the function of a bushing.

fiber. See *active optical fiber; all-glass fiber; all-plastic fiber; buffered fiber; compensated optical fiber; depressed-cladding fiber; dispersion-shifted fiber; dispersion-unshifted fiber; doubly clad fiber; drawn glass fiber; fully filled fiber; graded-index (GI) fiber; hard-clad silica fiber; launching fiber; Lorentzian fiber; medium-loss fiber; multimode fiber; optical fiber; overcompensated optical fiber; plastic-clad silica (PCS) fiber; polarization-maintaining optical fiber; quadruply clad fiber; single-mode fiber; step-index fiber; superfiber; tapered fiber; transition fiber; triangular-cored optical fiber; ultralow-loss fiber; undercompensated optical fiber; weakly guiding fiber; W-type fiber.*

fiber acoustic sensor. See *optical fiber acoustic sensor.*

fiber active connector. See *optical fiber active connector.*

fiber amplifier. An *optooptic* (optically pumped) *fiber laser* used as an amplifier to boost *optical signals.* The fiber amplifier can be used in lieu of an *optoelectronic repeater* in *local loops, LANs,* and *long-haul systems,* thus allowing more sequential switching stages between repeaters and an increased number of *nodes* for a single *transmitter* in a *network.* The fiber amplifier is a high gain, low noise, *wideband,* relatively low-cost device for *optical* amplification. The amplifier can be used for *signal* processing in the logic devices of *integrated optical circuits (IOCs)*.

fiber-and-splice organizer. A device that provides for (1) accommodation of *optical fibers* and *splices* in an orderly manner; (2) protection of all types of fibers and splices, such as mechanical, fusion, and multiple-fiber arrays; (3) fiber splice identification; (4) fiber rearrangement; (5) storage of excess fibers, *optical fiber cable cores,* and *fiber ribbons;* and (6) modularity to allow for additional capacity for fibers and splices. Synonymous with *fiber organizer; optical fiber organizer.*

fiber axial-alignment sensor. A *fiber optic sensor* in which the amount of *light coupled* from a *source optical fiber* mounted on one platform to a *photodetector* optical fiber mounted on another platform decreases as the *angular misalignment* of the axes of the two fibers increases. *Axial alignment* is considered to have occurred when the output signal of the photodetector becomes a maximum as the alignment is varied. The sensor can be used to measure angular alignment of the two platforms, one attached to each fiber. When the angle between the axes is greater than half the apex angle of the *acceptance cone,* light is no longer coupled from the source fiber to the detector fiber, at which point the output of the photodetector is reduced to zero.

fiber axial-displacement sensor. A *fiber optic sensor* in which the amount of *light coupled* from a *source optical fiber* mounted on one platform to a *photodetector* fiber mounted on another platform decreases as the axes of the two fibers are moved laterally, that is, transversely farther apart, even though the axes remain parallel, thus causing increasing *lateral offset loss.* The output *signal* from the photodetector is a function of the amount of displacement, varying form zero to a maximum. The time rate of change of the signal is proportional to the velocity at which the fibers are displaced laterally relative to each other. If the fibers oscillate with respect to each other, an oscillating signal will be produced by the photodetector.

fiber bandwidth. That value numerically equal to the lowest *frequency* at which the magnitude of the *modulation transfer function* of an *optical fiber* decreases to a specified function, generally to one-half of the zero frequency value. It is not a good measure of the information-carrying capacity of an optical fiber at a specified *optical wavelength.* Fiber *bandwidth* for an optical fiber should not be considered in the same terms as for other types of communication systems, such as electronic systems. It is a function of the modulation transfer characteristic of the fiber and not the optical frequency transfer function. The bandwidth is limited by several mechanisms: (a) in *multimode fibers*—primarily by *modal distortion* and *material dispersion;* (b) in *single-mode fibers*—primarily by material and *waveguide dispersion.* A better measure of the information-carrying capacity of an optical fiber is the *bit-rate-length product (BRLP),* which is limited by the modulation scheme and the desired *bit-error ratio (BER).* The BER is limited by the tolerable level of *intersymbol interference* caused by all forms of *dispersion, attenuation,* and *jitter;* the *dynamic range; photodetector sensitivity;* and other factors. The dispersion is a function of the *spectral width* of the *source,* the *refractive indices* of the *core* and *cladding,* and the length of the fiber. *Step-index fiber* can be made with *zero dispersion* at specific *wavelengths.* With the proper *refractive-index profile* and wavelength, *graded-index fibers* can also be made with zero dispersion. Another measure of the *data signaling rate (DSR)* or information-carrying capacity of an optical fiber is the *full-wave-half-power point.*

fiber blank. See *optical fiber blank.*

fiber buffer. A material that is used to protect an *optical fiber* from physical damage and provide mechanical strength and protection. Some fiber fabrication techniques result in firm contact between the fiber and its protective *buffering* material. Other techniques result in a loose fit, permitting the fiber to slip in the buffer tube. Added protection may be obtained by using several buffer layers.

fiber bundle. See *optical fiber bundle.*

fiber cable component See *optical fiber cable component (OFCC).*

fiber cladding. A *transparent* material that surrounds the *core* of an *optical fiber* and that has a lower *refractive index* than the core material.

fiber coating See *optical fiber coating.*

fiber concentricity error. See *optical fiber concentricity error.*

fiber connector set. See *optical fiber connector set.*

fiber core. The central portion of an *optical fiber.* The *core* has a higher *refractive index* than the *cladding* that surrounds it. The bulk of *lightwave* energy of an *electromagnetic wave propagating* in the fiber is confined to and propagates in the core. Fibers can be made to operate in *single mode.* Their *core diameters* range form 2 to 11 μ *(microns),* depending on the *numerical aperture,* core radius, and *wavelength* of the *incident light.*

fiber crosstalk. In an *optical fiber,* the exchange of *lightwave* energy between the *core* and the *cladding,* between the cladding and external surroundings, or between layers with different *refractive indices.* Fiber *crosstalk* is usually undesirable, because differences in *optical path length* and *propagation* time can result in *dispersion* and consequent *distortion,* thus limiting the *data signaling rate (DSR)* for a given *bit-error ratio (BER).* *Attenuation* may be deliberately introduced into the cladding to suppress the crosstalk by making it a *lossy medium* and thereby reducing dispersion.

fiber cutoff wavelength. For a short, uncabled, *single-mode optical fiber* with a specified large radius of curvature, that is, *macrobend,* the *wavelength* at which the presence of a second-order *mode* introduces a significant *attenuation* increase when compared with a fiber whose *differential mode attenuation* is not changing at that wavelength. Because the *cutoff wavelength* of a fiber is dependent upon length, bend, and cabling, the cabled fiber cutoff wavelength may be a more useful value for cutoff wavelength from a systems point of view. In general, the *cable cutoff wavelength* is less than the fiber cutoff wavelength. Also see *cable cutoff wavelength*.

fiber-cutting tool. A special cutting tool designed to prepare the ends of *optical fibers* for *splicing* or connecting. When using the tool, the fiber is bent, and a small groove is made with the tool at the spot under tension. This causes the fiber to fracture, producing a smooth cleaved end-surface along a molecular lattice.

fiber delay line. See *optical fiber delay line.*

fiber diameter. The nominal diameter of a round *optical fiber,* normally including the *core,* the *cladding,* and any *coating* not normally removed when making a connection, such as hermetic seals and special *buffers.*

fiber dispersion. The lengthening of the duration, that is, the width, of an *electromagnetic pulse* as it *propagates* along an *optical fiber,* caused by *material dispersion* which is caused by the *wavelength*-dependence of the *refractive index; modal dispersion* that is caused by different *group velocities* of the different *modes;* and *waveguide dispersion* that is caused by *frequency*-dependence of the *propagation constant* for a given mode.

fiber distributed-data-interface (FDDI) standard. A *fiber optic* communication *network* standard, developed by the American National Standards Institute (ANSI), in which a single-bit-stream format is used, rather than a set of *optical data signaling rates (DSR)* and formats using a byte format. Also see *synchronous optical network (Sonet) standard.*

fiber drawing. The fabrication of an *optical fiber* from the controlled pulling of the fiber, in a melted or softened state, through an *aperture,* causing it to elongate and reduce its diameter as it cools, adding *buffers* and other *coatings,* and winding the fiber on a spool, in an inside-payout or outside-payout ball or cone-shaped winding called a *bundpack.*

fiber end-to-end separation sensor. A *fiber optic sensor* in which the amount of *light coupled* from a *source fiber,* mounted on one platform, to a *photodetector* fiber mounted on another platform, is a function of the distance of separation of the end faces of the two fibers, that is, as the separation distance increases, the amount of coupled light decreases from a maximum to zero. Operation is dependent upon *exit* and *acceptance cone angles,* that is, *numerical aperture, core diameters,* and *wavelength.* If the end-to-end separation distance oscillates, the photodetector output *signal* will oscillate.

fiber faceplate. A plate made by cutting a transverse slice from a boule, that is, from a *bundle* of fused *optical fibers.*

fiber global attenuation-rate characteristic. A plot of the *attenuation rate* as a function of *transmitter wavelength* for an *optical fiber.* The plot should show the peaks and valleys caused by different *absorption, scattering,* and *reflection* levels at different wavelengths. The plot should also show (1) the useful wavelength regions for the fiber and (2) typical values rather than worst-case values so that the plot can be used to determine how the fiber will respond to new applications and to changes made in existing systems. Designers ensure that fibers operate at a wavelength at which the attenuation rate is the lowest, such as 1.31 μ *(microns)* for *silica glass* fibers. Another trough occurs at 1.55 μ. Global values are typical values because the only worst-case end-of-life values are those specified at time of purchase. Attenuation rates in dB/km are usually plotted at 23°C for wavelengths about 20% on either side of the *transmitter nominal central wavelength.* When designing an *optical station/regenerator section,* allowance must be made for increases in typical attenuation rates because of transmitter *spectral width* and the effect of temperature at the worst-case temperature conditions. Synonymous with *fiber global-loss characteristic.* Also see *loss-budget constraint; optical cable facility loss.*

fiber global-dispersion characteristic. A plot that shows the *dispersion coefficient* (ps/nm-km) for an *optical fiber* as a function of wavelength. If the *refractive index profile* is properly shaped, the fiber dispersion coefficient can be reduced to zero at a given wavelength. This wavelength should also coincide with the wavelength at which the *attenuation rate* is a minimum on the *fiber global-attenuation characteristic.* The plot should show (1) the useful wavelength regions for the fiber and (2) typical values rather than worst-case values, so that the plot can be used to determine how the fiber will respond to new applications and to changes made in existing systems.

fiber global-loss characteristic. See *fiber global-attenuation-rate characteristic.*

fiber hazard. See *optical fiber hazard.*

fiber identifier. A device that indicates the operational status of an *optical fiber,* such as whether there is data traffic or whether a tone is being *transmitted.* The identifier is used to help ensure that service is not interrupted when repairs or connections are made to *fiber optic cables.* It can be pocketed, hand-held, and taken anywhere.

fiber jacket. See *optical fiber jacket.*

fiber junction. See *optical fiber junction.*

fiber laser. An optically pumped, solid-state *laser* in which an *optical fiber* is the *active laser medium.* The *core* of the fiber is doped, usually with one of the standard rare earth ions, such as neodymium (Nd) or erbium (Er), used for bulk crystalline or glass lasers. The fiber laser is similar to a bulk laser, such as an Nd:YAG laser, except that an optical fiber is used as the gain medium instead of a rod or a slab. The high degree of confinement of the *optical (electromagnetic) fields* within the core and the long interaction distance in the fiber combine to provide excellent operational characteristics, including low lasing thresholds and high *quantum efficiencies,* that is, relatively high electrical to *optical power* conversion ratios. One fiber laser, developed by GTE Laboratories, is a neodymium-doped fluorozirconate heavy metal fluoride glass fiber laser *emitting* at a *wavelength* between 1.33 and 1.40 μ *(microns).* Fiber lasers exhibit a *spectral (line) width* of only a few MHz compared to 10 MHz or more for the distributed feedback laser. Only the external cavity semiconductor laser has a narrower spectral (line) width. The laser may be used in *local (subscriber) loops, local-area networks (LANs), metropolitan-area networks (MANs),* and *long-haul communication networks.* Synonymous with *self-lasing fiber.*

fiber longitudinal compression sensor. A *fiber optic sensor* in which a force applied to a length of *optical fiber* causes the fiber to shorten. By using *interferometric* techniques, the variation *in monochromatic light irradiance* at a *photodetector* can be made a function of the applied compressive force.

fiber-loop multiplexer. A *fiber optic multiplexer* for providing multi*channel* (multiplexed) capability for *local loops* to user end-instruments in offices and homes.

fiber mandrel. See *optical fiber mandrel.*

fiber merit figure. See *optical fiber merit figure.*

fiber net. A communication *network* in which *optical fibers* are used in the cables, rather than copper wires and coaxial cables. An all-fiber net has fiber in trunk lines between *switching centers* and central offices, as well as in *local loops* and *local-area networks (LANs).*

fiber optic. 1. Pertaining to *optical fibers* and the systems in which they are used. 2. Pertaining to the branch of science and technology devoted to combining the features and use of *optical fibers* and other *dielectric waveguides,* with other components, such as electronic components, to fabricate communication, sensing, telemetry, endoscopic, illumination, display, and other types of devices, components, and systems.

fiber optic attenuator. A device that operates upon its input *optical signal power level* in such a way that its output signal power level is less than the input level, the reduction caused by such means as *absorption, reflection, diffusion, scattering,* deflection, *diffraction,* and *dispersion,* and usually not a result of *geometric spreading,* for example, a length of *high-loss optical fiber.* Also see *optical attenuator.*

fiber optic borescope. A device in which *optical fibers* are used to view the interior of a device or machine. The borescope may also illuminate the object and obtain a *reflected* image. For example, a fiber, or an *aligned bundle* of fibers, in a flexible or rigid cable with appropriate *terminations,* or a bundle of fibers embedded in the housing of a turbine to view the blade clearance. Also see *flexible borescope; rigid borescope.*

fiber optic branching device. An *optical waveguide* that combines *light* from two or more inputs or distributes light to one or more outputs, such as a *splitter, combiner,* or *star coupler.*

fiber optic bundle. See *optical fiber bundle.*

fiber optic cable. A cable in which one or more *optical fibers* are used as the *propagation medium.* The optical fibers are usually surrounded by *buffers,* strength members, and *jackets* for protection, stiffness, and strength. Synonymous with *optical cable; optical fiber cable.* See *Tempest-proofed fiber optic cable.*

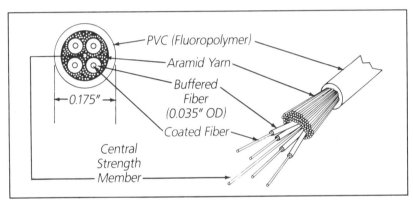

F-1. The Lightguide Building Cable LGBC–004A–LRX (LPX), a **fiber optic cable** designed for use in high-rise buildings, campus, and other *LAN* applications where *point-to-point data links* or *network* distribution systems are needed. Types include 50/125-μ *(micron)* and 62.5/125-μ *multimode,* and 8.3/125-μ *single-mode fibers,* with fiber counts from 2–12. (Courtesy AT&T.)

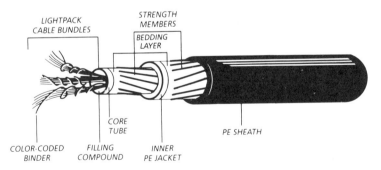

F-2. The Lightpack **fiber optic cable** can be factory-*connectorized* for single-fiber and/or mass (array) *splicing* in a single, lightweight compact cable. Many fiber counts are available. (Courtesy AT&T.)

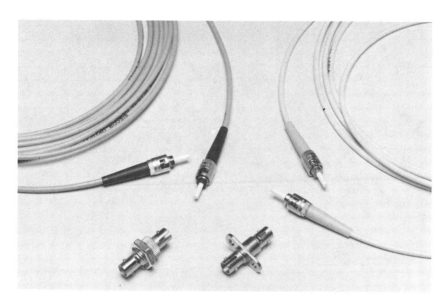

F-3. Dorran™ ST Compatible Cable Assemblies and Couplings for *single-mode* and *multimode* **fiber optic cable** connections. (Courtesy 3M Fiber Optic Products, EOTec Corporation.)

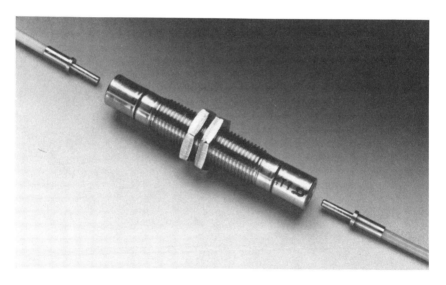

F-4. This **fiber optic cable** in-line adapter is designed to accept MIL-C-38999 contacts in *optical fiber*-to-fiber, single-*channel* avionics applications. (Courtesy ITT Cannon, Military/Aerospace Division, Fiber Optics Products Group).

F–5. The 12-fiber B-series Breakout **Fiber Optic Cable** featuring the Core-Locked™ PVC pressure extruded outer *jacket*. Full rated tensile load can be applied to the cable with wire mesh-type pulling grips for installation in indoor or outdoor environments. (Courtesy Optical Cable Corporation).

F–6. A **fiber optic cable** assembly for *single-mode fiber* connection. (Courtesy 3M Fiber Optic Products, EOTec Corporation).

fiber optic cable assembly. **1.** A *fiber optic cable* that is *terminated* with *fiber optic connectors,* that is, with one part of a two-part connector on each end. **2.** A *fiber optic cable* that is *terminated* and ready for installation. Synonymous with *optical cable assembly; optical fiber cable assembly.*

fiber optic cable core. The portion of a *fiber optic cable* that serves as the *waveguide* proper, such as an *optical fiber,* a *bundle* of optical fibers, or a *ribbon* holding a group of optical fibers. Usually the cable core is protected from environmental effects by various other components, such as additional *buffers,* strength members, gels, tubes, *jackets,* overall sheaths, or *overarmor.*

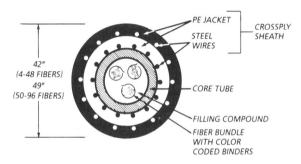

F–7. The Lightpack **fiber optic cable core** consists of *bundles* of *fibers* wrapped with color-coded binders. The core is filled with a water-blocking compound that, along with the large core tube clearance, virtually eliminates *microbending losses.* (Courtesy AT&T.)

fiber optic cable facility. See *optical cable facility.*

fiber optic cable facility loss. See *optical cable facility loss.*

fiber optic cable feed-through. A mechanism that provides strain relief to a *fiber optic cable* entering an *interconnection box* and that may also be used to seal around the cable.

fiber optic connector. A device that simply and easily permits *coupling,* decoupling, and recoupling of *optical signals* or *power* from each *optical fiber* in a cable to corresponding fibers in another cable, usually without the use of any tool. The connector usually consists of two mateable and demateable parts, one attached to each end of a cable, or to a piece of equipment, for the purpose of providing connection and disconnection of *fiber optic cables.* Synonymous with *optical connector.* (See Figs. F–8 and F–9 p. 92)

fiber optic connector variation. See *optical connector variation.*

fiber optic coupler. A device that transfers *optical signals* from one *propagation medium* to another, usually without using a *fiber optic splice* or *connector,* for example, in *fiber optic transmission* systems, a device that transfers optical signals

F-8. A Keyed Bayonet Biconic **fiber optic connector.** (Courtesy 3M Fiber Optic Products, EOTec Corporation).

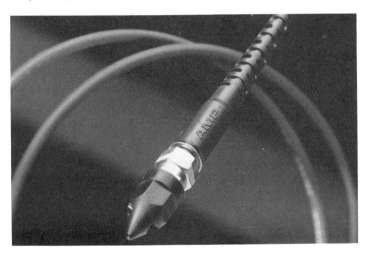

F-9. The FSMA Connector, a SMA style **fiber optic connector** provides repeatable *fiber optic termination* with industry standard *interface.* (Courtesy AMP Incorporated).

from one *optical fiber* to one or more other fibers, such as by placing their *cores* in proximity. Thus, the coupler transfers *optical power* to a number of ports in a predetermined manner. The ports may be connected to fiber optic components, such as fiber optic *light sources, waveguides,* and *photodetectors.* Synonymous with *optical coupler; optical fiber coupler.*

fiber optic cross-connection. Connection of two *optical fibers* by means of a third optical fiber, that is, a jumper, which serves as a *link* between the two fibers. Although two connections have to be made rather than the one required for a direct connection between the two fibers, the jumpers can be mounted on a panel so connections can be made conveniently at one place. Also see *fiber optic interconnection.*

fiber optic data link. A *link,* usually consisting of an electronically *modulated light source,* a *fiber optic cable,* a *photodetector,* and associated *fiber optic connectors,* capable of *transmitting data* in the form of *optical signals* from one location to another.

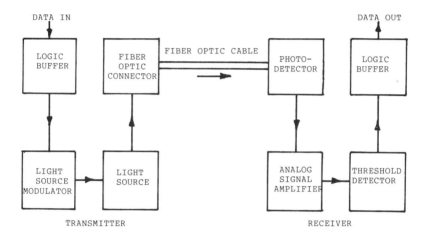

F-10. Components of a **fiber optic data link.**

fiber optic data-transfer network. A *data-transfer network (DTN)* in which *fiber optic* modules and components are used to sense, transmit, display, or otherwise process *data.*

fiber optic demultiplexer (active). An electronically operated *optical device,* with input and output *optical fibers,* that accepts information in the form of *modulated lightwaves* that have resulted from combining two or more *optical signal* streams in a *multiplexer* and places each of the original streams on a separate output channel, for example, a device that uses the *electrooptic effect* to disperse a single-*channel* input *polychromatic lightwave* into its separate constituent *colors,* each of which may have been separately modulated. The active optical demultiplexer may have two or more output fibers, each operating at a different *wavelength,* and one input fiber bearing all the wavelengths. Electrical energy, or other form of energy other than optical energy, is required to operate the active optical demultiplexer. Synonymous with *optical fiber demultiplexer (active).* Also see *fiber optic multiplexer (active); optical demultiplexer (active); optical multiplexer (active).*

fiber optic demultiplexer (passive). A purely passive *optical device* that accepts information in the form of *modulated lightwaves* that have resulted from combining two or more *optical signal* streams in a multiplexer and places each of the original streams on a separate output *channel,* for example, a prism that disperses a single-*channel* input *polychromatic lightwave* into separate constituent *colors,* each of which may have been separately modulated. For example, a passive optical demultiplexer may consist of two or more output *optical fibers,* each operating at a different *wavelength,* and one input fiber bearing all the wavelengths. No energy is required to operate the passive optical demultiplexer other than that contained in the input lightwave. Synonymous with *optical fiber demultiplexer (passive).* Also see *fiber optic multiplexer (passive); optical demultiplexer (passive); optical multiplexer (passive).*

fiber optic display device. A device used to display pictorial, alphanumeric, or graphical *data,* usually consisting of a *fiber faceplate* made of the end faces of *optical fibers* in an *aligned bundle,* the aligned bundle of fibers being used to transfer an image to the display surface, and perhaps a lens to enlarge or project the image.

fiber optic distribution frame. A structure with *optical waveguide terminations* for interconnecting permanently installed *fiber optic cable* in such a way that interconnection by cross-connections may be made. The frame may hold *connectors, splice trays,* or both. The fiber optic distribution frame is usually located at a central office, *switching center,* or *node* in a communication *network* and considered part of the inside plant. Synonymous with *fiber optic interconnection frame.* Also see *fiber optic interconnection box.*

fiber optic dosimeter. An instrument for measuring cumulative exposure to high-energy *radiation* or bombardment, such as *x-rays, gamma radiation,* cosmic rays, and high-energy neutrons, by measuring the *induced attenuation* in an *optical fiber* from exposure to the radiation. Also see *induced attenuation.*

fiber optic drop. A *fiber optic transmission* line that carries multiplexed *signals* from a central-office inside plant to a distribution point where lines lead to taps with *user end-instruments,* such as telephones and *data* terminals, perhaps in individual offices and homes. Current trends are to run *optical fiber* all the way to the taps for video, telephone, and *data channels.* Satellite dishes, coaxial cable, and *fiber optic cables* are competing *signal*-transport media for television, telephone, and data systems into home and office.

fiber optic endoscope. An *optical device* used to view and to take pictures of the internal parts of a system, such as the internal cavities and organs of the human body or the interior of machines. In medicine, fiber optic endoscopes, such as bronchoscopes, gastroscopes, colonoscopes, cholodochoscopes, cystoscopes, laparoscopes, and arthroscopes are used for in vivo examining internal body cavities, performing biopsies, and performing polypectomies. Industrial counterparts of the endoscope are used for remote nondestructive inspection of internal parts and cavities of machine systems while they are operating. Also see *fiberscope.*

fiber optic filter. A device that operates on the various *frequencies* in an input *optical signal* in such a way that certain frequencies are allowed to pass while others are blocked, for example, a piece of blue glass that will *transmit* only the blue *wavelengths* contained in *incident white light* and block all the other wavelengths, or a device that screens out all the wavelengths above or below a given value.

fiber optic flood illumination. Illumination of a relatively large area with a wide light beam emanating from the end of an *optical fiber* or *bundle* of fibers.

fiber optic gyroscope (FOG). A gyroscope in which *optical fibers,* such as *single-mode* or *polarization-preserving fibers,* and *monochromatic light,* such as that produced by *lasers,* are used to measure changes in rotational displacement, velocity, and acceleration, and hence changes in orientation and direction of the platform holding the sensor. *Interferometric* techniques, such as are used in the *Sagnac fiber optic sensor,* are used to detect the changes in *phase* or *polarization* of the *lightwaves.*

fiber optic illumination detection. The use of an *optical fiber,* or *bundle* of fibers, to determine whether *light* is emanating from a given *source.*

fiber optic illuminator. A device in which an *optical fiber,* or *bundle* of fibers, is used to convey *light* to and illuminate an area remote from the *source* of light. Illumination devices may be used to shed light on specific and often inaccessible surfaces, such as surgical fields, instrument panels, and auxiliary equipment, particularly in atmospheres where electrical illumination might be hazardous.

fiber optic interconnection. Connection of two *optical fibers* by means of a direct connection from one to the other, thus requiring no *fiber optic jumpers.* Also see *fiber optic cross-connection.*

fiber optic interconnection box. A housing for holding *fiber optic splices, connectors,* and *couplers* used to distribute *signals* on incoming cables to outgoing cables by means of connections. The connections may be made by interconnection or cross-connection. It is usually a local box or housing at a user premises for holding connections to *fiber optic cables* from *distribution frames* and fiber optic splices, or *splice trays* that hold splices, to loops going to *optical transceivers* or other *user end-instruments.* Synonymous with *fiber optic distribution box; optical fiber interconnection box.* Also see *fiber optic distribution frame.* (See Fig. F-11, p. 96)

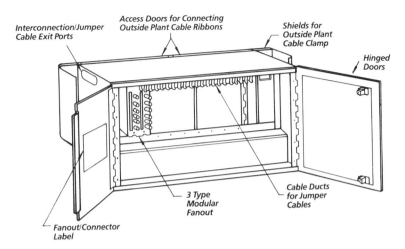

F-11. The ED8C005-50 Lightguide Cable Interconnection Terminal (Modular LCIT) serves as a **fiber optic interconnection box** or *distribution frame* for *termination, interconnection,* and *splicing* of *connectorized ribbon*-based outside plant or riser lightguide *cables*. The LCIT may be used in central offices, minihuts, maxihuts, remote terminals, customer premises, and 80-type cabinets. Measuring 12 × 23 in., it features 72 fiber termination/connection *ports,* 12 splice-through connection capacity, and top or bottom cable entry. (Courtesy AT&T.)

fiber optic isolater. An *optical link* inserted into a communication system to provide electrical isolation between two or more parts of the system. The fiber optic isolator is often used to provide nonmetallic entry and exit from secure areas. The fiber optic isolator serves as an *optical transmission link,* consisting of an *optical transmitter,* and *optical waveguide,* and a *photodetector,* to provide electrical isolation between two communication systems or components. The optical waveguide serves as an electrical insulator as well as an optical transmission line. The fiber optic isolator also perfoms impedance transformation. Also see *optical isolator.*

fiber optic jumper. A *jacketed,* short *optical fiber* with a *fiber optic connector* on both ends used for effecting a *fiber optic cross-connection.* Jumpers are usually mounted on panels for access convenience.

fiber optic light source. A device that *emits lightwaves,* that is, *radiation* in or near the *visible region* of the *electromagnetic spectrum,* for example, *light-emitting diodes (LEDs), lasers,* and lamps, for providing *optical power* to *optical* systems. Some fiber optic light sources are capable of being *modulated* by an electronic input *signal* and have an integral *optical fiber* as a *pigtail,* in which case they can serve as a component in *fiber optic transmitters* for command, communication, control, and telemetry systems. Others simply supply *light* that can be *modulated* by means other than by operating on the source, such as by a sensing mechanism. Fiber optic light sources usually have an optical fiber as an output lead and one or more wires for input electrical power and modulation signals.

fiber optic link. A communication *link* that *transmits signals* by means of *modulated light propagated* in an *optical fiber.* The link usually consists of a *fiber optic light source* as a *transmitter,* a *fiber optic cable,* and a *photodetector* as a receiver. Light from the *fiber optic transmitter* is usually *modulated* by an intelligence-bearing signal. The intelligence-bearing signal may be in the form of an electrical signal imposed on the *light source,* thus varying the *radiance,* that is, the *light intensity,* emanating from the source, or in the form of a modulator that modulates the light after it emanates from a constant *optical power* source, such as by a sensing device. The receiver demodulates the signal to obtain the original modulating signal containing the original intelligence. Synonymous with *optical link.* See *TV fiber optic link.*

fiber optic mixer. A device that accepts *wavelength-division multiplexed signals* from two or more input *optical fibers,* mixes the signals, and *transmits* the mixed signal containing all the input *wavelengths,* that is, transmits the resulting *dichromatic* or *polychromatic light.* For example, a fiber optic mixer can be a rod of transparent material with optical fiber input *ports* and internal mirrors for multiple *reflection* everywhere except at the output ports, at which optical fibers are attached.

fiber optic mode stripper. Material applied to the *cladding* of an *optical fiber, slab-dielectric waveguide,* or *integrated optical circuit (IOC)* that allows *optical energy propagating* in the *cladding* to leave the cladding, that supports only certain propagating *modes,* that usually does not support the modes that are propagating in the cladding, and that does not disturb the modes that are propagating in the *core.* *Optical power* removed from a *waveguide* by a stripper usually is not used. Power removed by a *coupler* usually is used.

fiber optic modulator. A device capable of accepting a nonintelligence-bearing input *signal,* that is, a *carrier,* and an intelligence-bearing input signal, and mixing them in such a manner that the output signal consists of a parameter of the carrier varied in accordance with the intelligence-bearing input signal. At least one of the input or output leads is an *optical fiber.*

fiber optic multimode dispersion. See *multimode dispersion.*

fiber optic multiplexer (active). An *optoelectronic* device that accepts information in the form of *modulated lightwaves* in *optical fibers,* that is, *optical signal* streams, on two or more *channels,* and places all the input information on a single output channel in such a way that the information on the input channels can be recovered at the end of the single output channel. For example, an active fiber optic multiplexer may consist of two or more input optical fibers, each operating at a different *wavelength,* and one output fiber containing all the wavelengths. The multiplexing component may or may not be constructed with optical fibers. Electrical energy or other forms of energy, other than that contained in the input lightwaves, are required to operate active fiber optic multiplexers and *optical multiplexers.* Synonymous with *optical fiber multiplexer (active)*. Also see *fiber optic demultiplexer (active); optical demultiplexer (active); optical multiplexer (active)*.

fiber optic multiplexer (passive). A purely *passive optical device* that accepts information in the form of *modulated lightwaves* in *optical fibers,* such as streams of *optical pulses,* on two or more *channels,* and places all the input information on a single output channel in such a way that the information on the input channels can be recovered at the end of the single output channel. For example, the passive fiber optic multiplexer may consist of two or more input optical fibers, each operating at a different *wavelength,* and one output fiber containing all the wavelengths. The multiplexing component may or may not be constructed with optical fibers. For example, they may be *optical fiber couplers.* They may be constructed of *optical mixing rods* or *boxes* or other mixing devices. No energy is required to operate the passive fiber optic multiplexer or the *optical multiplexer (passive)* other than that contained in the input lightwaves. Synonymous with *optical fiber multiplexer (passive).* Also see *optical demultiplexer (passive); optical multiplexer (passive).*

fiber optic multiport coupler. A *passive optical device* with two or more output or input *ports* that can be used to *couple* various sources to various receivers. Optical fibers are attached to the coupler at the ports. If there is only one input and one output it is a *fiber optic connector* or *splice.* Most fiber optic multiport couplers consist of a piece of *transparent* material. Operation depends on *reflection, transmission, scattering,* and *diffusion.*

fiber optic patch panel. A panel on which *fiber optic connectors* and *couplers* are mounted in an organized array for easy access. Usually both sides are accessible with sufficient spacing between components to facilitate handling individual components. Individual connectors should be easily identifiable. An individual connectorization and organization plan should accompany each panel and the *interconnection box* in which it is installed.

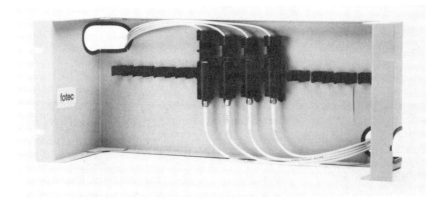

F-12. The A430P Attenuation Patch Panel, a **fiber optic patch panel** that includes *variable attenuators* for handling *optical power budgets* (Courtesy fotec incorporated).

fiber optic penetrator. A device that allows an *optical fiber* or *fiber optic cable* to pass from one compartment to another, such as through a wall, partition, bulkhead, hull, fuselage, or cabinet.

fiber optic phase modulator. See *optical phase modulator.*

fiber optic photodetector. A device capable of converting input *optical signals* into equivalent electronic output signals, usually consisting at least of an *optical fiber* signal-input *pigtail;* a light-sensitive element, such as a *photodiode;* an electrical signal output wire; and an electrical power input wire.

fiber optic polarizer. See *optical fiber polarizer.*

fiber optic preform. See *optical fiber preform.*

fiber optic receiver. An *optoelectronic* device that accepts *optical signals* from a *transmitter* or from a *propagation medium* and converts them into electronic signals for further processing.

fiber optic ribbon. *Optical fibers* arrayed side by side and held in place by tapes, adhesives, or plasitc strips, for example, a cable consisting of optical fibers laminated within a flat plastic strip. Synonymous with *fiber ribbon; optical fiber ribbon.*

fiber optic rotary joint. A *passive optical device* made of *optical fibers* and *connectors* that allows signals from one or more fixed platforms to be *coupled* to one or more platforms that are rotating relative to the fixed platforms. Also see *optical rotor.*

fiber optics. **1.** The branch of science and technology devoted to the *transmission* of *radiation,* that is, *radiant power,* through *optical fibers* made of *transparent* materials, such as glass, *fused silica,* and plastic. Optical fibers in fiber optic cables may be used for data transmission and for sensing, illumination, endoscopic, control, and display purposes, depending on their use in various geometric configurations, modes of excitation, and environmental conditions. The fibers usually may be wound and bound in various shapes and distributions singly or in bundles, which may be *aligned bundles* or unaligned bundles. The aligned bundles are used to transmit and display images as well as to scramble images. **2.** In 1956, Kapany first defined fiber optics as the art of active and passive guidance of *light* as *rays* and *waveguide modes* in the *ultraviolet, visible,* and *infrared spectra* in *transparent* fibers along predetermined *paths.* See *woven fiber optics; ultraviolet fiber optics.*

fiber optic scrambler. Similar to a *fiberscope,* except that the center section of the *aligned bundle* has its *optical fibers* randomly distributed. When the randomized fibers are potted and sawed, each half is capable of coding a picture that can be readily decoded only by the other half.

fiber optic sensor. A device that responds to a physical, chemical, or *radiant field* input stimulus, such as heat, *light,* sound, pressure, temperature, strain, *magnetic* *field,* and *electric field,* and *transmits* on an *optical fiber* a *pulse, signal,* or other representation indicating one or more characteristics of the stimulus, for example, a thermometer, tachometer, hydrophone, microphone, or barometer whose output lead is an optical fiber. See *Fabry-Perot fiber optic sensor; Mach-Zehnder fiber optic sensor; Michelson fiber optic sensor; Sagnac fiber optic sensor.* Also see *optical fiber sensor.*

fiber optic splice. An inseparable, that is, permanent, joint between one *dielectric waveguide,* such as an *optical fiber,* and another, with minimal power *loss* at the junction. The purpose of the splice is to *couple optical power* between the two waveguides. Synonymous with *optical splice.* Also see *optical fiber splice.*

fiber optic splice tray. A flat, usually rectangular, pan-shaped device for holding and protecting the individual *optical fibers* from two *fiber optic cables* that have been *spliced.* The tray holds the fibers even though they may not be sheathed, *jacketed,* or otherwise covered as a group, although the fiber optic splices may be individually covered.

fiber optic splitter. A device that extracts a portion of the *optical signal power propagating* in a *dielectric waveguide,* such as an *optical tap* or a *partial mirror,* and diverts that power to a destination different from that of the remaining portion.

fiber optic spot illumination. Illumination of a relatively small area with a narrow beam of *light* emanating from the end of an *optical fiber* or *bundle* of fibers.

fiber optic station cable. See *optical station cable.*

fiber optic station facility. See *optical station facility.*

fiber optic station/regenerator section. See optical *station/regenerator section.*

fiber optic supporting structures. Structures that are used to fasten, guide, and hold *fiber optic* components in place in operational systems, for example, *bend limiters,* raceways, harnesses, hangars, ducts, and housings, without which a fiber optic system cannot be installed, held in place, and operated.

fiber optic switch. A device that can selectively transfer *optical signals* from one *optical fiber* to another depending upon the application of external stimuli, such as a force or an applied *electric, magnetic, electromagnetic,* acoustic, gravitational, or other force *field.* The switch may be *latching* or *nonlatching.*

fiber optic system. See *integrated fiber optic system.*

fiber optic telemetry system. A telemetry system in which *fiber optic* components, such as *fiber optic sensors, data links, light sources, photodetectors,* and related components, are used to gather and distribute *data* originating from a sensor.

fiber optic terminus. 1. A device used to *terminate* a *dielectric waveguide,* such as an *optical fiber,* that provides a means of locating and holding the conductor or *waveguide* within a connector. **2.** The part of a *fiber optic connector* into which the *optical fiber* is inserted and attached mechanically or by adhesive.

fiber optic test method (FOTM). A general description of the overall approach to the testing of *fiber optic* systems and components, identifying the system or component to be tested, the type and nature of the test, the parameters to be measured, and the environmental conditions under which the test is to be conducted. The FOTM does not explicitly describe in every detail the exact steps to be taken during the conduct of the test. Engineering analysis is required and one or more *fiber optic test procedures (FOTPs)* must be prepared before an FOTM can be executed. Also see *fiber optic test procedure (FOTP).*

fiber optic test procedure (FOTP). A detailed, highly explicit, unambiguous, stand-alone, step-by-step sequence of actions, or the document describing them, for testing *fiber optic* systems and components, each step precisely described, with all dimensions, instrument settings, apparatus, structures, parameters to be measured, environmental conditions, exact action to be taken at each step, and data to be taken, clearly described for each part or run of the test. An FOTP can be executed without additional engineering analysis. FOTPs are usually prepared from or based on previously prepared *fiber optic test methods (FOTMs).* Also see *fiber optic test method (FOTM).*

fiber optic transmitter. An *optoelectronic* device capable of accepting electrical *signals,* converting them into *optical signals,* and *launching* the optical signals into one or more *optical waveguides.*

fiber optic waveguide. See *optical waveguide.*

fiberoptronics. 1. An abbreviation of *fiber optics, optoelectronics,* and *electrooptics.* **2.** The branch of science and technology devoted to exploiting the advantages obtained by combining *optical fibers,* other *active* and *passive optical devices,* and electronic circuits into useful systems, for example electronically *modulating* a *light source* in order to impress information on its output light beam, guiding the beam through a *fiber optic cable* to a *photodetector,* and recovering the impressed information in electronic form.

fiber organizer. See *fiber-and-splice organizer.*

fiber parameter. See *global fiber parameter.*

fiber pattern. In the *fiber faceplate* of a *fiberscope,* the arrangement of the ends of the individual *optical fibers* that comprise the faceplate. Inspection of an image on the end surface might reveal that the fiber pattern makes the image appear as a mosaic of individual picture elements.

fiber pigtail. See *optical fiber pigtail.*

fiber polarizer. See *optical fiber polarizer.*

fiber preform. See *optical fiber preform.*

fiber-pulling machine. A device capable of *drawing* an *optical fiber* from a *preform* heated to fusion temperature. The increased temperature provides for ductility and fusion of *dopants* to control the *refractive-index profile.* The cross section decreases and the length increases as the fiber is pulled from the preform. Further steps in the fiber-drawing process include application of *buffers* or other *coatings* and packaging, such as winding on spools in readiness for the cabling process.

fiber pulse compression. See *optical fiber pulse compression.*

fiber radiation damage. See *optical fiber radiation damage.*

fiber retention. A measure of the ability of a *fiber optic* component with an *optical fiber* for access, such as a *fiber splice* or a *fiber optic coupler* with an *optical fiber pigtail,* to retain or hold the fiber when an attempt is made to pull the fiber from the component with a known or measured tensile force.

fiber ribbon. See *fiber optic ribbon.*

fiber ringer. See *optical fiber ringer.*

fiber scattering. See *configuration scattering.*

fiberscope. A device consisting of an *aligned bundle* of *optical fibers* with terminating *fiber faceplates,* an objective lens on one end for focusing the image of an object, and an eyepiece for viewing the *transmitted* image in full color on the faceplate at the opposite end. The fiberscope is used for viewing objects that are otherwise inaccessible for direct viewing, such as the interior of machines and the internal organs of the human body. See *hypodermic fiberscope.* Also see *fiber optic endoscope.*

fiberscopic recording A recording method in which *light* emerging from the *fiber faceplate* of an *aligned bundle* of *optical fibers,* that is, a *fiberscope,* is used to expose film.

fiber sensor. See *coated-fiber sensor.*

fiber source. See *optical fiber source.*

fiber splice. See *optical fiber splice.*

fiber splice housing. A housing, such as tapes, *jackets,* coatings, and other components necessary for attaching, supporting, and aligning fibers, applied over a *fiber optic splice* and proximate fiber for their protection against the environment, for some mechanical strength, and for preservation of the integrity of the *fiber coating, buffer, cladding,* and *core.* The housing includes a means for mounting the aligned assembly in an *interconnection box* or cable splice closure.

fiber strain-induced sensor. A *fiber optic sensor* in which *optical time-domain reflectometry (OTDR), interferometry, birefringence, polarization* and *polarization-plane rotation, wave coherency,* special *coatings, microbending,* and other types of schemes are used together with the application of a transverse, longitudinal, or torsional stress that produces a strain on an *optical fiber,* causing a change in *irradiance,* that is, *light intensity,* at a *photodetector.*

fiber tensile strength. See *optical fiber tensile strength.*

fiber tension sensor. A *fiber optic sensor* in which an increase in tension applied to an *optical fiber* results in an elongation. When a reference fiber not subject to the increase in tension is used and *interferometric* techniques are applied, the force producing the tension can be measured. If the force is applied directly, it can be measured directly. If the fiber is wound on a bobbin and pressure is applied to the inside of the bobbin to cause it to expand, the fiber will be subjected to tension and thereby elongated.

fiber transfer function. See *optical fiber transfer function.*

fiber transverse compression sensor. A *fiber optic sensor* in which the *optical attenuation* is a function of the longitudinally applied pressure, that is, the pressure is uniformly applied radially against the outside of an *optical fiber.* A length of cabled fiber, a *light source,* not necessarily *coherent* or *monochromatic,* and a *photodetector* are usually required.

fiber trap. See *optical fiber trap.*

field. See *borescope view field; electric field; electromagnetic field; evanescent field; reference-surface tolerance field.*

field component. See *magnetic field component.*

field coupling. See *evanescent-field coupling.*

field diameter. See *mode field diameter.*

field diffraction pattern. See *near-field diffraction pattern.*

field intensity. See *field strength.*

field radiation pattern. See *near-field radiation pattern.*

field scanning technique. See *near-field scanning technique.*

field splicing. The joining of the ends of two *fiber optic cables* in a field environment, such as on open terrain, aboard ships, and under water, without the use of
detachable *connectors* and in such a manner that continuity through the *splice* is
preserved in all elements of the cable, such as optical continuity of *optical fibers,*
electrical continuity, and continuity of *jacketing,* sheathing, *buffers,* stuffing, insu-
lation, strength members, and waterproofing.

field strength. The intensity, that is, the amplitude or magnitude, of an *electric,
magnetic, electromagnetic,* gravitational, or other force *field* at a given point. The
term is normally used to refer to the rms value of the electric field, expressed in
volts per meter, or of the magnetic field, expressed in amperes per meter. Field
strength is usually given as a gradient and therefore is a vector quantity, because it
has both magnitude and direction. For example, the units of *electric field strength*
may be volts per meter, kilograms per coulomb, or lines per square meter divided
by the *electric permittivity* of the medium at the point it is measured; *magnetic field
strength* might be given as amperes per meter, oersteds, or lines per square meter
divided by the magnetic permeability of the medium at the point it is measured;
and gravitational field strength as newtons per kilogram. See *electric field strength;
magnetic field strength.* Synonymous with *field intensity.*

field template. See *four-concentric-circle near-field template.*

field vector. See *electric field vector; magnetic field vector.*

figure. See *optical fiber merit figure.*

filled cable. A cable that has a nonhygroscopic material, usually a gel, inside the
sheath or *jacket.* The gel is used to prevent moisture from entering minor leaks.

filled fiber. See *fully filled fiber.*

film. See *optical thin film; photoconductive film.*

film optical modulator. See *thin-film optical modulator.*

film optical multiplexer. See *thin-film optical multiplexer.*

film optical switch. See *thin-film optical switch.*

film optical waveguide. See *thin-film optical waveguide.*

film waveguide. See *periodically distributed thin-film waveguide.*

filter. In *optics,* a device used to modify the *spectral* composition of *light,* such as *absorb* certain *wavelengths* or *transmit* certain wavelengths. See *core mode filter; dichroic filter; fiber optic filter; high-pass filter; interference filter; low-pass filter; modefilter; optical filter.*

finish. See *end finish.*

fixed attenuator. An *attenuator* in which the amount of *attenuation* produced cannot be varied. The attenuation is expressed in *dB*. The operating *wavelength* for *optical attenuators* and *fiber optic attenuators* should be specified for the rated attenuation, because *optical attenuation* of a material varies with wavelength.

fixed connector. In *fiber optics,* a *connector* that has one of its mating halves mounted on board, panel, casing, bulkhead, or other rigid object or surface. Also see *free connector.*

fixed optics. The development and use of *optical* components whose characteristics cannot be changed or controlled during their operational use, except perhaps for minor adjustments in position, such as in focusing telescopes, binoculars, and microscopes, to accommodate the human eye. Also see *active optics.*

flat-coil fiber sensor. A *distributed-fiber sensor* in which the *optical fiber* is distributed in a plane in rows and columns or in a flat spiral, so that the distance, usually measured by *optical time-domain reflectometry (OTDR),* to a stimulated point on the plane, such as a pressure point or a hot point, can be used to calculate the cartesian or polar coordinates of the point.

flexible borescope. A portable *fiber optic borescope* that has an *aligned bundle* (coherent bundle) in a *fiber optic cable* with sufficient flexibility to reach an inspection area via a multiturn *path* and sufficient rigidity to cross unsupported gaps in the path. There is usually an eyepiece or a *fiber faceplate* for viewing the image and a controllable articulating tip on the objective end to enable inspection at various angles, such as up to 90° from the cable *optical axis.* The borescope is hand-held and energized via a fiber optic cable from a *light source.* The flexible borescope usually consists of a basic borescope unit and ancillary equipment required for operation. Also see *rigid borescope.*

flood illumination. See *fiber optic flood illumination.*

fluorescence. The *emission* of *electromagnetic radiation* by a material during *absorption* of *electromagnetic radiation* from another *source.* Also see *luminescence; phosphorescence.*

flux. See *radiant power.*

FOG. *Fiber optic gyroscope.*

foot-candle. A unit of illuminance equal to 1 lumen (of light flux) incident per square foot. It is the illuminance of a curved surface, all points of which are placed 1 foot from a light source having a luminous intensity (candlepower) of 1 candle, that is, a candela. Thus, 4π lumens emanate from 1 candela.

format. See *signal format.*

forward-scatter. **1.** The deflection by *reflection* or *refraction* of an *electromagnetic wave* or *signal* in such a manner that at least a component of the wave is deflected in the direction of *propagation* of the *incident wave* or signal. **2.** The component of an *electromagnetic wave* or *signal* that is deflected by *relection* or *refraction* in the direction of *propagation* of the *incident wave* or *signal.* **3.** To deflect, by *reflection* or *refraction,* an *electromagnetic wave* or *signal* in such a manner that a component of the wave or signal is deflected in the direction of *propagation* of the *incident wave* or signal. The term *scatter* can be applied to reflection or refraction by relatively uniform media, but it is usually taken to mean propagation in which the *wavefront* and direction are modified in a relatively disorderly fashion. In some instances, particularly in radio transmission, the term "forward" refers to the resolution of field components when resolved along a line drawn from the source of radiation to the point of scatter. Also see *backscatter.*

FOTM. *Fiber optic test method.*

FOTP. *Fiber optic test procedure.*

four-concentric-circle near-field template. A template comprising four concentric circles applied to a *near-field radiation pattern radiating* from the exit face of a round *optical fiber.* The template is used as an overall check of various geometrical properties of the fiber all at once.

four-concentric-circle refractive-index template. A template comprising four concentric circles applied to a complete *refractive-index profile* of a round *optical fiber.* The template is used as an overall check of various geometrical properties of the fiber all at once.

fraction. See *packing fraction.*

fraction loss. See *packing fraction loss.*

frame. See *fiber optic distribution frame.*

Fraunhofer diffraction pattern. See *far-field diffraction pattern.*

free connector. In *fiber optics,* a *connector* that is not attached or associated with any other object or surface, that is, neither of its mating halves is mounted on board, panel, casing, bulkhead, or other rigid object or surface. Also see *fixed connector.*

free space. A theoretical concept of space devoid of all matter. In *electromagnetic wave propagation,* the term implies remoteness from material objects that could influence the propagation of electromagnetic waves, freedom from the influence of extraneous fields other than the fields of interest, and thus usually free of electric charges, except when the charge itself is of interest. Thus, the interior of an *optical fiber* is not considered free space, but there might be some free space between fibers, fibers and *light sources,* and fibers and *photodetectors.*

free-space coupling. Energy transfer via *electromagnetic fields* that are not in a conductor, that is, the fields are in *free space.*

free-space loss. The *signal attenuation* that would result if all obstructing, *scattering,* or *reflecting* influences were sufficiently removed so as to have no effect on *propagation. Free-space* loss is primarily caused by beam divergence, that is, geometric spreading of signal energy over larger areas at increased distances from the *source.*

frequency. For a periodic function, the number of cycles or events per unit of time. When the unit of time is 1 second (s), the measurement unit is the *hertz* (Hz). For example, in an *electromagnetic wave,* the *frequency* is the number of times per second the *electric field vector* reaches its peak value in a given direction. See *cutoff frequency; normalized frequency; pump frequency; spectrum designation of frequency; threshold frequency.*

frequency band. A continuous and contiguous group of *electromagnetic wave frequencies,* usually defined by a lower and an upper limit of frequency.

frequency cutoff. See *low-frequency cutoff.*

frequency-division multiplexing (FDM). A multiplexing scheme in which the available *transmission frequency* range is divided into narrow *frequency bands,* each used as a separate *channel.* Because an *optical fiber* can *transmit* more than one *wavelength* of *light* at the same time, each wavelength can be separately *modulated* and used as a separate transmission channel as long as a combination of dispersive components, such as prisms, and *photodetectors* on the receiving end are wavelength-sensitive for *demultiplexing.* In *fiber optics,* FDM is usually called *wavelength-division multiplexing (WDM),* because *lightwaves* and *optical* components are best and more often described in terms of wavelength and it avoids confusion with FDM that also may be in use in the same system. (See Fig. F–13, p. 108)

frequency-response function. See *transfer function.*

Fresnel diffraction pattern. See *near-field diffraction pattern.*

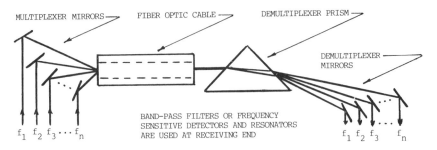

F-13. A method of performing **frequency-division multiplexing (FDM)**, *optical transmission, and demultiplexing.*

Fresnel equations. The equations that define the *reflection* and *transmission coefficient* at an *optical interface* when an *electromagnetic wave* is *incident* upon the interface surface. The coefficients are functions of the *refractive indices* of the *transmission media* on both sides of the interface, the *incidence angle,* and the direction of *polarization* with respect to the interface surface.

Fresnel reflection. The *reflection* of a portion of *incident light* at an *interface* between two homogeneous *transmission media* having different *refractive indices.* Fresnel reflection occurs at the air-glass interfaces at the entrance and exit ends of an *optical waveguide.* Resultant transmission *losses,* on the order of 4% per interface, can be eliminated by use of *antireflection coatings* or *index-matching materials.* Fresnel reflection depends on the refractive index difference across an interface surface and on the *incidence angle.* In *optical* elements, a thin *transparent* film may be used to cause an additional Fresnel reflection that cancels the original one by interference. This is called an antireflection coating. For incident *electromagnetic waves* with the *magnetic field vector* parallel to the interface, there is no Fresnel reflection at the *Brewster angle.* The *reflection coefficient* is zero and the *transmission coefficient* is unity.

Fresnel reflection method. The method for determining the *refractive-index profile* of an *optical waveguide* by measuring the *reflectance* as a function of position on an end-face.

Fresnel reflective loss. *Loss* caused by *refractive index* differences at a *terminus interface.* The loss is the *optical power* in the *Fresnel reflection.*

front-emitting LED. See *surface-emitting LED.*

front velocity. See *phase-front velocity.*

frustrated total (internal) reflection sensor. A *fiber optic sensor* consisting of two *optical fibers* with an end of each separated by an air gap produced by cutting a fiber and polishing the cut ends at a precise angle to produce *total (internal) reflection* in the *source* fiber when the fibers are widely separated. When the distance between the ends, that is, the air gap, is sufficiently small, a large portion of the *light* in the source fiber is *coupled* to the fiber connected to a *photodetector*. As the gap is widened, say by moving the fiber axes laterally away from one another, by moving the fiber end faces longitudinally away from one another, or by increasing the *angular misalignment* of the fibers, the light coupled across the gap decreases. The *irradiance* that is, the *light intensity,* at the photodetector will be a function of the *lateral offset, longitudinal offset,* or angular misalignment of the fibers.

full-duration-half-maximum (FDHM). Pertaining to the time interval during which a characteristic or property of a variable, such as the amplitude of a *pulse,* has a value greater than 50% of its maximum value.

full-wave-half-power point. In *fiber optics,* a measure of the information-carrying capacity of an *optical fiber* of given length in which the power or amplitude of an *optical pulse* of a full wave, that is, a 50% duty cycle, is reduced to half of the steady-state level. *Dispersion* and *attenuation* contribute to the reduction in pulse amplitude or *power level.*

full-width-half-maximum (FWHM). Pertaining to the range over which a characteristic or property of a variable has a value greater than 50% of its maximum value. FWHM is applied to such characteristics as *radiation patterns, optical pulse widths, spectral linewidths, beam diameters,* and *beam divergences;* and to such variables as *wavelength,* voltage, power, and current.

fully connected mesh network. In *network* topology, a network configuration in which every *node* is directly connected to every other node. Thus, each node is connected to $n-1$ *branches.* The number of branches in a fully connected mesh network is given by the relation:

$$N_b = n(n-1)/2$$

where n is the number of nodes in the network.

fully filled fiber. An *optical fiber* in which an *electromagnetic wave* is *propagating* in such a manner that the *optical power* is distributed among all the *modes* the fiber is capable of supporting in the *equilibrium modal power distribution* condition. *Leaky,* that is, *high-order, modes* are lost in long fibers. These modes are filtered by being *stripped* off. Some *light sources,* such as lensed *light-emitting diodes (LEDs)* and *edge-emitting LEDs, couple* most of their optical power into the *low-order modes.* Fiber optic *connectors* do some mode *mixing* by adding power back into the high-order modes.

function. See *impulse-response function; optical fiber transfer function (OFTF); transfer function; work function.*

fundamental mode. 1. In a *multimode fiber,* the *lowest order mode* a *waveguide* with a given *numerical aperture* and set of *launch conditions* is capable of supporting. It corresponds to the *cutoff frequency,* that is, the longest *wavelength* the guide is capable of supporting. **2.** In a *single-mode fiber,* the one *mode* is the fundamental mode. If the *frequency* of the *source* is increased, and consequently the *wavelength* decreased, so that the same *optical fiber* can support additional modes, the fundamental mode still remains the same mode, even though the same fiber is supporting more modes with wavelengths shorter than that of the fundamental mode.

fused quartz. Glass that is made by melting crystals of natural quartz, resulting in a glass that is not as pure as the *fused (vitreous) silica* glass.

fused silica. Glass that consists of pure silicon dioxide (SiO_2). Synonymous with *vitreous silica.*

fusion splice. A *fiber optic splice* made by applying sufficient heat to melt, fuse, and thus join an end from each of two lengths of *optical fiber* in order to form a single optical fiber with near-zero *attenuation* at the splice.

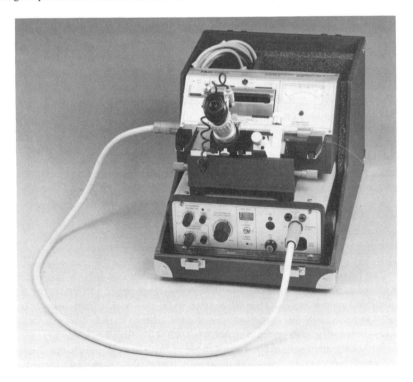

F–14. The OSG-15 Portable Fusion Splicing Kit for **fusion splicing** *optical fibers.* (Courtesy Philips Telecom Equipment Company, a division of North American Philips Corporation).

fusion splicing. In *optical transmission* systems using solid *dielectric propagation media,* the joining of two media by butting them, forming an *interface* between them and then removing the common surfaces by a melting process so that there is no interface between them. Thus, a *lightwave* will be propagated from one propagation medium, such as an *optical fiber,* to another without *reflection* at the joint and therefore without an *insertion loss.* However, the joint is never perfect, and therefore some small loss will occur at the *fiber optic splice.* (See Fig. F-14)

FWHM. *Full-width-half-maximum.*

G

g. In *fiber optics,* the symbol used for the *refractive-index profile parameter.* Also see *Power-law refractive-index profile.*

gain. See *statistical terminal/regenerator system gain; terminal/regenerator system gain.*

gamma photon. A fundamental particle or *quantum* of *gamma radiation,* with an energy equal to *hf,* where *h* is *Planck's constant* and *f* is the *frequency* of the *radiation,* that is, of the *photon.* Gamma-radiation frequency is higher than that of *x-rays* and *optical spectra,* which includes *visible, infrared,* and *ultraviolet spectra.* Gamma radiation occupies the 10^{10} − to 10^{12} − GHz *band.*

gamma radiation. *Electromagnetic waves,* of a *frequency* higher than that of *light-waves* and *x-rays,* that can be stopped or *absorbed* only by dense materials, such as lead and depleted uranium. They cannot be confined to or guided in *optical fibers* and are usually produced by nuclear reaction and high-powered *lasers* when transitions occur from very high to very low electron energy levels in molecules. Frequencies of gamma rays lie in the 10^{10} − to 10^{12} − GHz *band.* At these frequencies even single *photons* can be destructive to human tissues.

gap energy. See *band-gap energy.*

gap loss. In *fiber optics,* the *radiant power loss* caused by a space between *optical fiber* end faces at a joint, such as in a connector, even though the fiber axes are angularly aligned. The loss increases as the spacing departs from optimum. The loss is extrinsic to the fibers. For *waveguide*-to-waveguide *coupling,* it is called *longitudinal offset loss.*

gas laser. See *mixed-gas laser.*

Gaussian beam. A beam whose *irradiance,* that is, *radiant power density* or *light intensity* across its diameter, is somewhat proportional to a Gaussian distribution curve, that is, the bell curve, in which the maximum *electromagnetic field strength* occurs at the center and diminishes toward the edges, the cross section normally being symmetrical in field strength about the center. When such a beam is circular in cross section, the *electric field strength* is given by the relation:

$$E_r = E_0 e^{-(r/w)}$$

where E_r is the electric field strength at the distance r from the center, E_0 is the electric field strength at the beam center, r is the distance from the center for which E_r is determined, and w is the beam width, that is, the radius at which the field strength is $1/e$ of its value at the center.

Gbps. *Gigabits per second.*

geometric optics. 1. The *optics* of *light rays* that follow mathematically defined *paths* when passing through *optical* elements, such as lenses, prisms, and other *optical transmission media* that *refract, reflect,* or *transmit electromagnetic radiation.* **2.** The branch of science and technology that treats *lightwave propagation* in terms of *rays* considered as straight or curved lines in homogeneous, nonhomogeneous, *isotropic,* and *anisotropic media,* rather than as *wavefronts* as in *physical optics.* Light rays and wavefronts are always and everywhere orthogonal.

geometric spreading. In a *wave* propagating in a *propagation medium* in which there are no *sources,* the decrease in *irradiance,* that is, the *power density,* as a function of distance, in the direction of propagation, that is caused only by the divergence of rays rather than by *absorption, scattering, diffusion,* or transverse *radiation.* For example, a point source, or a source with a nonzero *exit angle,* will have its energy spread over larger and larger areas as the distance from the source increases. Geometric spreading is not considered to take place in a *collimated beam,* though some spreading will occur because collimation is never perfect and some *light* is scattered transverse to the beam by discontinuities in the *transmission* media, such as air molecules, smoke particles, or water vapor.

GI fiber. See *graded-index (GI) fiber.*

gigabit per second. A unit of *data signaling rate* equal to 10^9 bits/s.

gigahertz. A unit of *frequency* equal to 1 billion (U.S.) *hertz,* or 10^9 Hz.

glass. See *bulk optical glass; dry glass; halide glass.*

glass fiber. See *all-glass fiber; drawn glass fiber.*

glass powder. Pulverized glass produced from raw materials by industrial processing. When in the molten state, the glass is purified and dried for making *optical fiber preforms.*

global attenuation-rate characteristic. See *fiber global attenuation-rate characteristic.*

global-dispersion characteristic. A plot of *dispersion* (ps/nm-km) as a function of *wavelength* for a dielectric waveguide. Also see *global-dispersion coefficient; zero-dispersion slope.*

global-dispersion coefficient. In an *optical station/regenerator section,* a value of the *optical fiber chromatic dispersion coefficient* that lies between the upper and lower limit of the optical fiber *global-dispersion characteristic.* The upper limit of the global-dispersion coefficient (ps/(nm-km)) is given by the relation:

$$D_1(\lambda) = S_{0\,max}\,(\lambda - \lambda_{0\,min}{}^4/\lambda^3)/4$$

and the lower limit is given by the relation:

$$D_2(\lambda) = S_{0\,max}\,(\lambda - \lambda_{0\,max}{}^4/\lambda^3)/4$$

where λ is the operating wavelength, $S_{0\,max}$ is the maximum worst-case zero-dispersion slope (ps/(nm²-km)), $\lambda_{0\,min}$ is the minimum value of the zero-dispersion wavelength, and $\lambda_{0\,max}$ is the maximum value of the zero-dispersion wavelength, the variation in values being caused by manufacturing tolerances, temperature, humidity, and aging.

global fiber parameter. An *optical fiber* parameter, such as fiber *attenuation rate, dispersion,* and *splice insertion loss,* that is likely to vary as a result of temperature, humidity, and aging and for which an allowance should be made when designing a *terminal/regenerator section,* so as to permit unforseen conditions and events, such as upgrading of terminal equipment, rerouting of traffic, restoration of services, and prevention of traffic saturation. If a known upgrade is planned in the design of an *optical station/regenerator section,* the required (global) parameters should be specified at the outset.

Goos-Haenchen shift. The *phase shift* that occurs in a *lightwave* when it is *reflected,* the magnitude of the shift being a function of the *incidence angle* and the *refractive-index* gradient at the reflecting *interface* surface. This shift occurs to some degree with every internal reflection that occurs in an *optical fiber.* When *total (internal) reflection* occurs, the amount of phase shift in the direction of *propagation* can be precisely determined. The *ray* trajectory is changed from what would be predicted from simple ray theory because of the penetration of the ray into the lower refractive-index material of the *cladding* a short distance before it is turned completely from the *incidence angle* to the *reflection angle.* The ray trajectory is shifted in the direction of the fiber *optical axis.* If the refractive indices at the *core*-cladding interface were a perfect step function, that is, a step-change in refractive index over zero distance, there would be no Goos-Haenchen shift. There would simply be *Fresnel spectral reflection.* Also see *reflected ray.*

graded-index (GI) fiber. An *optical fiber* with a *core refractive index* that is a function of the radial distance from the fiber *optical axis,* the refractive index becoming progressively lower as distance from the axis increases. This characteristic causes the *light rays* tending to leave the *core* to be continuously *refracted,* that is, redirected back into the core, thus forcing them to remain in the core.

graded-index profile. 1. In an *optical fiber,* a *refractive-index profile* in which the *refractive index* varies with the radial distance from the *optical axis* of the *core.* The refractive index decreases as the distance from the center increases. **2.** The condition of having the *refractive index* vary continuously and smoothly between two extremes. **3.** Any *refractive-index profile* that varies with radial distance. It is distinguished from a *step-index profile,* in which there is no variation of refractive index from the *optical axis* to the *core-cladding interface.*

grating. See *diffraction grating.*

grating sensor. See *moving grating sensor.*

Green interferometer. See *Twyman-Green interferometer.*

grooved cable. A *fiber optic cable* in which the *optical fibers* are fitted into grooves in a cylindrical element. Fiber optic cables with a large number of fibers can be produced by stranding two or more cylindrical elements together and sheathing, or *jacketing,* the whole assembly.

group delay. In *optics,* the time required for *optical power propagating* in a given *mode* at the modal *group velocity* to travel a given distance. For measurement purposes, the unit of interest is the *group delay* per unit length, which is the reciprocal of the group velocity of the particular mode. The measured group delay per unit length of an *optical signal* through an *optical fiber* exhibits a *wavelength* dependence caused by the various *dispersion* mechanisms in the fiber. Also see *Sellmeier equation.* See *multimode group delay.*

group-delay time. The rate of change of the total *phase shift* with angular *frequency,* that is, angular velocity, through a device or *transmission* system. Group-delay time is the time interval that is required for the crest of a group of *waves* to *propagate* through a device or transmission facility or system where the component wave trains have slightly different individual frequencies. Any other *significant instant* or point in the wave may be chosen instead of the crest. Thus, there is a rate of change of the total phase shift with angular velocity through the device, facility, system, or component. The group delay time is expressed by the relation:

$$t_g = \Delta\theta/\Delta\omega \qquad \text{or } t_g = d\theta/d\omega$$

where θ is the phase shift angle, and ω is the angular frequency, that is, the angular velocity. The group-delay time has the units of time because θ is normally expressed in radians and ω in radians/second. The angular velocity is given as $\omega = 2\pi f$, where f is the frequency. The changes occur over the length of the *propagation medium* in the direction of propagation. Thus, for a wave of a certain frequency in a group of waves with different frequencies, the group-delay time is the first derivative with respect to frequency of the phase shift between two given points in the propagating wave. In a transmission system where phase shift is proportional to frequency, the group-delay time is constant.

group index (N). In *fiber optics,* for a given mode in an *electromagnetic wave propagating* in a given *propagation medium,* the group index is given by the relation:

$$N = c/v_g$$

where c is the velocity of *light* in vacuum, and v_g is the *group velocity* of the *mode* in the medium. For a *plane wave,* the group index for a given mode is given by the relation:

$$N = n\lambda(dn/d\lambda)$$

where n is the *refractive index* of the medium in which the wave is propagating, and λ is the *wavelength.*

group velocity. The velocity of *propagation* of an envelope produced when an *electromagnetic wave* is *modulated* by or mixed with other waves of different *frequencies.* Ideally, the group velocity would be the velocity of a *signal* represented by two superimposed sinusoidal waves of equal amplitude and slightly different frequencies approaching a common limiting value. It is the velocity of information propagation and, loosely, of energy propagation. The modulating signal is considered the information-bearing signal, and the modulated signal is considered the *carrier.* Thus, the group velocity is the velocity of *transmission* of energy associated with a progressing wave consisting of a group of sinusoidal components, that is, the velocity of a certain feature of the wave envelope, such as the crest. The group velocity differs from the *phase velocity.* Only the latter varies with frequency. In general, the group velocity, represented by the relation:

$$v_g = d\omega/d\beta$$

is less than the phase velocity, represented by the relation:

$$v_p = \omega/\beta$$

where ω is the carrier angular velocity, $d\omega$ is the *signal* angular velocity, and β is the *phase constant* at the carrier frequency. βL is the carrier phase delay for a line L meters long. The angular velocity, that is the angular frequency is $2\pi f$, where f is the frequency. The group velocity is also equal to the rate at which frequency changes with respect to the reciprocal of the *wavelength.* In *nondispersive media,* the group velocity and the phase velocity are equal if the phase constant is a linear function of the angular velocity. In a *waveguide,* each mode has its own group velocity. Also see *phase velocity.*

guided mode. See *bound mode.*

guided ray. In an *optical waveguide,* a *ray* that is confined to the *core* or is *bound* to a *mode* in the core. Specifically for an *optical fiber,* a ray at a given radial position having a direction that satisfies the relation:

$$0 \leq \sin \theta_r \leq [n_r^2 - n_a^2]^{1/2}$$

where *r* is the radial distance of the ray from the *optical axis,* θ_r is the angle that the ray makes with the *optical axis* at the radial position *r,* n_r is the *refractive index* at the radial position *r,* and n_a is the refractive index at the core radius, that is, the radius at the core-cladding interface. Synonymous with *bound ray; trapped ray.*

guided wave. A *wave* whose energy is confined between surfaces or in the vicinity of surfaces of materials because of specific properties of the materials. In guided *electromagnetic waves,* wave energy is concentrated near *interfaces,* i.e., between substantially parallel boundaries separating materials of different properties and whose direction of *propagation* is effectively parallel to these boundaries. A guided electromagnetic wave may consist of several *bound modes.* Also see *bound mode.*

guiding fiber. See *weakly guiding fiber.*

gyroscope. See *fiber optic gyroscope (FOG).*

H

h. See *Planck's constant.*

half-duration. See *pulse half-duration.*

half-maximum. See *full-duration-half-maximum (FDHM); full-width-half-maximum (FWHM).*

half-power point. See *full-wave-half-power point.*

halide glass. A special glass, containing halogens, such as fluorine, in combination with heavy metals, such as zirconium, barium, or hafnium, that transmits *light* in the *visible* and *infrared spectra* of the *electromagnetic spectrum,* such as *wavelengths* up to 7 μ *(microns).* Applications include remote-sensing systems, night-vision devices, and fiber optic systems. Halide glasses have a potential *ultralow loss fiber optical attenuation rate* of 10^{-3} dB/km, making them particularly attractive for long-distance, repeaterless communication *links.* (See Fig. H–1)

hard-clad silica (HCS) fiber. An *optical fiber* with a *silica (glass) core* and a hard polymer (plastic) *cladding* bonded to the core.

hardness. See *nuclear hardness; radiation hardness.*

harmonic distortion. See *total harmonic distortion.*

harness. See *optical harness.*

harness assembly. See *optical harness assembly.*

hazard. See *optical fiber hazard.*

H-bend. A gradual change in the direction of the axis of a *waveguide* throughout which the axis remains in a plane parallel to the direction of the *magnetic vector (H) transverse polarization.* Synonymous with *H-plane bend.* Also see *E-bend.*

HCS fiber. See *hard-clad silica (HCS) fiber.*

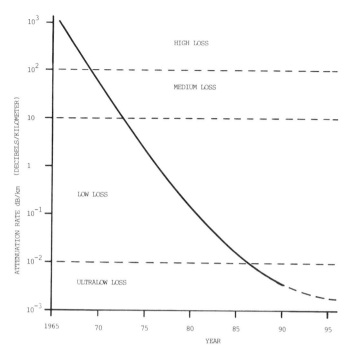

H-1. Reduction in *optical attenuation* in *optical fibers* achieved by industry and research laboratories by improved *bulk optical glass* purification processes and forming methods and by production of special glasses, such as **halide glasses.** The lower limit is caused by *Rayleigh scattering, infrared absorption, overtone absorption, ultraviolet absorption,* and other causes.

HDTV system. *High-definition television system.*

heavy-duty connector. In *fiber optics,* a *connector* that is designed and intended for use outside an *interconnection box* (distribution box) or cabinet.

helical polarization. See *left-hand helical polarization; right-hand helical polarization.*

Helmholtz equations. The set of equations that describe *uniform plane-polarized electromagnetic waves* in an unbounded, lossless, that is, non*absorptive, source*-free *propagation medium,* such as an *optical fiber,* when (1) the *wave equations* are expressed in cartesian, that is, rectangular, coordinates, (2) the components of the *electric* and *magnetic fields* in the direction of *propagation* are identically equal to zero, and (3) the transverse first derivatives d/dx and d/dy when the wave is propagating in the z direction are also identically zero. The Helmholtz equations are given by the relations:

$$d^2E_x/dz^2 - \Gamma^2E_x = 0 \qquad d^2E_y/dz^2 - \Gamma^2E_y = 0$$
$$d^2H_x/dz^2 - \Gamma^2H_x = 0 \qquad d^2H_y/dz^2 - \Gamma^2H_y = 0$$

where E is the scalar *electric field strength* component, and H is the scalar *magnetic field strength* component in the direction indicated by the subscript for a z-direction propagating *wave*. $\Gamma^2 = j\omega\mu(\sigma + j\omega\epsilon) = -\omega^2\mu\epsilon + j\omega\mu\sigma$, where j is $-1^{1/2}$, ω is the angular velocity, μ is the *magnetic permeability*, ϵ is the *electric permittivity*, and σ is the *electrical conductivity*. Thus, Γ^2 is a complex function. For *dielectric* materials, such as the glass used in *dielectric waveguides*, $\mu = 1$ and $\sigma = 0$. These equations specifically apply to *propagation* in *optical* elements under certain geometrical conditions. These equations are solutions of the vector Helmholtz equations given by the relations:

$$\nabla^2E - \Gamma^2E = 0 \qquad \nabla^2H - \Gamma^2H = 0$$

where ∇ is the spatial derivative used in the vector algebra, and Γ is as above.

hertz (Hz). The Systemè International (SI) unit for *frequency,* equivalent to 1 cycle of a recurring phenomenon per second. The equivalent term cycles per second is no longer used. For *electromagnetic waves* the frequency in hertz is given by their speed in meters per second divided by their wavelength in meters.

Hertzian wave. An *electromagnetic wave* that has a *frequency* lower than 3000 GHz, that is, 3×10^{12} *hertz,* and is *propagated* in *free space,* that is, without a *waveguide,* for example, radio *waves. Lightwaves* are of the order of 3×10^{14} hertz. Hertzian waves and lightwaves are electromagnetic waves.

heterodyne. The process of combining two *electromagnetic waves* of different *frequency* in a nonlinear device to produce frequencies that are equal to the sum and difference of the combining frequencies. See *self-heterodyne.* Also see *homodyne.*

heterojunction. A semiconductor *p-n junction* in which the two regions differ in conductivity, because of their different *dopant* levels, and differ in their atomic compositions. In heterojunction transistors, diodes, and *lasers,* a sudden transition occurs in material composition across the *junction* boundary, such as a sudden change in *refractive index* or a transition from *p-type,* that is positively doped, to *n-type,* that is, negatively doped, material. See *single heterojunction.* Also see *homojunction.*

high-definition television (HDTV) system. A television system with more than two times the *resolution* of the existing National Television Standards Code (NTSC) published by the National Television Standards Committee (NTSC) for North America, with 1125 horizontal scan lines and a wider screen. *Fiber optic cables* have the broader *bandwidth* required for over 100 HDTV station *channels,* with *bandwidth* to spare for an *integrated-services data network (ISDN),* including *optical fiber* in the *local loop.* HDTV systems are not compatible with conventional systems and would render them obsolete. See *advanced television system; enhanced-definition television (EDTV) system; improved-definition television (IDTV) system.*

high-order mode. In an *electromagnetic wave propagating* in a *waveguide*, a *mode* corresponding to the larger eigenvalue solutions to the *wave equations*, regardless of whether it is a mode in *transverse electric (TE), transverse magnetic (TM),* or *transverse electromagnetic (TEM) waves.* Thus, it is a mode in which more than just a few whole *wavelengths* of the wave fit transversely in the guide and therefore can be supported by the guide. A *multimode fiber* supports as many modes as the *core diameter, numerical aperture,* and wavelength permit. For the high-order modes, the *normalized frequency* must be above 4 in order that the number of modes be sufficiently high for there to be high-order modes. In a fiber supporting a *single mode,* there are no high-order modes. Conceptually, however, if there are only two modes, one could be considered the *low-order mode* and the other the high-order mode. Also see *low-order mode.*

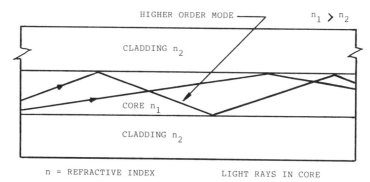

HIGHER ORDER MODE

$n_1 > n_2$

CLADDING n_2

CORE n_1

CLADDING n_2

n = REFRACTIVE INDEX   LIGHT RAYS IN CORE

H–2. **High-order modes** in an *optical fiber* have shorter *wavelengths* and higher *frequencies* than *low-order modes.* The higher order modes experience more *reflections* and hence are more apt to leak out of the *core* and *radiate* away.

high-pass filter. A *filter* that passes *frequencies* above a given frequency and *attenuates* all others.

homodyne. Pertaining to the process of combining two *waves,* such as *electromagnetic waves,* of the same *frequency.* For example in *fiber optics,* pertaining to *detection* based on the use of only one frequency, such as *monochromatic light,* to achieve detection. The *light* is *launched* into a length of *optical fiber.* The *baseband signal,* that is, the parameter being detected, varies the length of the fiber, thus shifting the *phase* of the output wave. The output wave is combined with the original wave with a *beam splitter* and sent to a *photodetector.* Shifting the phase of one wave relative to the other causes the combined wave intensity at the photodetector to vary as the amounts of cancellation and reinforcement of the two waves vary. Thus, the *photocurrent* amplitude will be a function of the phase shift, which in turn is a function of the baseband signal that is *modulating* the length of the fiber. If the relative phase shift varies through many cycles, the photodetector photocurrent frequency will be proportional to the rate of change of phase shift. Calibration enables measurement of any physical parameter that changes the length of the fiber, such as a sound wave, pressure, force, *magnetostriction,* or displacement. Another example of homodyning pertains to a system of suppressed *carrier* radio reception in which the receiver generates a voltage having the original carrier frequency and combines it with the incoming signal (zero-beat reception). Also see *heterodyne;* *homodyne detection.*

homodyne detection. *Detection,* that is, *demodulation,* using techniques that depend on mixing two *signals* of the same *frequency* to obtain the *baseband signal.* Also see *homodyne.*

homogeneous cladding. *Cladding* in which the *refractive index* has the same value everywhere. An *optical fiber* may have several homogeneous claddings, each having a different *refractive index.*

homojunction. A semiconductor *p-n junction* in which the two regions differ in *electrical conductivity* because of their different *dopant* levels, while their atomic compositions are the same. Also see *heterojunction.*

hopping. See *mode hopping.*

housing. See fiber *splice housing.*

H-plane bend. See *H-bend.*

hybrid. In electrical, electronic, *fiber optic,* or communications engineering, pertaining to a device, circuit, apparatus, or system made up of two or more components, each of which is normally used in a different application and not usually combined in a given requirement. Some examples of hybrid systems include an electronic circuit having both vacuum tubes and transistors; a mixture of thin-film and discrete integrated circuits; a computer that has both analog and digital capabilities; a transformer, or a combination of transformers and resistors, affording paths to three branches, that is, circuits A, B, and C, so arranged that A can send to C, B can receive from C, but A and B are effectively isolated; or a cable containing *optical fibers* and electrical wires.

hybrid cable. In *fiber optics,* a cable that contains *optical fibers* and electrical conductors.

hybrid connector. In *fiber optics,* a *connector* that contains *optical fibers* and electrical conductors.

hybrid mode. In an *electromagnetic wave propagating* in a *waveguide,* a *mode* that has components of both the *electric* and *magnetic field* in the direction of *propagation.* In *fiber optics,* hybrid modes correspond to *skew,* that is, non*meridional, rays.*

hydroxyl ion absorption. The *absorption* of *electromagnetic waves,* particularly *lightwaves* in *optical* glass, caused by the presence of trapped hydroxyl ions remaining from water as a contaminant. The hydroxyl ion can penetrate glass during and after product fabrication, such as while an *optical fiber* is being drawn, after it has been wound on a spool, after it has been cabled, during storage, and during operational use.

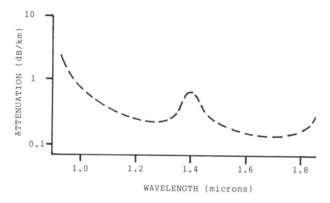

H–3. *Attenuation rate* versus *nominal central wavelength* for a typical *single-mode optical fiber.* The peak at about 1.4μ *(microns)* is due primarily to **hydroxyl ion absorption.** The attenuation rate is typically about 0.4 dB/km *(decibles/*kilometer).

hypodermic fiberscope. A *fiberscope* consisting of a *light source* for illumination; one or more *optical fibers,* say of the order of a total of 4 to 8 μ *(microns)* in diameter for forward *transmission* of the illumination for the object field and backward *transmission* of the light *reflected* from the object; and a device for displaying the reflected image from subcutaneous tissue.

Hz. Abbreviation for *hertz.*

I

identification. See *equipment identification.*

IDTV system. *Improved-definition television system.*

ILD. *Injection laser diode.*

illumination. See *fiber optic flood illumination; fiber optic spot illumination.*

illuminator. See *fiber optic illuminator.*

imagery. The representation of objects reproduced electronically or by *optical* means on film, electronic display devices, *fiber optic* display devices, or other media.

imaging. The sensing of real objects or representations of objects, followed by their representation by *optical,* electronic, or other means, on a film, screen, plotter, cathode ray tube, thin-film electroluminescent (TFEL) display panel, or other device. *Aligned bundles* of *optical fibers* may be used for imaging. Imaging *photon*-counting detector systems can produce images of objects by scanning and counting photons over a shorter period than even the fastest photographic films, such as 20 photons/ms for 2 min, compared with 14 h of exposure on the ASA 400 photographic film to produce the same quality image of the same object illuminated at an extremely low *irradiance* level.

imaging system. A display system capable of forming images of objects. Examples of imaging systems include lens systems, *optical fiber* systems, computer-controlled CRT or *fiberscope* terminals, film systems, video tape systems, and thin-film electroluminescent (TFEL) systems.

impedance. See *characteristic impedance; wave impedance.*

improved-definition television (IDTV) system. A television system in which improvements in conventional-system performance and picture definition are made that do not make existing receivers and other equipment obsolete. These improvements are considered as compatible improvements. See *advanced television system; enhanced-definition television (EDTV) system; high-definition television (HDTV) system.*

impulse response. The function that describes the response of a device after a delta-function, such as a step voltage or *optical signal,* is applied to its input. The impulse response is the inverse Laplace or Fourier transform of the *transfer function,* that is, the output function is determined by the transfer function operating upon the input function.

impulse-response function. The function that describes the response of an initially relaxed system after a Dirac-delta (step) function is applied at time $t = 0$. The root-mean-square (rms) duration of the *impulse response* may be used to characterize a component or system by means of a single parameter rather than a function. The rms value is given by the relation:

$$\sigma_{rms} = [(1/M_0)\int_{-\infty}^{+\infty} (T - t)^2 h(t) \, dt]^{1/2}$$

where:

$$M_0 = \int_{-\infty}^{+\infty} h(t) \, dt \text{ and } T = (1/M_0)\int_{-\infty}^{+\infty} t h(t) \, dt$$

The impulse response may be obtained by deconvolving (deriving) the input waveform from the output waveform, or as the inverse Fourier transform of the *transfer function.* Also see *root-mean-square (rms) pulse broadening; root-mean-square (rms) pulse duration; root-mean-square (rms) deviation; spectral width.*

incandescence. The *emission* of *light* by thermal excitation, which brings about energy-level transitions that produce sufficient quantities of *photons* with sufficient energies to render the *source* of *radiation* or the radiation itself visible.

incidence. See *normal incidence.*

incidence angle. The angle between a ray striking a surface, such as a *reflecting* or *refracting surface,* and the normal to the surface at the point where the ray strikes the surface.

incident wave. A *wave* that impinges upon or intersects an *interface* surface.

inclusion. In *optical* glass, a particle of foreign or extraneous matter, such as an air bubble or speck of dust, including undesirable impurities, such as hydroxyl or iron ions; usually an impurity. *Dopants* used to achieve desired *refractive-index profiles* are normally not considered inclusions.

incoherent bundle. See *unaligned bundle.*

incoherent radiation. *Radiation* that has a low *coherence degree.*

index. See *absorption index; group index (N); refractive index; relative refractive index.*

index contrast. See *refractive-index contrast.*

index difference. See *ESI refractive-index difference.*

index dip. See *refractive-index dip.*

index-matching material. *Transparent* material, that is, *light-transmitting* material, with a *refractive index* such that when used in intimate contact with other transparent materials, *radiant power loss* is reduced at *interfaces* by reducing *reflection,* increasing transmission, avoiding *scattering,* and reducing *dispersion.* For example, a liquid or cement whose refractive index is nearly equal to that of the core of an *optical fiber,* used to reduce *Fresnel reflections* from the fiber end face. Also see *antireflection coating.*

index of refraction. See *refractive index.*

index profile. See *graded-index profile; linear refractive-index profile; parabolic refractive-index profile; power-law refractive-index profile; radial refractive-index profile; refractive-index profile; step-index profile.*

index-profile parameter. See *refractive-index-profile parameter.*

index template. See *four-concentric-circle refractive-index template.*

indirect wave. A *wave,* such as a *lightwave,* radio wave, or sound wave, that arrives at a point in a *propagation medium* by *reflection* or *scattering* from surrounding objects or surfaces, such as a mountain for sound waves, a steel structure for radio waves, and a *refractive-index interface* surface for lightwaves.

induced attenuation. The increase in *attenuation* in a component or system or the increase in *attenuation rate* caused by external factors, such as environmental changes, mechanical treatment, and immersion in fluids over and above the *attenuation* or attenuation rate measured before the external factors were applied to the component or system. When *fiber optic* components, such as *optical fibers, connectors, splices,* and *couplers* are tested, the induced attenuation or attenuation rate is determined by measuring the attenuation or attenuation rate before and after the test. Synonymous with *change in transmittance; transmittance change.* Also see *fiber optic dosimeter.*

induced modulation. See *mechanically induced modulation.*

information descriptor. See *receiver information descriptor; transmitter information descriptor.*

information payload capacity. See *payload rate.*

information rate. See *system information rate.*

infrared (IR). Pertaining to *electromagnetic radiation* in the range of *frequencies* that extends from the longest *wavelength* of the *visible red region* of the *spectrum* to the *radio* wave region, the frequency being lower and the wavelength longer than visible red. The *band* of IR lies between the longest wavelength of the visible spectrum, about 0.8 μ *(microns)* and the shortest radio waves, about 100 μ. The IR region of the spectrum is often divided into *near*-IR, 0.8-3; *middle*-IR, 3-30; and *far*-IR, 30 to 100 μ. See *far-infrared; middle-infrared; near-infrared.*

injection fiber. See *launching fiber.*

injection laser diode (ILD). See *laser diode.*

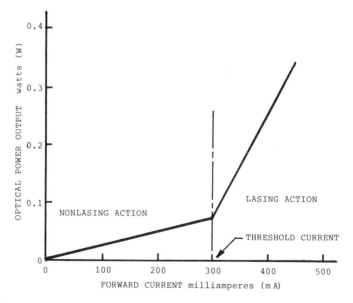

I-1. *Optical power output* versus forward electrical current input for a typical **injection laser diode** *light source.*

injection-locked laser. A *laser* whose *peak spectral wavelength,* that is, peak intensity emission wavelength, is controlled by the injection of a separate *optical signal* from a different source or from an optical signal *reflected* from an external mirror.

input. See *maximum receiver input.*

input power. See *optical receiver maximum input power.*

insert. In a *fiber optic connector,* the portion of the connector that contains the precision alignment components. See *aerial insert.*

insertion loss. 1. In an *optical system,* the *radiant power loss* caused by *absorption, scattering, diffusion, leaky waves, dispersion, microbends, macrobends, reflection, radiation,* or other causes when an *optical* component is inserted into the system, such as when a *fiber optic connector* or *splice* is inserted in a *fiber optic cable.* **2.** In a *fiber optic transmitter* for an *optical link,* the *radiant power* lost at the point where the *light source* output is *coupled* into the *optical fiber,* that is, coupled into the *optical fiber pigtail.* Also see *launch loss; transmission loss.*

inside vapor deposition (IVD). In the production of *optical fiber preforms,* the deposition of *dopants* on the inside of a rotating glass tube, called a *bait tube,* which is then sintered to produce a doped layer of higher *refractive-index* glass on the inside. A solid *optical fiber* is then pulled from the tube as part of the *fiber drawing* process.

inside vapor-phase-oxidation (IVPO) process. A *chemical vapor-phase-oxidation (CVPO) process* for production of *optical fibers* in which *dopants* are burned with dry oxygen mixed with a fuel gas to form an oxide stream, that is, a soot stream. The stream is deposited on the inside of a rotating glass tube, called a *bait tube,* and then sintered to produce a doped layer of higher *refractive-index* glass on the inside for forming the *core.* A solid optical fiber is then pulled from the tube as part of the *fiber-drawing* process.

inspection lot. A group of units of a given product, taken at random from a given production run in which all units were produced under the same conditions, from which a sample, consisting of one or more units, is to be taken and inspected to determine conformance with acceptability criteria. The number of units in the inspection lot and the number of units to be randomly selected from the inspection lot for testing depend on the number of units being acquired or purchased.

instant. See *significant instant.*

integrated fiber optic communication system. A communication system or *network* consisting of *fiber optic links* connected to other types of links, such as wire and microwave links.

integrated optical circuit (IOC). A circuit or group of interconnected circuits either monolithic or hybrid, consisting of miniature solid-state active and *passive optical,* electrical, *electrooptic, or optoelectronic components,* such as *light-emitting diodes (LED); optical filters; photodetectors; memories;* differential amplifiers; logic gates, such as AND, OR, NAND, NOR, NEGATION gates; and *thin-film optical wave-guides,* usually on *semiconductor* or *dielectric* substrates, used for *signal* processing.

integrated optics. 1. The branch of science and technology devoted to the interconnection of miniature *optical components* via *optical waveguides* on *transparent dielectric* substrates, such as lithium niobate and titanium-doped lithium niobate, to make *integrated optical circuits (IOCs).* **2.** The design, development, and operation of circuits that apply the technology of integrated electronic circuits produced by planar masking, etching, evaporation, and crystal film growth techniques to microoptic circuits on a single planar *dielectric* substrate. Thus, a combination of electronic circuitry and *optical waveguides* is produced for performing various communication, switching, and logic functions, including amplification, gating, *modulating, light* generation, *photodetecting, filtering, multiplexing, signal* processing, *coupling,* and storing.

integrated receiver. See *PIN-FET integrated receiver.*

integrated-services digital (or data) network (ISDN). A communication *network* capable of providing video, voice, *digital,* and *analog data transmission* and interconnection services over a single *transmission medium,* such as a network in which *optical fiber* is used for *local loops, local-area networks (LANs), metropolitan-area networks (MANs),* and long-haul trunks between *switching centers.* For example, one *single-mode optical fiber* in the local loop to home or office could handle all forms of *signals* and many different *user end-instruments* because of the *wideband* capability of the fiber, such as 60 analog video *channels,* or, for 100-Mbit/s FSK *digital data* signaling with 200-MHz spacing, a *carrier/noise* ratio of 16 dB, a BER of 10^{-9}, and *modulation* of a *laser source* with a 4-GHz signal at *frequencies* between 2 and 6 GHz, integrated services on the single fiber can be all of (1) 50 analog broadcast-quality video channels: 2 GHz; (2) 4 *HDTV* channels, FM modulation: 800 MHz; (3) 4 switched-video channels, digital: 800 MHz; (4) 25 digital audio channels: 200 MHz; and (5) voice, data, and other services channels: 200 MHz. The *bandwidths* needed for each service group sum to 4 GHz.

intensity. See *radiant intensity.*

interconnect feature. See *optical cable interconnect feature.*

interconnection. See *fiber optic interconnection; Open Systems Interconnection (OSI).*

interconnection frame. See *fiber optic interconnection box.*

interface. 1. A concept involving the interconnection between two devices or systems. The definition of the interface includes the type, quantity, and function of the interconnecting circuits and the type and form of *signals* to be interchanged via those circuits. Mechanical details, such as details of plugs, sockets, and pin numbers, may be included in the interface definition. **2.** A shared boundary, such as the boundary between two subsystems or two devices. **3.** A boundary or point common to two or more similar or dissimilar command and control systems, subsystems, or other entities at, against, or through which information flow takes place, **4.** A boundary or point common to two or more systems or other entities across which useful information flow requires the definition of the interconnection of the systems that enables them to interoperate. **5.** The process of interrelating two or more dissimilar circuits or systems. **6.** In telephone *networks,* the point of interconnection between terminal equipment and telephone company communication facilities.

interface device. See *network-interface device.*

interface rate. The gross bit rate of an *interface* after all processing is completed, for example, the actual bit rate at the boundary between the physical layer and the physical medium.

interface standard. See *fiber distributed-data interface (FDDI) standard.*

interference. 1. The phenomenon that occurs when two or more *coherent waves* are superimposed so that they form beats in time or patterns in space. **2.** Noise induced in a system by another system. See *intersymbol interference.*

interference filter. An *optical filter* that consists of one or more thin layers of *dielectric* or metallic material and whose operation is based on *interference* effects. Synonymous with *multilayer filter.*

interferometer. An instrument in which the *interference* of *lightwaves* is used for making measurements. See *Twyman-Green interferometer.*

interferometric sensor. In *fiber optics,* a sensor in which a single *coherent* narrow-*spectral-width* beam is divided into two or more beams. The individual beams are subjected to the various environments or stimuli that are being sensed, then recombined with the original beam so that cancellation and enhancement vary as the *phases* of the waves are shifted relative to each other, resulting in a *modulation* of *irradiance,* that is, *light intensity, incident* upon a *photodetector. Optical fibers* are used to convey the light beams. Usually there is a sensing leg and a reference leg, the outputs of which are combined to develop the intensity variation. The sensing leg is exposed to the physical variable being sensed, while the reference leg is protected from the stimulating physical variable. The sensor will still perform properly if both legs are equally exposed to another physical variable not being sensed.

interferometry. The branch of science and technology devoted to measurements involving the interaction of *waves,* such as the interaction between *electromagnetic waves,* such as *lightwaves.* Lightwaves can be made to interact so as to produce various spatial, time-domain, and frequency-domain *light* energy distribution patterns. *Modulation* of these variations can be used in *fiber optic sensors.* See *slab interferometry; transverse interferometry.*

intermediate-field region. Relative to a source of *electromagnetic radiation,* the region between the *near-field* and the *far-field regions.*

intermodal dispersion. See *modal dispersion.*

intermodal distortion. See *multimode distortion.*

internal photoeffect detector. A *photodetector* in which *incident photons* raise electrons from a lower to a higher energy state, resulting in an altered state of the electrons, holes, or electron-hole pairs generated by the transition, which is then detected. The effect is used in photodetectors to detect *signals* at the end of an *optical fiber.*

internal photoelectric effect. The changes in characteristics of a material that occur when *incident photons* are *absorbed* by the material and excite the electrons in the various energy bands in the molecules composing the material. Characteristic changes include changes in *electrical conductivity,* photosensitivity, and electric potential development. For example, electrons may move from a valence band to a conduction band when the material is exposed to *radiation.*

internal reflection. In an *optical* element in which an *electromagnetic wave* is *propagating,* a *reflection* at an outside surface toward the inside, such that a wave that is *incident* upon the surface is reflected wholly or partially back into the element itself. *Optical fibers* depend on *total (internal) reflection* at the *core-cladding interface* for successful *transmission* of *lightwaves* by confining enough of the transmitted energy to the core all the way to the end of the fiber.

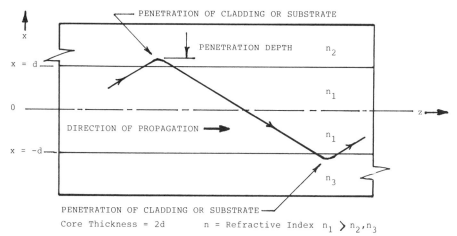

I-2. **Internal reflection** of a *ray* of an *electromagnetic wave* in a *slab-dielectric (planar) waveguide* in which n_2 and n_3 are each less than n_1.

internal reflection sensor. See *frustrated total internal reflection sensor.*

intersymbol interference. Extraneous energy from a *signal* in one or more keying intervals, that is, temporal or spatial separation beteen corresponding points on two consecutive *pulses,* that tends to interfere with the reception of the signal in another keying interval, or the disturbance that results therefrom. In *fiber optics,* intersymbol *interference* can occur when *dispersion* causes *pulse broadening,* that is, *spreading* in time and space along the fiber as it *propagates,* resulting in *pulse* overlap that might be so large that a *photodetector* could no longer distinguish clear boundaries between pulses. When this occurs to a sufficient degree in *digital data* systems, the *bit-error ratio (BER)* may become excessive. If the spread is too much, the *data signaling rate (DSR)* has to be reduced so as to provide more time, or space, between pulses.

interval. See *Nyquist interval.*

intramodal dispersion. See *material dispersion.*

intramodal distortion. *Distortion* caused by *dispersion,* such as *material* or *profile dispersion,* of a specific *propagating mode,* for example, distortion resulting from dispersion of *group velocity* of a propagating mode. Intramodal distortion is the only form of modal distortion that can occur in a *single-mode fiber.*

intrinsic coupling loss. In *fiber optics,* the *coupling insertion loss* caused by *optical fiber* parameter mismatches when two dissimilar fibers are joined. Synonymous with *intrinsic joint loss.* Also see *extrinsic coupling loss.*

intrinsic joint loss. See *intrinsic coupling loss.*

intrinsic quality factor (IQF). In *fiber optics,* a factor that indicates the quality of a *multimode fiber* referenced to the intrinsic *splice loss.* The IQF can be used as an alternative to specifying precise requirements of *refractive-index profile* characteristics, such as *core diameter, numerical aperture, concentricity errors,* and *core noncircularity* by simply requiring that fibers meet an average intrinsic splice loss. This approach allows simultaneous parameter deviations to compensate for each other in terms of the measured intrinsic splice loss.

inversion. See *population inversion.*

IOC. *Integrated optical circuit.*

ion absorption. See *hydroxyl ion absorption.*

ion-exchange process. In *fiber optics,* a *graded-index optical fiber* manufacturing process in which the fiber is fabricated by replacing undesirable ions with desirable ions to achieve the desired composition and structure of the material used to fabricate the fiber.

ion laser. A *laser* using ionized gases, such as argon, krypton, and xenon, to produce *lasing* action.

IQF. See *intrinsic quality factor (IQF).*

IR. *Infrared.*

irradiance. 1. The *optical* energy per unit of time *transmitted* by a *light* beam through a unit area normal to the direction of *propagation* or the direction of maximum power gradient, expressed in watts per square meter or joules per second-(square meter), joules per cubic meter, or lumens per square meter, that is, it is the *electromagnetic radiant power* per unit area *incident* upon a surface. If the *radiant power* is *incident* upon a surface, the amount of radiant power distributed per unit area of the surface is the irradiance. Thus, the radiant power in a beam is the total power obtained by integrating the irradiance over the cross-sectional area of the beam. If the radiant power in a beam is distributed uniformly over the cross-sectional area of the beam, the irradiance is the radiant power divided by the cross-sectional area, which is also the average irradiance. For an *electromagnetic wave propagating* in a vacuum or a material medium, the radiant power per unit transverse area propagating in the direction of maximum power gradient is also the irradiance, that is, the power density, which is usually expressed in watts per square meter. Because the wave is propagating with a velocity given by the ralation:

$$v = c/n$$

where c is the speed of light in a vacuum, and n is the *refractive index* of the material at the point at which the speed is being considered, the irradiance can also be expressed in terms of the energy per unit volume of vacuum or *propagation medium,* expressed as joules per cubic meter, hence giving rise to the term "energy density." **2.** *Radiant power* per unit area passing through *free space* or a material medium in a direction perpendicular to the unit area, that is, in the same direction as the unit area vector. In an *electromagnetic wave,* the power-flow rate, a vector quantity, propagates in a specific direction at a particular point in a *propagation medium.* Therefore, for an electromagnetic wave, the irradiance may be expressed (1) as the energy per unit time, that is, the power, passing through a unit cross-sectional area that is perpendicular to the direction of propagation of the wave, (2) as the scalar energy per unit volume contained in an electromagnetic wave propagating in a propagation medium, and (3) as *visible radiation,* that is, as the lumens per square meter of area perpendicular to the direction of propagation. All three dimensional concepts give rise to the term "power density." Irradiance, or power density, usually diminishes with distance from the *source,* due to *absorption, dispersion, diffusion,* deflection, *reflection, scattering, refraction, diffraction,* and *geometric spreading.*

Irradiance is a point function. It can be increased at a point by causing convergence of waves, such as might be accomplished by a convex lens or concave mirror or reflecting antenna, that is, a dish. A convex lens only a few centimeters in diameter and less than a centimeter thick can increase the irradiance of solar radiation, that is, sunlight, when focused at a point on the surface of a combustible material, to a level high enough to ignite the material, but the radiant power entering the lens and incident on the surface is the same. Synonymous (colloquial) with *power density.* See *spectral irradiance.* Also see *radiance; radiant power.*

irradiation. The product of *irradiance* and time, that is, the *radiant energy* received per unit area during a given time interval.

ISDN. *Integrated-services digital network.*

isochrone. A line on a map or chart joining all points associated with a constant *propagation* time from the *transmitter* or *source.*

isolation. In a *fiber optic coupler,* the extent to which *optical power* from one *signal path* is prevented from reaching another signal path. For example, path A is considered to be isolated from path B if the amount of optical power from path B coupled into path A is zero or extremely small compared with the power levels involved.

isolator. See *fiber optic isolator; optical isolator; waveguide isolator.*

isotropic. Pertaining to material whose properties, such as *electric permittivity, magnetic permeability,* and *electrical conductivity,* are the same regardless of the direction in which they are measured, that is, their spatial derivatives are 0 in all directions. Therefore, an *electromagnetic wave* propagating in the material will be affected in the same way no matter what the direction of *propagation* and direction or type of *polarization* of the wave. Also see *anisotropic; birefringent medium.*

isotropic propagation medium. Pertaining to a material medium in which properties and characteristics of the medium are everywhere the same in reference to a specific phenomenon, such as a medium whose *electromagnetic* properties, such as the *refractive index,* at each point are independent of the direction of *propagation* and *polarization* of a wave propagating in the medium.

isotropic source. A theoretical antenna, such as a *light source* or radio antenna, that *radiates* with equal *irradiance,* that is, with equal *power density* or equal *field intensity,* in all directions. It can only be approximated in actual practive. However, it is a convenient conceptual reference for comparing and expressing the directional properties of actual sources.

IVD. *Inside vapor deposition.*

IVPO. *Inside vapor-phase oxidation.*

IVPO process. See *inside vapor-phase-oxidation (IVPO) process.*

J

jacket. See *optical fiber jacket.*

jacket leak. A measure of the imperviousness of a *fiber optic cable jacket,* determined by such techniques as the water-submerge-and-vacuum method and the gas-detection-and-vacuum method.

jacketed cable. See *tight-jacketed cable.*

jitter. **1.** The variation in time or space of a received *signal* compared to the instant or point of its *transmission* or to a fixed time frame or point at the receiver. Sources of jitter include signal-pattern-dependent *laser* turn-on delay, noise induced at a gating turn-on point, gating hysteresis, and variations in signal delay that accumulate on a *data link* caused by cable vibration. The signal variations are usually abrupt and spurious. They include time and spatial variations in length, amplitude, spacing, or phase of successive pulses. Because the jitter may occur in time, amplitude, *frequency, phase,* or other signal parameter, there should be an indication of the measure of jitter, such as average, RMS, peak-to-peak, or maximum jitter. **2.** Rapid or jumpy undesired movement of display elements, display groups, or display images about their normal or mean positions on the display surface of a display device. For example, movement of a display image on the *fiber faceplate* or a *fiberscope* when the image *source* at the objective end of the *aligned bundle* or *optical fibers* is vibrated relative to the aligned bundle, or the movement of an image on the screen of a CRT when the bias voltage on the deflecting plates oscillates at, say, 5 to 10 Hz. If the jitter rate is above 15 Hz, or if the persistence of the screen is long, the jitter will cause the image to appear blurred because of the persistence of vision. See *longitudinal jitter; phase jitter; pulse jitter; time jitter.* Also see *phase perturbation; swim.*

Johnson noise. See *thermal noise.*

joint. See *fiber optic rotary joint; multifiber joint.*

jumper. See *fiber optic jumper.*

junction. **1.** In a semiconducting material, such as that used in a transistor, diode, or laser, the contact *interface* between two regions of different semiconducting material, for example, the interface between *n-type material,* which is negatively doped and therefore donor material, and *p-type material,* which is positively doped and therefore acceptor material; or between n-type material and intrinsic material. At a *junction,* a voltage, that is, an electrical potential gradient or energy barrier, occurs because of the migration of electric charges (holes or electrons) to the junction and the accumulation of unlike charges on either side of the junction. **2.** An *optical interface.* See *heterojunction; homojunction; optical fiber junction; optical junction; p-n junction; single heterojunction.*

K

k. *Boltzmann's constant; wave number.*

Kerr cell. A substance, usually a liquid, whose *refractive index* changes in direct proportion to the square of the applied *electric field strength*. The substance is configured so as to be part of another system, such as part of the *optical path* of another system. The cell thus provides a means of *modulating* the *light* in the optical path. The device is used to modulate light passing through a material. The modulation depends on the *rotation* of the *polarization plane* caused by the applied electric field. The amount of rotation determines how much of a beam can pass through a fixed polarizing *filter.* Also see *Pockels cell.*

Kerr effect. The creation of *birefringence* in a material that is not normally birefringent by subjecting the material to an *electric field.* The degree of birefringence, that is, the difference in *refractive indices* for *light* of orthogonal *polarizations,* is directly proportional to the square of the applied *electric field strength.* Thus, the effect is a *polarization-plane rotation* that can be used with *polarizers* to *modulate light,* such as to modulate the *irradiance* of light *incident* upon a *photodetector.* Also see *Pockels effect.*

Kerr-effect sensor. A *fiber optic birefringent sensor* in which the *phase* of the induced *ordinary ray* can be advanced or retarded relative to the phase of the *extraordinary ray* when an *electric field* is applied. The electric field causes *double refraction,* that is, an *incident* beam of *light* is divided into two beams that *propagate* at different *phase* velocities, resulting in a *rotation* of the *polarization plane.* The amount of phase shift is directly proportional to the square of the applied voltage, and the direction of shift depends on the polarity of the applied voltage.

knife-edge effect. The *transmission* of electromagnetic waves into the line-of-sight shadow region caused by the *diffraction* that occurs because of an obstacle in the *path* of the waves, such as a sharply defined mountain top. The deflection of a light beam caused by a *diffraction grating* is an example of a multiplicity of knife-edge effects.

L

lambert. A unit of luminance equal to $10^4/\pi$ candles per square meter. The SI unit of luminance is the lumen per square meter, where 4π lumens of light flux emanate from 1 candela.

Lambertian radiator. A radiator or *source* of *radiation* in which the radiation is distributed angularly according to *Lambert's cosine law*. Synonymous with *Lambertian source*.

Lambertian reflector. A *reflector* of *radiation* in which the radiation is distributed angularly according to *Lambert's cosine law*.

Lambertian source. See *Lambertian radiator*.

Lambert's cosine law. The *radiance* of certain idealized surfaces is independent of the angle from which the surface is viewed. According to this law, the energy emitted in any direction by an *electromagnetic* radiator is proportional to the cosine of the angle that the selected direction makes with the normal to the *emitting* surface, given by the relation:

$$N = N_0 \cos A$$

where N is the *radiance,* N_0 is the radiance normal to the emitting surface, and A is the angle between the viewing direction and the normal to the emitting surface. Emitters that radiate according to this law are certain highly diffuse radiators or *scatterers* called *Lambertian radiators* or *sources*. Synonymous with *cosine emission law; Lambert's emission law*.

Lambert's emission law. See *Lambert's cosine law*.

Lambert's law. In the *transmission* of *electromagnetic radiation,* such as *light,* when *propagating* in a *scattering* or *absorptive* medium, the internal *transmittance* is given by the relationship:

$$T_2 = T_1{}^{d_2/d_1}$$

where T_2 is the unknown internal transmittance of a given thickness d_2, and T_1 is the known transmittance of a given thickness d_1.

LAN. See *local-area network (LAN)*.

laser. A device that produces a high *irradiance*, that is, a high-*intensity*, narrow-*spectral-width, coherent,* highly directional, that is, near-zero-divergence, beam of *light* by *stimulating* electronic, ionic, or molecular transitions to higher energy levels and allowing them to fall to lower energy levels, thus producing the *photon* stream. The *lasing* action is produced by *population inversion*. See *atmosphere laser; atomic laser; fiber laser; injection-locked laser; ion laser; liquid laser; mixed-gas laser; molecular laser; multiline laser; multimode laser; tunable laser*. Also see *light amplification by stimulated emission of radiation (laser)*.

L-1. The S600, a ruggedized HeNe **laser** source for locating faults in *fiber optic cables* with *single-mode* or *multimode optical fibers,* particularly for use where an *OTDR* is not appropriate, such as in short *jumper cables* or *pigtails* often used in buildings and telco central offices. (Courtesy fotec incorporated.)

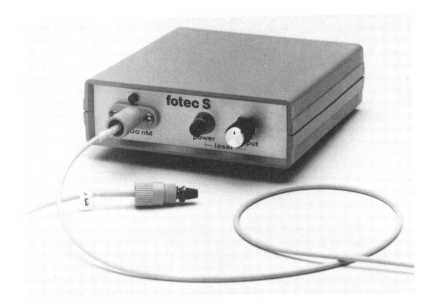

L-2. The S380 Portable Fiber Optic Laser Source, a lightweight, small, economical **laser** *light source* for *fiber optic* testing. (Courtesy fotec incorporated.)

laser beam. A *collimated,* highly directional beam of narrow-*spectral-width light,* that is *monochromatic light,* with near-zero divergence and exceptionally high *irradiance,* that is, *power density,* emitted from materials that are undergoing *lasing* action.

laser chirp. A form of *laser* instability in which the *nominal central wavelength* of the *spectral emission* of the laser is shifted during a single *light pulse.*

laser diode. A *laser* in which stimulated *emissions* of *coherent light* occur at a *p-n junction.* The laser diode operates as a laser producing a *monochromatic light modulated* by insertion of carriers across a *p-n junction* of semiconductor materials. It has a narrower spatial and *wavelength* emission characteristic needed for the higher *data-signaling-rate (DSR)* systems than the *light emitting diodes (LED).* The LEDs are more useful for large-diameter, large-*numerical-aperture optical fibers* that operate at lower frequencies. Typical characteristics of semiconductor laser diodes for fiber optic communication applications are (1) for a GaAlAs type, 0.80 μ, 1–40 mW (optical), 10–500 mA, 2 V, 1–20% eff., 1 gm, $10^4 - 10^7$ h, 10 × 35° divergence, and modulated directly by the drive current; and (2) for an InGaAsP type, 1.1–1.6 μ, 1 − 10 mW (optical), 20–200 mA, 1.5 V, 1–20% eff., 1 gm, up to 10^5 h, 10 × 30° to 20 × 40° divergence, and modulated directly by the drive current. Synonymous with *diode laser; injection laser diode; semiconductor diode laser; semiconductor laser.* See *stripe laser diode.*

laser medium. See *active laser medium.*

lasing. A phenomenon occurring when resonant *frequency*-controlled energy is *coupled* to a specially prepared material, such as uniformly doped semiconductor crystals that have free-moving or highly mobile loosely coupled electrons. As a result of resonance and the imparting of energy by collision or close approach, electrons are excited to high energy levels. When the electrons move to lower levels, *quanta* of high-energy *radiation* are released as *coherent lightwaves.* The action takes place in a *laser.* In addition to certain semiconductors and crystals, gases, such as helium, neon, argon, krypton, and carbon dioxide; and liquids, such as organic and inorganic dyes, can also be made to undergo lasing action.

lasing condition. See *prelasing condition.*

lasing threshold. The lowest excitation input *power level* at which the output of a *laser* is primarily caused by *stimulated emission* rather than *spontaneous emission.*

latching. A switch actuation method that requires a specific *signal* to place the switch in a position where it remains until another actuating force or signal is applied. Also see *nonlatching.*

latching switch. In *fiber optics,* a switch that will selectively transfer *optical signals* from one *optical fiber* to another when an actuating force or signal is applied and will continue to transfer signals after the actuating force or signal is removed until another actuating force or signal is applied. Also see *nonlatching switch.*

lateral offset loss. In *fiber optics,* an *optical power loss* caused by transverse or lateral deviation from optimum alignment of the *optical axes,* that is, the sidewise offset distance between the axes, of *source* to *optical waveguide,* waveguide to waveguide, or waveguide to *photodetector.* Synonymous with *transverse offset loss.* (See Figs. L-3 – L-4, p. 142)

launch angle. 1. In an *optical fiber* or *fiber bundle,* the angle between the input *radiation* vector, that is, the input *light chief ray,* and the *optical axis* of the fiber or fiber bundle. If the end surface of the fiber is perpendicular to the axis of the fiber, the launch angle is equal to the *incidence angle* when the *ray* is external to the fiber and equal to the *refraction angle* when initially inside the fiber, because when inside the fiber, the ray might be considered as having been "launched" by the face of the fiber. **2.** The beam divergence from any *emitting* surface, such as that of a *light-emitting diode (LED), laser,* lens, prism, or *optical fiber* end face. **3.** The angle at which a *light ray* emerges from a surface. Synonymous with *departure angle.* Also see *exit angle.*

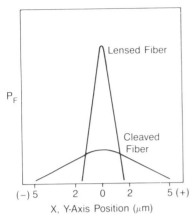
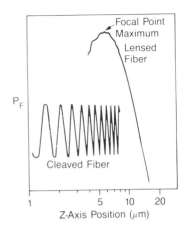

X, Y-Axis Position (μm)

Z-Axis Position (μm)

L-3. The curves show the precision positioning required for lateral alignment for *source*-to-*optical fiber coupling* to minimize **lateral offset loss** (*x,y*-axes displacement, left). For semispherical and conical lens coupling, a lateral displacement of 0.25 μ *(microns)* causes a decrease in coupled *optical power*, P_f, of 10%, or 0.5 *dB*; 0.75 μ results in a 50%, or 3-dB decrease. Coupling for a cleaved fiber is relatively insensitive to lateral offsets of these magnitudes. For *gap loss* (*z*-axis displacement, right) in source-to-fiber coupling, or *longitudinal offset loss* for fiber-to-fiber coupling, *Fabry-Perot* half-wave *reflections* cause about 15% fluctuations in coupled power. Thus, for optimum coupling, precision positioning must be obtained and maintained. (Courtesy Tektronix.)

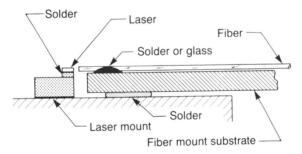

L-4. Precise alignment of *source* and *optical fiber* for optimum *coupling* of *optical power* must be obtained and then maintained. The structure shown maintains source and fiber positioning for minimizing **lateral offset loss**, *angular misalignment loss*, and *gap loss (longitudinal offset loss)*. (Courtesy Tektronix.)

launch condition. When *light rays* exit from the surface of a material, such as the surface of a *light source* or the exit end face of an *optical fiber,* a parameter that describes the geometric relationships between the ray and the surface, the optical characteristics of the *interfaces* and paths, the characteristics of the rays, and other related parameters, for example, *exit angle, launch angle, wavelength, refractive-index contrast, launch numerical aperture, launch loss,* and *spectral width*.

launching. See *single-mode launching.*

launching fiber. An *optical fiber pigtail* attached to an *optical source* used to enable *splicing* to another component, such as a *coupler* or *optical fiber.* The launching fiber may be used in conjunction with a *light source* to excite the *modes* of another fiber in a particular fashion. They are often used in test systems to improve the precision of measurements. Synonymous with *injection fiber.*

launch loss. The *radiant power loss* at the point where the *radiant power* output from a *light source* is *coupled* into an *optical waveguide,* such as an *optical fiber pigtail* attached to the light source. Launch loss can be caused by *aperture* mismatch, *angular misalignment, longitudinal offset, lateral offset,* and *Fresnel reflection.* Also see *insertion loss.*

launch numerical aperture (LNA). The *numerical aperture* of an *optical* system used to *couple,* that is, launch, *power* into an *optical waveguide.* The LNA may differ from the stated NA of a final focusing element if, for example, that element is underfilled or the focus is other than that for which the element is specified. The LNA is one of the parameters that determine the initial distribution of power among the modes of an *electromagnetic wave propagating* in an *optical waveguide.*

launch spot. The spot of *light* produced by a light beam from an *optical* element, such as from a lens or the exit end face of an *optical fiber,* when the beam is *incident* upon a transverse surface, that is, a screen perpendicular to the *optical axis* of the element. The diameter of the spot is dependent upon the departure angle, that is, the *launch angle,* from the element and the distance between the transverse surface and the element. The launch spot size can be used to determine the *numerical aperture* of the optical element. The optical element is aligned to provide maximum *irradiance* on the screen. The numerical aperture is given by the relation:

$$NA = D/(D + 4d^2)^{1/2}$$

where D is the spot diameter and d is the distance from the element, such as the fiber exit end face, to the screen. If $D/2d$ is less than 0.25, the numerical aperture of the element is given by the approximate relation:

$$NA \approx D/2d = r/d$$

where r is the spot radius and r/d is the tangent of the *launch angle.* Also see *numerical aperture* and *xx/yy restricted launch* in the appendix.

law. See *Bouger's law; Brewster's law; Lambert's cosine law; Lambert's law; Planck's law; radiance conservation law; reflection law; Snell's law.*

layer. See *application layer; barrier layer; data-link layer; network layer; physical layer; presentation layer; session layer; transport layer.*

leak. See *jacket leak.*

leaky mode. 1. In an *optical waveguide,* a *mode* that has an *evanescent field* that decays monotonically for a finite distance in the transverse direction outside the *core,* but becomes oscillatory and conveys power in the transverse direction everywhere beyond that finite distance. Leaky modes in *physical optics* are equivalent to *leaky rays* in *geometric optics.* **2.** An *electromagnetic wave propagation mode* in a *waveguide* that *couples* significant energy into *leaky waves.* Leaky modes are usually *high-order modes.* Specifically, a leaky mode is a mode that satisfies the relation:

$$[(n_a k)^2 - (m/a)^2]^{1/2} \le \beta \le n_a k$$

where n_a is the *refractive index* at the *core-cladding interface* surface, that is, at the distance a from the *optical axis* where a is the *core* radius, k is the *free-space wave number,* given by the relation $k = 2\pi/\lambda$, where λ is the *wavelength,* m is the azimuthal index of the mode, and β is the imaginary part *(phase term)* of the *axial propagation constant.* Leaky modes experience *attenuation* even if the waveguide is perfect in every respect. Synonymous with *tunneling mode.* Also see *bound mode; cladding mode; leaky ray; unbound mode.*

leaky ray. In an *optical waveguide,* a *ray* that escapes from the guide even though the theory of *geometric optics* predicts that *total (internal) reflection* should be taking place at the *core-cladding* boundary, perhaps caused by *microbends* and curvatures in the boundary. Leaky rays in geometric optics are equivalent to *leaky modes* in *physical optics.* Calculations in geometric optics for the leaky ray would predict total (internal) reflection at the core-cladding *interface,* but there is a *loss* because the core-cladding interface surface is curved. Specifically, a leaky ray is a ray at radial distance r from the *optical axis* that satisfies the relation:

$$(n_r^2 - n_a^2) \le \sin^2 \theta_r$$

and the relation:

$$\sin^2 \theta_r \le [n_r^2 - n_a^2]/[1 - (r/a)^2 \cos^2 \Phi_r]$$

where θ_r is the angle the ray makes with the optical axis of the waveguide, n_r is the *refractive index* at the radial distance r from the optical axis, a is the *core* radius, and Φ_r is the azimuthal angle of the projection of the ray on the transverse plane. Leaky rays correspond to leaky modes in the terminology of mode descriptors. Also see *bound mode; cladding mode; leaky mode; unbound mode.* Synonymous with *tunneling ray.*

leaky wave. In a *waveguide,* an *electromagnetic wave* that is *coupled* to *propagation media* outside the waveguide, such as inside and outside the *cladding* of *optical fibers.* Certain *leaky modes* are no longer guided because they are no longer coupled to modes inside the waveguide. They normally stem from *incident waves* that have a large skew component at entry into the fiber. They become detached and *radiate* from the guide after a short distance, such as a few *wavelengths,* within the fiber. These *modes* will have escaped by the time the *equilibrium length,* that is, equilibrium modal power distribution length, is reached. They are usually *high-order modes* and are in the *cutoff frequency* region. In *optical fiber* waveguides, *low-order modes* are usually bound to the *core* and therefore are not leaky modes. Unless coupling to other guides is desired, efforts are usually made to reduce leaky waves to a minimum.

LED. *Light-emitting diode.* See *edge-emitting LED (ELED); front-emitting LED; superluminescent LED (SRD); surface emitting LED.*

left-hand circular polarization. *Circular polarization* of an *electromagnetic wave* in which the *electric field vector* rotates in a counterclockwise direction, as seen by an observer looking in the direction of *propagation* of the wave. Also see *right-hand circular polarization.*

left-hand helical polarization. *Helical polarization* of an *electromagnetic wave* in which the *electric field vector* rotates in a counterclockwise direction as it advances in the direction of *propagation,* as seen by an observer looking in the direction of propagation. The tip of the electric field vector advances like a point on the thread of a left-hand screw, the normal screw being a right-hand screw, when entering a fixed nut or tapped hole.

left-hand polarized electromagnetic wave. An *elliptically* or *circularly polarized electromagnetic wave* in which the direction of rotation of the *electric field vector* is counterclockwise as seen by an observer looking in the direction of *propagation* of the wave. Also see *right-hand polarized electromagnetic wave.*

length. See *borescope overall length; borescope tip length; borescope working length; coherence length; dispersion-limited length; electrical length; equilibrium length; nonequilibrium-modal-power-distribution length; optical path length; wavelength.*

lensed connector. See *expanded-beam connector.*

level. See *power level.*

lever. See *optical lever.*

light. The region of the *electromagnetic spectrum* that can be perceived by human vision, designated the *visible spectrum,* and nominally including the *wavelength* range of 0.4 to 0.8 μ *(microns),* including all the *colors,* from red to violet, to which the normal human retina responds. In the *laser* and *optical* communication fields, custom has extended usage of the term to include a somewhat broader portion of the electromagnetic spectrum that can be handled by the basic *optical* techniques and components used for the visible spectrum. This region has not been clearly defined, but, as employed by most workers in the field, *light* may be considered to extend from the near-*ultraviolet region* of approximately 0.3 μ, through the *visible region,* and into the *midinfrared region* to 3 μ, designated by what has been called the near-visible spectrum. The near-visible spectrum extends part way into the ultraviolet and *infrared* regions of the *optical spectrum.* For example, the 1.31-μ wavelength used in *fiber optic* systems is considered to be within the realm of *lightwaves,* though it is not within the visible spectrum and hence cannot be detected by the human eye but can cause damage to the retina. The optical spectrum is considered to extend even farther, that is, from the beginning of vacuum ultraviolet at 0.001 μ to the end of *far-infrared* at 100 μ. See *coherent light; convergent light; divergent light; monochromatic light; time-coherent light; ultraviolet light; white light.* Also see *optical spectrum.*

light amplification by stimulated emission of radiation (laser). The generation of *coherent light* by having molecules of certain substances *absorb incident* electrical or *electromagnetic* energy at specific *frequencies,* store the energy for short periods in higher electron energy-band levels, and then release the energy upon return to the lower levels in the form of light, that is, *photons* or *lightwaves,* at particular frequencies in extremely narrow *bands.* The release of *radiated* energy can be controlled in time and direction so as to generate an intense, highly directional, narrow beam of coherent electromagnetic energy, that is, the electromagnetic fields at every point in the beam are uniquely and specifically definable and predictable. Also see *laser.*

light attenuation. The conversion of *lightwave* energy into other forms of energy or directions when *propagating* a given distance in a *propagation medium,* thus reducing the *transmitted* lightwave energy, that is, by reducing the *transmittance.* When lightwaves propagate in the medium for a given distance, they undergo *transmission, reflection, absorption, scattering,* and *radiation.* The law of conservation of energy is obeyed, as given by the relationship:

$$T + R + A + S + D = 1$$

where *T, R, A, S,* and *D* are the coefficients of energy *transmittance, reflectance, absorption, scattering,* and *radiation,* respectively. All the coefficients are normalized to the input energy. All the coefficients, except transmittance, contribute to the *attenuation.*

light beam. A bundle of *covergent, divergent,* or parallel *(collimated) optical rays.* Also see *light ray; ray.*

light current. See *photocurrent.*

light-duty connector. In *fiber optics,* a *connector* that is designed and intended for use inside an *interconnection box* (distribution box) or cabinet.

light-emitting diode (LED). A *p-n junction* semiconductor device that produces *incoherent radiation* by *spontaneous emission* under suitable operational conditions. The LED operates in a manner similar to a *laser diode,* that is, by injecting electrons and holes across the junction. It has about the same total *optical power output,* limited *data signaling rate (DSR),* and operational electric current densities as the laser diode, but is simpler and cheaper, has a lower tolerance requirement, and is more rugged. However, its *spectral width* is about 10 times that of a laser and its *launch angle* is greater.

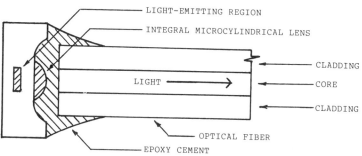

L-5. *Butt-coupling* an *optical fiber* and a **light-emitting diode (LED)** *source,* showing the integral microcylindrical lens for *coupling optical power* from the *emitting area* into the *optical fiber core.*

light pipe. 1. A *passive optical device* that *transmits light* from one location to another, such as an *optical fiber* or *slab-dielectric waveguide.* **2.** A hollow tube with a *reflecting* inner wall that guides *lightwaves* in its hollow center. *Aligned* and *unaligned bundles* of *optical fibers* may be used as light pipes.

light ray. 1. In *geometric optics,* the *path* described by a succession of tangents at each point in the direction of *propagation* of *light* energy. The light ray is perpendicular to the *wavefront* of a *lightwave propagating* in an *isotropic propagation medium.* It represents the lightwave itself. For *plane-polarized light,* the ray is in the same direction as the *Poynting vector* **2.** A *ray* in the *optical spectrum.* Also see *light beam; ray.*

light source. See *fiber optic light source.*

light susceptibility. See *ambient light susceptibility.*

lightwave. An *electromagnetic wave* with a *wavelength* within the *optical spectrum,* namely a wavelength within or near the *visible spectrum,* usually including the *ultra-*

violet and *near-infrared.* Its wavelength ranges from about 0.3 to 3 μ *(microns),* that is, about 300 to 3000 nm *(nanometers).*

lightwave-spectrum analyzer. A device capable of determining the existence and measuring the energy levels at various *wavelengths* in a *light* beam. The device is essentially a tunable *filter* that can examine each portion of the *optical spectrum* in the beam. For example, one type of analyzer has a slit for admitting the *incident* beam and excluding stray ambient light, some lenses, a *diffraction grating* that is tuned by a stepper motor, more lenses, and another slit. A *Fabry-Perot interfero-meter* at the output of the diffraction grating will improve the *resolution.* Further improvements can be made using a *lightwave* synthesizer, that is, a tunable local oscillator with a sufficiently narrow *bandwidth.*

limited. See *diffraction-limited.*

limited operation. See *attenuation-limited operation; quantum-limited operation; quantum-noise-limited operation.*

limiter. See *bend limiter.*

line. See *optical fiber delay line; spectral line.*

linear device. A device whose output is a linear function of its input. If the output is represented by *y* and the input by *x,* then the output is related to the input by the relation:

$$y = mx + b$$

where *m* and *b* are constants. For example, a device in which the output *electric field,* voltage, or current is linearly proportional to the input electric field, voltage, or current, and no new *wavelengths* or *modulation frequencies* are generated by the device. The behavior of a linear device can be described by a *transfer function* or an *impulse-response function.* Also see *nonlinear device.*

linearly polarized (LP) mode. A *mode* of a *linearly polarized electromagnetic wave, propagating* in a *weakly guiding propagation medium,* such as an *optical fiber,* whose *electric* and *magnetic field* components in the direction of propagation are small compared with the transverse components.

linear polarization. *Electromagnetic wave polarization* in which the *electric field vector* maintains a fixed spatial direction and varying magnitude.

linear refractive-index profile. A *refractive-index profile* of the *core* of a *dielectric waveguide,* such as an *optical fiber,* in which the *refractive index* varies uniformly with distance, from a given value at the center to a given value at the *core-cladding interface* surface. Thus, the linear refractive-index profile is produced when the *refractive-index profile parameter* is unity, that is, precisely equal to 1. Synonymous with *uniform refractive-index profile.* Also see *parabolic refractive-index profile; power-law refractive-index profile; profile parameter; radial refractive-index profile.*

linear scattering. See *nonlinear scattering.*

line code. A sequence of symbols that represent binary *data* for *transmission* purposes, e.g., Manchester, return-to-zero, and block codes. Line codes are used to recover precise timing and may be used to detect errors.

line rate. See *optical line rate.*

line source. 1. An *optical source* that *emits* one or more *spectral lines* of narrow *spectral linewidth,* that is, one or more *monochromatic* beams, as compared with a more or less *continuous spectrum* with many *wavelengths.* 2. An *optical source* that *emits* a spatially very narrow beam, spatially narrow in one dimension, usually emitted by an optical source whose active area, that is emitting area, forms a spatially narrow line of light. Also see *coherent; spectral width.*

line spectrum. The various *wavelengths* of *radiation emitted* by a *source.* It consists of one or more *spectral lines* rather than a *continuous spectrum.*

linewidth. See *spectral linewidth.*

link. 1. In a communication *network,* the facilities between adjacent *nodes.* 2. A portion of a circuit connected in tandem. See *control link; data link; dedicated data link; fiber optic link; fiber optic data link; optical link; repeatered optical link; repeaterless optical link; satellite optical link; TV fiber optic link.* (See Fig. L-6, p. 150)

link layer. See *data-link layer.*

liquid laser. A *laser* whose active medium is in liquid form, such as organic dyes and inorganic solutions. Dye lasers are commercially available and are often called "organic dye" or "tunable dye" lasers.

LNA. *Launch numerical aperture.*

local-area network (LAN). A nonpublic telecommunication system, within a specified geographical area, designed to allow a number of *user end-instruments* and independent devices, such as telephones, *data* terminals, television sets, and computers, to communicate with each other over a common *transmission* system at fairly high *data signaling rates.* LANs are usually restricted to relatively small geographical areas, such as rooms, buildings, or clusters of buildings. An LAN is not subject to public telecommunications regulations. LANs are usually connected to a *switching center* or central office for outside communication via telephone and other *networks.*

L–6. This Fiber Optic Network **Link** allows RS-232C, RS-422, or TTL *data bus networking* using daisy chain or *star architecture*. (Courtesy 3M Fiber Optic Products, EOTec Corporation.)

local loop. 1. A communication *channel* from a *switching center* or an individual message or packet distribution point to a user terminal or *user end-instrument*. For example, in a telephone system, a pair of wires or a *fiber optic cable* to a user's telephone might constitute a local loop. **2.** In *fiber optics,* a *fiber optic cable* that connects *user end-instruments,* such as telephones, *data* terminals, computers, and television sets, to a communication *network,* usually via multiplexing equipment to a network central office or *switching center.* Synonymous with *subscriber loop.*

locked laser. See *injection-locked laser.*

long-haul communication network. A communication *network* designed for handling communication traffic over long distances, such as nationwide or worldwide traffic. Long-haul systems are characterized by long-distance trunks between towns and cities, large *switching centers* and central offices, high-quality equipment for high-fidelity and high-definition *analog* (voice) and *digital data transmission, integrated services digital (data) networks (ISDN),* high transmission capacity, and automatic switching for handling calls and messages without operator assistance. It is anticipated that by the year 2000 over 600 billion voice circuit-kilometers will be on fiber optic long- and short-haul networks and only 16 billion on microwave, 160 billion on satellite, and 3 billion on coaxial cable. Also see *short-haul communication network.*

longitudinal compression sensor. See *fiber longitudinal compression sensor.*

longitudinal jitter. In facsimile *transmission,* the effect caused by irregular scanning speed. The speed irregularity may occur from many causes, such as the irregular rotation of the drum or helix that causes slight waviness or breaks in the lines of the reproduced image, lines that were straight on the original object (document).

longitudinal offset loss. See *gap loss.*

long wavelength. In *fiber optics,* pertaining to *optical radiation* that has a *wavelength* greater than a nominal 1 μm *(micron).*

loop. See *local loop; subscriber loop.*

loop multiplexer. See *fiber loop multiplexer.*

loose cable structure. See *loose-tube cable.*

loose-tube cable. A *fiber optic cable* in which one or more *optical fibers* are fitted loosely in a tube. One or more such tubes may be used to form a single cable. The tubes may be filled with a seal or gel to protect the fiber. Within certain limits, stresses applied to the cable are not transferred to the fiber Synonymous with *loose cable structure.* Also see *tight-jacketed cable.*

loose-tube splicer. A tube with a square hole used to *splice* two *optical fibers.* The curved fibers are made to seek the same corner of the square hole, thus holding then in alignment until an *index-matching* epoxy already in the tube cures, thus forming a low-*loss* butted joint.

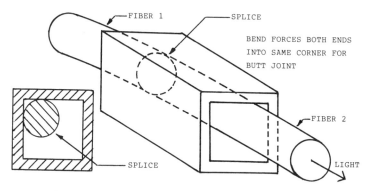

L-7. A **loose-tube splicer** for *butt-coupling optical fibers.*

Lorentzian fiber. An *optical fiber* that has a *refractive-index profile* defined by the relation:

$$n_r^2 = n_1^2/[1 + 2\Delta(n_1^2/n_2^2)(r/a)^2] \qquad \text{for } r \leq a$$

where Δ is given by the relation:

$$\Delta = (n_1^2 - n_2^2)/2\, n_1^2$$

n_1 is the maximum *refractive index* of the *core* at the *optical axis,* that is, at $r = 0$, n_2 is the refractive index of the *homogeneous cladding,* that is, the refractive index of the cladding is given by the relation:

$$n_r = n_2 \qquad \text{for } r > a$$

where r is the radial distance from the optical axis of the fiber, and a is the radius of the core, that is, one-half the *core diameter. Skew rays* entering the fiber at angles less than the *acceptance angle,* thus satisfying the *critical angle* criterion for *total* (internal) *reflection,* will travel a helical path within the fiber. A skew ray is a ray that is incident upon the end face of the fiber and that is not a *meridional ray* nor an *axial ray.* If the value of Δ is appropriately selected, all the helical rays will arrive at the end of the fiber at the same time. However, this profile will then not permit the meridional rays and the axial rays to arrive at the end at the same time as the helical rays.

loss. 1. The amount of electrical, *optical,* sound, or other form of *power* or energy consumed in a circuit or component. **2.** The energy dissipated without accomplishing useful work, usually expressed in decibels *(dB).* In directed *signal* and power *transmission,* the difference between power or energy that is dispatched and that which is received. For example, in a *fiber optic data link,* signal power that is *absorbed, reflected, scattered,* or *radiated* from a *fiber optic cable* between the *optical transmitter* and *receiver* is considered a loss. Thus, the loss is the transmitted power minus the received power. See *absorption loss; angular misalignment loss; bending loss; coupler loss; coupler excess loss; coupling loss; curvature loss; extrinsic coupling loss; free-space loss; gap loss; insertion loss; intrinsic coupling loss; lateral offset loss; microbend loss; modal loss; optical cable facility loss; packing fraction loss; reflection loss; return loss; scattering loss; splice loss; statistical optical cable facility loss; transmission loss.*

loss budget. In a *link,* such as a *fiber optic link,* the distribution of the total *loss* among the components of a system that can be tolerated in order that the received *signal* is above the threshold of *sensitivity* of the receiver, including allowance for a *power safety margin.* The loss is distributed among the components of the system, such as *cables, couplers,* and *splices,* in an optimum fashion, so as to design for minimum cost at tolerable *bit-error ratios* for *digital* systems. Required *transmitted power,* receiver sensitivity, intervening losses, and power margins are all considered in the distribution, that is, in the budgeting, of the losses.

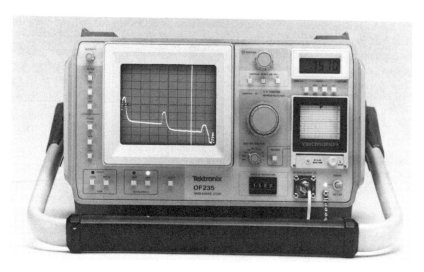

L-8. The OF235 Fiber Optic Time Domain Relectometer, capable of making quantitative cali-brated **loss** and distance measurements on *single-mode optical fiber* at 1.300 or 1.550μ *(mi-crons) wavelengths* at the touch of a button. (Courtesy Tektronix).

loss-budget constraint. In the design of an *optical station/regenerator section,* the *terminal/regenerator system gain* must be equal to or larger than the sum of all the *optical power losses* in the *optical fiber path* between terminal/regenerator *spliced interfaces.* The constraint is given by the relation:

$$G > L$$

where G is the *terminal/regenerator system gain* given by the relation:

$$G = P_T - P_R - P_D - R_P - M - U_{wdm} - \ell_{sm}U_{sm} - N_{con}U_{con}$$

where P_T is the *transmitter power,* P_R is the *receiver sensitivity* (power), P_D is the *dispersion power penalty,* R_P is the *reflection power penalty,* M is the overall *safety margin,* U_{wdm} is the worst-case value of all losses associated with *wavelength-division multiplexing* equipment at both ends, ℓ_{sm} is the fiber length in the stations, U_{sm} is the worst-case end-of-life loss (dB/km) of the *single-mode cable* at both stations, N_{con} is the number of *fiber optic connectors* within the stations (inside plants), and U_{con} is the *loss (dB)* per connector. G is used to overcome all the losses, L, in the *optical cable facility* (outside plant) of the station/regenerator section, given by the relation:

$$L = \ell_t(U_c + U_{cT} + U_\lambda) + N_S(U_S + U_{ST})$$

where ℓ_t is the total sheath length of *spliced-fiber cable* (km), U_c is the worst case end-of-life cable *attenuation rate* (dB/km) at the *transmitter nominal central wave-*

length U_λ is the largest increase in cable *attenuation rate* that occurs over the transmitter central wavelength range, U_{cT} is the effect of temperature on the end-of-life cable attenuation rate at the worst-case temperature conditions over the cable operating temperature range, N_s is the number of *splices* in the length of the cable in the optical cable facility, including the splice at the optical station facility on each end and allowances for cable repair splices, U_s is the loss (dB/splice) for each splice, and U_{ST} is the maximum additional loss (dB/splice) caused by temperature variation. *L* must be equal to or less than the *fiber optic terminal/regenerator system gain, G,* for the optical station/regenerator section to operate satisfactorily. Also see *statistical loss-budget constraint; fiber global attenuation-rate characteristic.*

loss characteristic. See *fiber global-loss characteristic.*

loss test set. See *optical loss test set (OLTS).*

lossy medium. A *propagation medium* in which significant amounts of energy in a *wave propagating* in the medium are *absorbed* per unit distance traveled by the wave. For example, in *optical fiber cladding,* a lossy medium may be used to *attenuate* waves by *absorbing* the energy in those that have leaked outside the *core.*

lot. See *inspection lot.*

lowest order transverse mode. The lowest *frequency* in an *electromagnetic wave* that can *propagate* in a given *waveguide* that can support more than one *transverse electric (TE) mode* or more than one *transverse magnetic (TM) mode,* the limitation in the number of *modes* being determined by the boundary conditions and the geometrical shape of the waveguide, as well as the *frequency* (or *wavelength*). Solution of *Maxwell equations* with the boundary conditions for a rectangular waveguide operating in the TM mode yields the relationship:

$$\omega^2 \mu\epsilon > [(m\pi/a)^2 + (n\pi/b)^2]$$

where ω is the angular velocity ($\omega = 2\pi f$ where *f* is the frequency), μ is the *magnetic permeability* of the material in the waveguide, ϵ is the *electric permittivity of the material* in the waveguide, *a* and *b* are the cross-sectional dimensions, and *m* and *n* are the whole numbers (eigenvalues) that satisfy and provide solutions to Maxwell's equations. The solutions to Maxwell's equations for transverse magnetic (TM) and transverse electric (TE) modes of *propagation* yield modes identified as TM_{mn} and TE_{mn}, except that for TM modes neither *m* nor *n* can be zero, since these conditions result in a zero *electric* and *magnetic field.* Therefore, the lowest order mode for TE is TE_{10} and the lowest order mode for TM is TM_{11}. The TE_{10} mode, called the dominant mode, is obtained by designing a waveguide with a width-to-depth ratio of about 2, using an operating frequency above the *cutoff frequency,* given by the relation:

$$f_c = v/2a$$

where v is the velocity of *light* in the medium of the guide, but below the next higher cutoff frequency. In *optical fiber* waveguides, the dimensional parameters relate to those of the fiber. For example, in an optical fiber with a circular cross section, the *core diameter* corresponds to the *a* and *b* for the rectangular waveguide. The variables for a rectangular electromagnetic metal waveguide relate one-for-one to the variables in the waveguide *normalized frequency* (*V*-value) equation for optical fibers, namely that the number of modes is a function of the number, that is, the *V*-value, determined by the relation:

$$V = (2\pi a/\lambda_0)(n_1^2 - n_2^2)^{1/2}$$

where *a* is half the *core diameter,* λ_0 is the wavelength, and n_1 and n_2 are *core* and *cladding refractive indices.* When both equations are solved for *wavelength,* it can be shown that the modes in both waveguides are functions of the dimensions of the guides, their refractive indices, the velocity of *light,* and the *launch conditions.* For the optical fiber having a circular cross section, the governing dimension is the diameter, or the radius, of the core. Thus, there is no fundamental difference between the equations for determining supportable modes in electromagnetic rectangular filled or hollow metal waveguides operating at radio, microwave, and video frequencies and the equations governing the number of supportable modes in optical fibers operating at *lightwave* frequencies. Also see *low-order mode.*

low-frequency cutoff. Because of dimensional and boundary conditions, the lowest *electromagnetic frequency* a *waveguide* with given dimensions and constructed of given materials is capable of supporting. The lowest frequency corresponds to the longest *wavelength* the guide can support. In *fiber optics,* the low-frequency cutoff is the frequency of the longest wavelength an *optical fiber* is capable of supporting. In essence, the *core diameter* has to be at least the order of magnitude of the wavelength of the *mode propagating* in the core or else the core cannot *transmit,* that is, cannot support, the *mode.*

low-loss fiber. See *ultralow-loss fiber.*

low-order mode. In an *electromagnetic wave propagating* in a *waveguide,* a *mode* corresponding to the lesser, that is, the first few, eigenvalue solutions to the *wave equations,* regardless of whether it is a mode in a *transverse electric (TE), transverse magnetic (TM),* or *transverse electromagnetic (TEM) wave.* Thus, it is a mode in which only a few whole *wavelengths* of the wave fit transversely in the guide and therefore can be supported by the guide. A *multimode fiber* supports as many modes as the *core diameter, numerical aperture,* and wavelength permit. For the waveguide to support only low-order modes, the *normalized frequency,* that is, the *V*-value, must be less than 4, in order that the number of modes be sufficiently small so there will be no *high-order modes.* In a *single-mode fiber* there are no high-order modes. Conceptually, however, if there are only two modes, one could be considered the low-order mode and the other the high-order mode. In addition, an *optical fiber* operating in single mode at say 1.31 μ *(microns)* may operate in multimode if the wavelength is reduced to say 0.85 μ. Also see *low-order mode; lowest order transverse mode.*

low-pass filter. A device that passes all *frequencies* below a specified frequency with little or no loss, but discriminates strongly against higher frequencies by removing, blocking, rejecting, *absorbing,* or otherwise heavily *attenuating* frequencies above the specified value.

LP. *Linearly polarized.*

LP mode. See *linearly polarized (LP) mode.*

luminance threshold. See *absolute luminance threshold.*

luminescence. The process by which certain materials, such as radium, *emit electromagnetic radiation,* that, for certain *wavelengths* or restricted regions of the *spectrum,* is in excess of that attributable to the thermal state of the material and the *emissivity* of its surface. The *radiation* is characteristic of the particular luminescent material and occurs without outside *stimulation.* See *electroluminescence.* Also see *fluorescence; phosphorescence.*

luminosity curve. See *absolute luminosity curve.*

M

m. *Meter.*

MAA. *Maximum acceptance angle.*

machine. See *fiber-pulling machine.*

Mach-Zehnder fiber optic sensor. An *interferometric sensor* in which an *electromagnetic wave,* such as a *lightwave,* is split, each half *propagating* around half a loop in opposite directions, one half via a *beam splitter* and a fixed mirror, the other via a movable mirror and a beam splitter, both halves recombining in an *optical fiber* or on the sensitive surface of a *photodetector* where their relative *phases* can enhance or cancel each other. The movable mirror can be used to *modulate* the resultant *irradiance,* that is, the *field intensity,* at the photodetector. Displacements as short as 10^{-13} m can be measured. Optical fibers may be used for the light paths.

macrobend attenuation. *Optical power radiated* at a *macrobend* in an *optical waveguide. Evanescent waves, cladding modes,* and *high-order leaky modes* in the *core* become decoupled from core *propagating* modes because of *phase* shifts and velocity differences at the bend. For macrobend radii less than the *critical radius* most of the propagating optical power is radiated laterally from the waveguide.

macrobending. In an *optical waveguide,* all macroscopic deviations of the *optical axis* from a straight line. Macrobending is distinguished from *microbending.* In *optical fibers,* macrobend radii must be greater than both the *minimum bend radius,* that is, the radius less than which the fiber will break; and the *critical radius,* that is, the radius less than which substantial *radiation* will occur. (See Fig. M-1, p. 158)

macrobend loss. See *curvature loss.*

magnetic field component. In an *electromagnetic wave,* the part of the wave that consists of a time-varying *magnetic field,* whose interaction with an *electric field* gives rise to the *propagation* of a field of force or energy in a direction perpendicular to both fields. *Reflection, refraction,* and *transmission* that occur at an *interface*

157

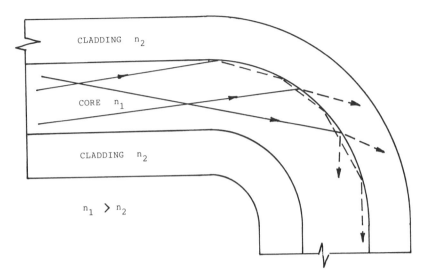

M-1. **Macrobending** *losses* occur to a *lightwave propagating* in an *optical fiber* when a bend occurs whose radius is less than the *critical radius,* because at the bend the *incidence angle* at the *core-cladding interface* surface is less than the *critical angle* or the velocity of the lightwave propagating along the outside of the bend exceeds the velocity of *light,* in which case *optical power* is *radiated* away.

between two different media depend on the direction of the magnetic field component and the electric field component relative to the interface surface, and on the *permittivities, permeabilities,* and *conductivities* of the *propagation media* on both sides of the interface.

magnetic field strength. The intensity, that is, the amplitude or magnitude, of a magnetic field at a given point. The term is normally used to refer to the rms value of the magnetic field, expressed in amperes per *meter.* Instantaneous magnetic field strength is usually given as a gradient and therefore is a vector quantity, because it has both magnitude and direction. For example, the units of magnetic field strength might be given as amperes per meter, oersteds, or lines per square meter divided by the *magnetic permeability* of the medium at the spatial position that the magnetic field strength is to be determined. Also see *electric field strength.*

magnetic field vector. In an *electromagnetic wave,* such as a *lightwave,* the vector that represents the instantaneous *magnetic field strength* at any point in a medium in which the wave is *propagating.*

magnetic permeability. A material medium parameter that defines the magnetic characteristic of the medium and serves as the constant of proportionality in the *constitutive relation* between the applied *magnetic field strength* and the resulting magnetic flux density, specifically, by the relation:

$$\mathbf{B} = \mu\mathbf{H}$$

where μ is the magnetic permeability, $\mathbf{B}$ is the magnetic flux density, and $\mathbf{H}$ is the magnetic field strength. If $\mathbf{B}$ is in gauss and $\mathbf{H}$ in oersteds, μ will be in SI (International System) units. The permeability is also the constant of proportionality in the equation expressing the force of attraction of unlike magnetic poles a given distance apart. Except for ferrous metals, which have a high magnetic permeabilty, and a few metals, which have a much lower permeability, but nevertheless greater than unity, the magnetic permeability of most materials, such as *dielectric* materials like glass and plastic, is unity. Thus, magnetic fields are practically unaffected by the glass and other dielectric materials from which *optical fibers* are made. Therefore, the magnetic permeability of glass does not contribute to the *refractive index* of glass. For electrically nonconducting media, that is, *propagation media* in which the *electrical conductivity* is zero, the relative velocities of *electromagnetic waves,* such as *lightwaves, propagating* within them are given by the relation:

$$v_1/v_2 = [(\mu_2\epsilon_2)/(\mu_1\epsilon_1)]^{1/2}$$

where v is the velocity and μ and ϵ are the magnetic permeability and electric permittivity, respectively, of media 1 and 2. If medium 1 is free space, the above equation becomes the relation:

$$c/v = [(\mu\epsilon)/(\mu_0\epsilon_0)]^{1/2}$$

where c is the velocity of light in a vacuum, v is the velocity in a material medium, the nonsubscripted values of the constitutive relations are for the material propagation medium, and the subscripted values are for free space. However, c/v is the definition of the refractive index of a medium relative to free space. From this is obtained the relation:

$$n = (\mu_r\epsilon_r)^{1/2}$$

where n is now a dimensionless number and relative to free space. The magnetic permeability and electric permittivity are also usually given relative to free space, or air. As stated above, for dielectric materials, μ_r is approximately 1, being very much greater than 1 only for ferrous substances and slightly greater than 1 for a few non-ferrous metals, such as nickel and cobalt. Hence, for the glass and plastics used to make optical fibers and other dielectric waveguides, the refractive index, relative to a vacuum, or air, is given by the relation:

$$n \approx \epsilon_r^{1/2}$$

where ϵ_r is the electric permittivity of the propagation medium. Unless otherwise stated, refractive indices are usually given relative to a vacuum, or relative to 1.0003 of air near the earth's surface and less as the air gets thinner in the upper atmosphere and outer space. In absolute units for free space, the electric permittivity is $\epsilon =$

8.854 × 10⁻¹² F/m (farads per meter), while the magnetic permeability is $4\pi \times 10^{-7}$ H/m (henries per meter). The reciprocal of the square root of the product of these values yields approximately 2.998×10^8 m/s (meters per second), the velocity of light in a vacuum. Because the refractive index for a propagation medium is defined as the ratio between the velocity of light in a vacuum and the velocity of light in the medium, the ratio becomes unity when the propagation medium is a vacuum. See *refractive index; relative magnetic permeability*.

magnetooptic. Pertaining to a change in a material's *refractive index* under the influence of a magnetic field. Magnetooptic materials generally are used to rotate the *polarization plane* of an *electromagnetic wave*.

magnetooptic effect. The rotation of the *polarization plane* of a *plane-polarized electromagnetic wave*, such as a *lightwave*, caused by subjecting the *propagation medium* to a magnetic field, that is, Faraday rotation. The effect can be used to *modulate* a *light beam* in a material because many coefficients, such as the *reflection* and *transmission coefficients* at *interfaces; acceptance angles; critical angles; refraction angles; reflection angles; transmission modes; propagation* velocities; and other properties are dependent upon the direction of propagation of the lightwave relative to the interface. The applied magnetic field adds to the *magnetic field component* of the electromagnetic wave. The amount of angular rotation, that is, the angular displacement, is given by the relation:

$$A = VHL$$

where V is a constant of proportionality, H is the applied *magnetic field strength*, and L is the distance the lightwave is in the field. The applied magnetic field is in the direction of propagation of the lightwave for polarization-plane rotation. In any case, the applied magnetic field affects the magnetic field of the electromagnetic wave, the resultant depending on their direction. Synonymous with *Faraday effect*. Also see *electrooptic effect*.

magnetooptics. The branch of science and technology devoted to the study and application of the influence of magnetic fields on *electromagnetic waves* in the *optical spectrum*. Also see *acoustooptics; electrooptics; optooptics; photonics*.

magnification. See *borescope magnification*.

magnitude. See *pulse magnitude*.

maintainability. A characteristic of design and installation of an item, expressed as the probability that the item will be retained in or restored to a specific condition within a given time, when the maintenance is performed in accordance with prescribed procedures and resources.

MAN. *Metropolitan-area network (MAN)*.

mandrel. See *optical fiber mandrel.*

margin. See *safety margin.*

maser. See *optical maser.*

matched cladding. In a *dielectric waveguide, cladding* composed of a single homogeneous layer of *dielectric* material.

material. See *index-matching material; nonlinear optical (NLO) material; optically active material; polymeric nonlinear optical material; third-order nonlinear material.*

material absorption. See *absorption.*

material dispersion. In *fiber optics,* the spreading, that is, the lengthening of the time and spatial width of a *light pulse* as it *propagates* in a *dielectric waveguide.* The spreading is caused by the different velocities of the different *wavelengths* in the pulse, that is, in the *spectral width* of the *light.* This should not be called spectral or intramodal *dispersion,* because all dispersion is spectral and material dispersion affects all modes. Material dispersion is that part of the total dispersion of an *electromagnetic* pulse in a waveguide, such as an *optical fiber,* due to the dependence on wavelength of the *refractive index* of the material used to make the waveguide. As the wavelength is increased, and *frequency* is decreased, material dispersion decreases. At high frequencies, the rapid interaction of the electromagnetic field with the waveguide renders the refractive index even more dependent upon frequency. Since each frequency experiences a different refractive index, some frequencies will propagate faster than others, so that they arrive at the end of the fiber at different times, making the pulse at the end of a path broader in space and time than at the beginning.

material-dispersion coefficient. See *material-dispersion parameter.*

material-dispersion parameter. A parameter that characterizes *material dispersion,* given by the relation:

$$M(\lambda) = -(1/c)(dN/d\lambda) = (\lambda/c)(d^2n/d\lambda^2)$$

where n is the *refractive index,* N is the *group index,* λ is the *wavelength,* and c is the velocity of light in a vacuum. $M = 0$ at a particular wavelength, λ_0, usually about 1.3 μ *(microns)* for many fiber optic *propagation media,* such as the silica glass used to make *optical fibers.* The minus sign is chosen so that M is positive for wavelengths less than λ_0 and negative for wavelengths greater than λ_0. The amount of *pulse broadening,* that is spreading or widening, per unit length of optical fiber is given by M times the *spectral width,* that is, the spectral linewidth, except at a wavelength near λ_0, where the square of the spectral width is more significant. Synonymous with *dispersion coefficient.*

material scattering. *Scattering* caused by the properties of the materials used in fabricating *electromagnetic wave propagation media.* For example, it is the part of the total scattering attributable to the properties of the materials used in fabricating an *optical fiber.*

matrix. See *transmittance matrix.*

maximum acceptance angle (MAA). In an *optical fiber,* the maximum angle between the longitudinal axis of the fiber and an *incident light ray* at the fiber end face, in order that that ray be *totally (internally) reflected* at the *core-cladding interface* inside the fiber. For an *optical fiber,* the sine of the maximum *acceptance angle* (MAA) is the *numerical aperture,* given by the relation:

$$MAA = \sin^{-1} NA, \text{ and } NA = (n_1^2 - n_2^2)^{1/2}$$

where NA is the numerical aperture, n_1 is the fiber core *refractive index* at the point on the fiber end face at which the NA is being determined, and n_2 is the cladding refractive index. For a *graded-index fiber,* because the NA depends on the refractive index, and the refractive index varies with distance from the center of the fiber, the NA varies with the distance from the center of the fiber, and therefore the true MAA also depends on the maximum refractive index found on the fiber end face, which is at the center.

maximum input power. See *optical receiver maximum input power.*

maximum optical reflection. In an *optical transmitter,* the percentage of total *optical power output reflected* back into the transmitter that it can accommodate and still maintain its stated performance.

maximum pulse rate. In *optical waveguides* and electrical conductors in which *pulse dispersion* limits the *pulse-repetition rate (PRR),* that is, the *data signaling rate (DSR),* the PRR that is just sufficient to create a specified *bit-error ratio (BER),* given by the relation:

$$PRR = AL^b$$

where A is determined by the line characteristics that also fix the value of b, and L is the length of the line. Normally, $b = 0.5$ for *single-mode optical fibers.* For *multimode fibers,* b is between 0.5 and 1.0. For wires, $b = 2$.

maximum receiver input. In an *optical receiver terminal/regenerator facility* of an *optical station/regenerator section,* the maximum value of the input *optical power (dBm)* to the receiver at the line side of the receiver module optical connector or *splice* when operated under standard or extended operating conditions that the receiver will accept and still maintain a specified *bit-error ratio (BER)*. If the receiver input power should exceed this maximum, the section is overdesigned or an *optical attenuator* must be inserted.

maximum theoretical numerical aperture. See *numerical aperture*.

maximum transceiver dispersion. The worst-case *dispersion* (ps/nm) caused by *optical fiber* length between transmitter and receiver pair that can be accommodated by the pair to meet the bit rate *(data signaling rate)* and *bit-error ratio (BER)* specified by the manufacturer, when operated under standard or extended operating conditions.

Maxwell's equations. A group of basic equations, in either integral or differential form, that (1) describe the relationships between the properties of magnetic and *electric fields,* their sources, and the behavior of these fields in *free space* and material *propagation media* and at *interfaces* between these media; (2) express the relations among electric and magnetic fields that vary in space and time in material media and free space; and (3) are fundamental to the *propagation* of *electromagnetic waves* in material media and free space. The equations are the basis for deriving the *wave equations* that express the *electric* and *magnetic field vectors* in a propagating electromagnetic wave in a propagation medium, such as a *lightwave* in an *optical fiber* or an electromagnetic wave in a hollow or filled metallic *waveguide.* Solutions to the equations are dependent upon boundary conditions. Maxwell's equations in differential form are given by the relations:

$$\nabla \times \mathbf{E} = - \partial \mathbf{B}/\partial t$$

$$\nabla \times \mathbf{H} = \mathbf{J} + \partial \mathbf{D}/\partial t$$

$$\nabla \cdot \mathbf{B} = 0$$

$$\nabla \cdot \mathbf{D} = \rho$$

where **E, H, B,** and **D** are the vectors of the electric field strength, that is, electric field intensity (volts per meter), the magnetic field strength, that is, magnetic field intensity (amperes per meter, oersteds), the magnetic flux density (webers per square meter or gauss), and the electric flux density or displacement (coulombs per square meter), respectively; **J** is the electric current density (amperes per square meter); and ρ is the electric charge density (coulombs per cubic meter). The ∇ is the "del" space derivative operator, expressing differentiation with respect to all distance coordinates, the $\nabla \times$ being the curl (a vector) and the $\nabla \cdot$ being the divergence (a scalar). The partial derivatives are with respect to time. These equations are used in conjunction with the *constitutive relations* to obtain useful practical results when actual sources of charge and current in free space and material media are known. The equations are valid only when the field and current vectors are single-valued, bounded, continuous functions of position and time and have continuous derivatives. Wave equations are the solutions of Maxwell's equations, given the boundary conditions. For optical fibers, the medium is considered charge-free and electrically nonconducting. Thus, $\rho = 0$ and $\mathbf{J} = 0$, which simplifies solving Maxwell's equations.

MCVD. *Modified chemical vapor deposition.*

MCVD process. See *modified chemical-vapor deposition (MCVD) process.*

mean-square pulse duration. See *root-mean-square (rms) pulse duration.*

mean time between failures (MTBF). For a particular measurement interval, the total functional life of a population of an item divided by the total number of failures within the population during the measurement interval. The definition holds for time, cycles, kilometers, events, or other measure-of-life units. It is the sum of all the operational times of all the items in the population, divided by the total number of failures within the population during the measurement time interval. For example, if kilometers are used, the result would be expressed as mean distance between failures.

mean time between outages (MTBO). The *mean time between failures* that result in loss of system continuity or unacceptable degradation of performance. The MTBO may be expressed by the relation:

$$MTBO = MTBF/(1 - FFAS)$$

where MTBF is the nonredundant mean time between failures, and FFAS is the fraction of failures for which the failed equipment is by-passed automatically.

mean time to failure (MTTF). In a population of similar items, the mean time an item functions successfully from the initial instant it is placed in operation to the instant of first failure, that is, the infant mortality period averaged over the population of items.

mean time to repair (MTTR). The total corrective maintenance time, that is, the total time devoted to maintenance, divided by the total number of maintenance actions, during a given period, such as a month, a year, or to date, since an item was placed in service.

mean time to service restoral (MTTSR). The mean time to restore service, that is, restore acceptable operational capability, following system failures that result in a service outage or unacceptable operational capability. The time to restore service is all the time from the occurrence of the failure until the restoral of service, including fault detection, fault location, and fault correction time, that is, the actual time to repair.

measurement range. In an *optical time-domain reflectometer (ODTR),* the length of *optical waveguide,* such as an *optical fiber,* that lies between the minimum length and the maximum length that the ODTR can make an *attenuation* measurement, such as measure the distance to, and the *insertion loss* caused by, a fiber optic connector. Measurement ranges are typically from several centimeters to over a hundred kilometers operating at typical *wavelengths* of 0.85, 1.3, and 1.55 μ *(microns)*. Also see *distance resolution; dynamic range*.

mechanically induced modulation. 1. In *fiber optics, modulation* of a *lightwave propagating* in an *optical fiber* in accordance with physical distortion of the fiber. The distortion may be caused by a force field deliberately applied to the fiber, such as by vibrating, bending, compressing, or stretching it, in order to measure or sense the force field. The modulation may appear as amplitude, *phase* shift, *wavelength, polarization rotation*, or other form of modulation. **2.** The introduction of undesirable *modulation* (noise) of a *lightwave* in an *optical fiber* caused by physical distortion of the fiber.

mechanical splice. A *fiber optic splice* made by mechanical fixtures or materials, such as *optical cements*, rather than by thermal fusion. (See Figs. M-2 –M-3)

medium. See *active laser medium; anisotropic propagation medium; birefringent medium; dispersive medium; isotropic propagation medium; lossy medium; nondispersive medium; propagation medium*.

medium-loss fiber. An *optical fiber* having a medium-level *optical signal* power loss per unit length of fiber, usually measured in decibels/kilometer at a specified *wavelength* and due to all intrinsic causes. In medium-loss fiber, *attenuation* in amplitude of a *propagating wave* is caused primarily by *scattering* caused by metal ions, *ab-*

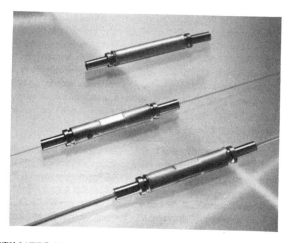

M-2. The OPTIMATE® **Mechanical Splice** for connecting *multimode* and *single-mode optical fibers* without fusion or gluing operations. Design and crimping methods compensate for fiber diameter variations for center *(optical axis)* alignment. *Insertion loss* is less than 0.25 dB. (Courtesy AMP Incorporated).

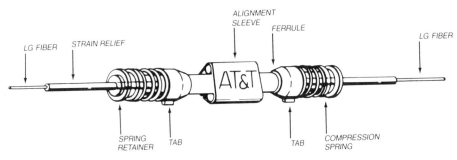

M-3. A mean *splice loss* of 0.20 *dB,* 0.10 dB, or less with a *test* set can be achieved with this enhanced rotary **mechanical splice.** It can be used in aerial, buried, underground, terminal, midspan, repair, and distribution applications. Mated glass plug assemblies *(termini)* are installed on each end of a *fiber.* These plugs are then joined together through an *alignment sleeve* while *index-matching* gel is used between the two *connectors* to minimize splice loss. The plugs are rotated in the alignment sleeve, and the tabs are aligned to obtain minimum loss (passive alignment). If extremely low loss is required, the splice can be tuned (active alignment). The Enhanced Rotary Mechanical Splice can be used for splicing *single-mode* or *multi-mode fiber.* (Courtesy AT&T.)

sorption caused by water in the form of the hydroxyl ion, and *Rayleigh scattering.* This does not include extrinsic losses, such as *high-order mode loss* caused by *launch conditions* or *fiber optic connector* losses at fiber-to-fiber *interfaces.*

megahertz (MHz). A unit of *frequency* denoting 1 million, that is, 10^6, Hz.

meridional ray. In *fiber optics,* any *ray* that passes through the axis of an *optical fiber.* Therefore, a meridional ray must lie in a plane that contains the fiber axis.

merit figure. See *optical fiber merit figure.*

mesh network. In network topology, a network configuration in which there is more than one *path* between any two *nodes* and thus there are no end-point nodes, that is, no nodes connected to only one *branch of the network.* See *fully connected mesh network.*

meter (m). The SI (International System) unit of length. The meter was originally established by Napoleonic scientists as 1 ten-millionth of the distance between a pole and the equator, that is, 10^{-7} of that distance along the surface of the earth. Later, the standard international meter was the distance between two fine lines engraved on a platinum bar held at the International Bureau of Weights and Measures near Paris. Now the meter is defined as 1,650,763.73 times the *wavelength* of the *spectral line* of orange *light emitted* when a gas consisting of the pure krypton isotope of mass number 86 is excited by an electrical discharge.

method. See *fiber optic test method (FOTM); Fresnel reflection method; reference test method (RTM); refracted-ray method; transverse-scattering method.*

metric system. A decimal system of weights and measures based on the *meter,* the kilogram, and the second. The modern version of this system employs "SI Units," or the International System of Units. See table: Metric System.

Metric System

QUANTITY	BASE UNITS	ABBREVIATION
length	meter	m
mass	kilogram	kg
time	second	s
electric current	ampere	A
temperature	kelvin	K
amount of substance	mole	mol
luminous intensity	candela	cd

SUPPLEMENTARY UNITS		
plane angle	radian	rad
solid angle	steradian	sr

Certain derived terms have also been standardized, such as the hertz (2π radians per second), equivalent to the obsolete cycle per second.

Prefixes used with metric units:

UNIT	ABBREVIATION	VALUE
exa	E	10^{18}
peta	P	10^{15}
tera	T	10^{12}
giga	G	10^9
mega	M	10^6
kilo	k	10^3
hecto	h	10^2
deka	da	10
deci	d	10^{-1}
centi	c	10^{-2}
milli	m	10^{-3}
micro	μ	10^{-6}
nano	n	10^{-9}
pico	p	10^{-12}
femto	f	10^{-15}
atto	a	10^{-18}

Examples:

TERM	ABBREVIATION	MEANING
megahertz	MHz	10^6 hertz
picofarad	pF	10^{-12} farads
nanosecond	ns	10^{-9} seconds
micrometer (micron)	μm (μ)	10^{-6} meters

metropolitan-area network (MAN). A communication network covering generally a larger geographical area than a *local-area network (LAN).* Typically, an MAN interconnects two or more local-area networks, may cross administrative boundaries, may use multiple access methods, and may operate at a higher data signaling rate (DSR) than an LAN.

MFD. *Mode field diameter.*

MHz. *Megahertz.*

Michelson fiber optic sensor. A high-resolution *interferometric sensor* in which an *electromagnetic wave,* such as *monochromatic light,* is split, one half *reflected* from a fixed mirror and back through the *splitter* to a *photodetector,* the other half passed directly through the splitter to a movable mirror that reflects it back to the splitter where it is reflected to the same photodetector. The two waves can enhance or cancel each other thereby *modulating* the *irradiance,* that is, the light intensity, at the photodetector in accordance with an input signal in the form of a displacement of the movable mirror, such as might be produced by a sound wave or a pressure, strain, or temperature variation. If an *optical fiber* is used, the ends of the fiber form the reflecting surfaces. Moving one end relative to the other produces the same effect as moving the mirror. Displacements less than 10^{-13} m can be measured when narrow spectral-width, that is, monochromatic, *laser sources* are used.

microbend. In *optical fibers,* a small deviation or dent in the *core-cladding interface* surface of a fiber. Undulated core-cladding interface surfaces with large numbers of microbends per unit length are often created during the cabling process when manufacturing fiber optic cables, when winding cables on spools and reels, and when passing cables around capstans. Microbends can be created in *fiber optic cables* under hydrostatic pressure, particularly when *buffering* is inadequate or *strength members* are improperly positioned. Simply bending a cable can introduce microbends because of stress differentials imposed on the fiber at the inside and outside radii of the bend. The size of microbends may be from a fraction of a *micron* to several microns in the direction of the *optical axis* or transverse distortion of the core-cladding interface. A waviness of the optical axis, up to several millimeters in *wavelength,* is also considered a microbend. Lightwaves striking a microbend may escape from the *core* because the *critical angle* at the microbend can become less than that required for *total (internal) reflection.*

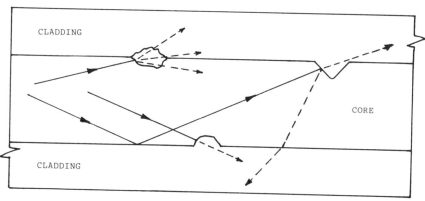

M-4. *Core-cladding interface* surface flaws and **microbends** in an *optical fiber* are often caused by stresses from forces applied during the *cabling process* and careless handling. The *losses* occur because the *incidence angle* is less than the *critical angle* at the microbends.

microbending. In an *optical waveguide,* the introduction of minute curves with small radii in the *optical axis* of the guide or the *core-cladding interface* surface, at which *incidence angles* of *light rays propagating* in the guide are or tend to become less than the *critical angle,* so that some of the *radiant power* in the ray escapes from the *fiber core* into the *cladding. Microbends* may result from manufacturing processes, such as application of *coatings, cabling processes,* spooling, packaging, shipping, storing, installing, and using. Microbends can cause *radiation loss* and *mode coupling.* The size of microbends may be from a fraction of a *micron* to several microns in the direction of the *optical axis* or transverse distortion of the core-cladding interface. A waviness of the optical axis, up to several millimeters in *wavelength,* is also considered a microbend. When microbends are large enough, they may be considered *macrobends,* such as will occur when the cable containing the fiber is bent around a curve in a raceway, duct, or trench. Care must be taken in installation that the macrobend radius be greater than both the *minimum bend radius* and the *critical radius.*

microbend loss. In an *optical fiber,* the loss caused by *microbends.*

microbend sensor. A *fiber optic sensor* in which an *optical fiber* is passed between toothed, or serrated, plates. When a force or pressure is applied to the plates, the fiber is squeezed between them, causing *microbends* in the fiber. An *electromagnetic wave,* such as a *lightwave, propagating* in the fiber will escape at the bends, ejecting *light* into the *cladding* at each point at which the *incidence angle* at the *core-cladding interface* is greater than the *critical angle* for *total (internal) reflection.* The increase in *attenuation* of the main *beam* in the fiber, caused by the loss of light into the cladding as pressure is applied to the plates, is detected by a *photodetector* whose output decreases as the pressure increases. Thus, the output signal of the sensor is a function of the applied force or pressure.

microcrack. In *optical fibers,* a minute crack or partial break, usually along the outer surface of the *cladding,* but possibly deep enough to enter the *core.* Micro-cracks may become enlarged by exposure to adverse environmental conditions, such as moisture, bending, twisting, tensile stress, thermal gradients, vibration, shock, and corrosive atmospheres. Light striking a microcrack will be *reflected, refracted,* and *scattered,* generally causing light to escape from the fiber. Proofing of fibers under tension enlarges microcracks. If the cracks become sufficiently large, the fiber ruptures under the applied tension and must be *spliced.* However, proofing is done to reduce the possibility of a fiber rupturing when it is not under tension.

micrometer (μm). See *micron (μ).*

micron (μ). One millionth of a *meter,* that is, 10^{-6} meters. The micron, instead of the *nanometer,* is widely used by the fiber optics community to express the *wavelengths* of *light* and the geometry of *optical fibers,* such as the *core, cladding,* and *mode field diameters* and radii, thus simplifying wavelength and geometric comparisons. The wavelengths of light used in *fiber optics* are of the order of 1 μ. Synonymous with *micrometer (μm).*

middle-infrared. Pertaining to the region of the *electromagnetic spectrum* that lies between the longer *wavelength* end of the *near-infrared* region and the shorter wavelength end of the *far-infrared* region, that is, from about 3 to 30 μ *(microns).* Thus, the near-infrared region is included in the near-*visible spectrum,* but the *middle-infrared* and far-infrared regions are not included in the near-visible region. The far infrared extends to the shorter wavelength end of the radio wave region, which is also the longer wavelength end of the *optical spectrum,* that is, to about 100 μ. None of the *infrared,* that is, the near, middle, or far, are included in the visible spectrum.

mirror. See *dichroic mirror; partial mirror.*

mirror sensor. See *optical cavity mirror sensor.*

misalignment loss. See *angular misalignment loss.*

mixed-gas laser. An *ion laser* in which a mixture of gases, such as argon and krypton, is used as the *active laser medium.*

mixer. See *fiber optic mixer.*

mixing. See *optical mixing.*

mixing box. See *optical mixing box.*

mixing rod. See *optical mixing rod.*

modal dispersion. In *fiber optics,* the spreading of a *light pulse* in an *optical waveguide* caused by the different paths taken by the various *modes* in the *lightwaves.* Synonymous with *intermodal dispersion; mode dispersion.*

modal distortion. See *multimode distortion.*

modal distribution. **1.** In an *optical waveguide,* such as an *optical fiber,* operating at a single *frequency* or narrow *spectral width,* the number and types of *modes* supported by the guide and their *propagation* time differences. **2.** In an *optical waveguide* operating at multiple *frequencies* simultaneously, the separation in frequency among the modes being supported by the guide.

modal loss. In an *open waveguide,* such as an *optical fiber,* a *loss* of power in one or more *modes* in an *electromagnetic wave propagating* in the guide. Modal loss is caused by anomalies inside and outside the guide, such as obstacles outside the waveguide, changes in dimensions of the guide, and sharp bends in the guide. A loss of power from a mode in an electromagnetic wave propagating in a waveguide also occurs when *radiant power* is transferred to *high-order modes* and *leaky modes* that enter the cladding where they may be *absorbed, scattered,* or *radiated* away from the waveguide.

modal noise. Noise generated in an *optical system* by the combination of *mode*-dependent *optical losses,* fluctuations in the distribution of *radiant power* among the modes or in the relative *phases* of the modes, and the effects of *differential mode attenuation.* Synonymous with *mode-partition noise; speckle noise.*

modal power distribution. See *nonequilibrium modal power distribution.*

modal-power-distribution length. See *nonequilibrium modal-power-distribution length.*

mode. One of the various possible patterns of *standing* or *propagating electromagnetic fields* in a *cavity* or *waveguide.* Modes are characterized by their *wavelength;* the spatial distribution and direction of their *electric* and *magnetic field components* relative to the boundaries of the waveguide; and the *field strengths* of these components. Any electromagnetic field distribution that satisfies *Maxwell's equations* and the boundary conditions, that is, is a solution to these equations, is a mode. The field *pattern* of a mode depends on the *wavelength, refractive index,* and *cavity* or *waveguide* geometry. In *guided electromagnetic waves,* each mode is a particular condition or arrangement of the waves in the waveguide. Each different orientation, that is, *polarization,* of the *electric* and *magnetic fields* of an electromagnetic wave in the guide corresponds to a different mode. Different *propagating wavelengths* correspond to different propagating modes. Different paths correspond to different modes. *Optical fibers* can support many modes, the number being given by the relation:

$$N = 2\pi^2 a^2 (n_1^2 - n_2^2)/\lambda^2$$

where a is the fiber radius (one-half the *core diameter*), n_1 and n_2 are the *refractive indices* of the *core* and *cladding*, and λ is the wavelength of the *lightwave launched* into the fiber by the *source*, not the wavelength inside the fiber. *Single-mode fiber* diameters range from 2 to 11 μ *(microns)*, but the number of modes that a given fiber can support depends on the core radius, the wavelength, and the *numerical aperture*, which depends on the core and *cladding* refractive indices. Thus, an optical fiber operating in single-mode can be made to operate in *multimode* if the wavelength is reduced. See *bound mode; cladding mode; coupled mode; cutoff mode; fundamental mode; high-order mode; hybrid mode; leaky mode; linearly polarized (LP) mode; lowest order transverse mode; low-order mode; propagation mode; radiation mode; transverse magnetic (TM) mode; transverse electromagnetic (TEM) mode; transverse electric (TE) mode; unbound mode.*

mode attenuation. See *differential mode attenuation (DMA)*.

mode conversion. In a *guided electromagnetic wave*, such as a *lightwave*, the transfer of some or all of the electromagnetic power in one *mode* to another mode. Mode conversion can be intentional, such as inducing *high-order cladding modes* for transferring energy to another fiber or for *microbend sensing;* or can be unintentional, such as when energy is transferred to high-order modes that escape from the guide and are radiated away causing *signal attenuation.*

mode coupling. In an *optical waveguide*, the exchange of *optical power* among modes. The exchange of power may reach statistical equilibrium after *propagation* over a finite distance in a waveguide, such as an *optical fiber*. This finite distance is designated as the equilibrium modal-power-distribution length, or simply the *equilibrium length.*

mode dispersion. See *modal dispersion.*

mode distortion. See *multimode distortion.*

mode field diameter. For Gaussian statistical distributions of power and energy among the *modes* of an *electromagnetic wave propagating* in a *single-mode optical fiber*, the diameter at which the *electric* and *magnetic field strengths* are reduced to $1/e$ of their maximum values. This is equivalent to the diameter at which the *radiant power* is reduced to $1/e^2$ of the maximum power, because the power is proportional to the square of the electric or magnetic field strength.

mode filter. An *optical device* that can accept, pass, reject, or *attenuate*, that is, reduce the power level of, a certain *mode* or modes in an *electromagnetic wave.* See *core mode filter.*

mode hopping. The transfer of *radiant power* from one *mode* to another. Mode hopping occurs in *lasers.* Synonymous with *mode jumping.*

mode jumping. See *mode hopping.*

mode mixer. See *mode scrambler.*

mode-partition noise. See *modal noise.*

mode scrambler. 1. In *fiber optics,* a device, usually consisting of one or more *optical fibers,* in which a desired distribution of *radiant power* among *modes propagating* in an optical fiber is accomplished by transferring power among the modes by means of induced *mode coupling.* The mode scrambler is frequently used to provide a *modal distribution* that is independent of *source* characteristics or that meets other specifications. **2.** A device that induces *mode coupling* in a *waveguide,* such as an *optical fiber.* Synonymous with *mode mixer.*

mode simulator. See *equilibrium mode simulator.*

mode stripper. See *cladding mode stripper; fiber optic mode stripper.*

mode volume. The number of *bound modes* that an *optical waveguide* is capable of supporting. The mode volume is approximately given by the relation:

$$M_{si} = V^2/2$$

for *step-index waveguides,* and by the relation:

$$M_{gi} = (V^2/2)(g/(g+2))$$

for *graded-index power-law profile waveguides,* where g is the *refractive-index profile parameter,* and V is the *normalized frequency,* that is, the V-parameter or V-value, when V is greater than 5. See *effective mode volume.*

modified chemical-vapor-deposition (MCVD) process. A *modified inside vapor-phase-oxidation (IVPO) process* for production of *optical fibers* in which the burner travels along the glass tube optical fiber *preform.* Soot particles are created inside the tubing rather than in the burner flame as in the *outside vapor-phase-oxidation (OVPO) process.* The chemical reactants, such as oxygen, *dopants,* and silicon tetrachloride, are caused to flow through the rotating glass tube at a pressure of about 1 atmosphere, gauge. The high temperature causes the formation of oxides, in the form of a soot, and a glassy deposit on the inside surface of the tube. This deposit increases the *refractive index* of the glass and forms the *core* when the tube is drawn into a solid fiber.

modulate. To vary a characteristic or parameter of an entity in accordance with a characteristic or *parameter* of another entity, for example, to vary the *irradiance,* that is, the *intensity,* of a *beam* from a *light source,* such as a *laser,* in accordance with an intelligence-bearing electronic *signal* applied to the source, or to vary the *radiant power* at a point in a *waveguide,* such as an *optical fiber,* in accordance with a physical variable being sensed or measured, such as in a *microbend* or *Sagnac sensor.* Also see *demodulate.*

modulation. The controlled variation of a *parameter,* such as amplitude, *phase, frequency,* or *pulse position* or *width,* of a *wave* usually for the purpose of transferring information. Modulation can be accomplished by superimposing another wave or by varying a physical parameter to which the wave is sensitive, such as by varying *attenuation* in an *optical fiber* or controlling the output of a *laser* by varying the driving voltage. Uncontrolled or random modulation is considered to be noise or interference. See *mechanically induced modulation; polarization modulation; pulse modulation; pulse-amplitude modulation (PAM).*

modulator. See *fiber optic modulator; optical phase modulator; thin-film optical modulator.*

molecular laser. A type of *gas laser* whose *active laser medium* is a molecular substance, that is, a compound, such as carbon dioxide, hydrogen cyanide, or water vapor. Also see *atomic laser.*

molecular stuffing process. A process of making *graded-index (GI) optical fibers* in perhaps five broad steps—glass melting, phase separation, leaching, *dopant* introduction, and consolidation.

monitor. In communications, to place *emissions,* such as radio, radar, video, *optical,* microwave, or sonar transmissions, under continuous surveillance, by detecting, measuring, recording, and interpreting them.

monochromatic. Pertaining to a single *wavelength* or pure *color.* In practice, a "single" wavelength *source* is at best a narrow *band* of wavelengths. Also see *coherent; line source; spectral width.*

monochromatic light. *Electromagnetic radiation,* in the *visible* or near-visible spectrum, that has only one *frequency* or *wavelength.* Because production of high-*radiance,* that is, high-intensity, high-energy *light* at a single wavelength, that is, with a zero *spectral linewidth,* is not practical at present, light with narrow *spectral widths,* such as that produced by *lasers,* is considered monochromatic light, that is, it consists of one *color.*

monochromator. An instrument for selecting narrow portions, that is, selecting specific *spectral lines,* from an *optical spectrum.*

monomode optical fiber. See *single-mode fiber.*

moving grating sensor. A *fiber optic sensor* in which two *optical fibers* are separated by a small gap in which a pair of gratings is placed in such a way that one is fixed and the other is movable. Both gratings consist of alternate opaque and *transparent* parallel elements of equal width. When the transparent elements of both gratings line up, *light* is *transmitted* from the *light source* fiber to the *photodetector* fiber. When the opaque elements of one grating line up with the transparent elements of the other, the light transmission from fiber to fiber is zero. Starting from a lineup of opaque elements covering transparent elements, and consequently zero transmission, as the movable grating starts to uncover the transparent elements, light begins to pass through the gratings, causing the transmitted amount of light to increase to the maximum when the transparent elements line up. As the movable grating continues to be moved in the same direction, the amount of light decreases to zero, rising again as the grating continues to move in the same direction, until the end of its movement. The number of times peak transmission is reached is proportional to the distance moved. The *frequency* of the output *signal* is proportional to the speed of movement. The acceleration of the movable grating is proportional to the time rate of change of output signal frequency. For example, a sound wave impinging on the movable grating will be reproduced as an electronic output signal of the photodetector.

MTBF. *Mean time between failures.*

MTBO. *Mean time between outages.*

MTTF. *Mean time to failure.*

MTTR. *Mean time to repair.*

MTTSR. *Mean time to service restoral.*

multifiber cable. A *fiber optic cable* that contains two or more *optical fibers.* Each fiber provides a separate *transmission channel.*

multifiber joint. An *optical fiber splice* or *connector* designed to mate two *multifiber cables.* The joint provides simultaneous *optical* alignment of all the individual fibers. *Optical coupling* between the aligned *waveguides* may be achieved by various techniques, such as proximity-butting, the use of lenses, and the use of *index-matching materials.*

multilayer filter. See *interference filter.*

multiline laser. A *laser* that *emits radiation* at two or more *wavelengths.*

multimode dispersion. See *optical multimode dispersion.*

multimode distortion. In an *optical waveguide, distortion* resulting from *multimode group delay* and *differential mode attenuation.* In an *optical fiber* in which more than one mode is *propagating,* multimode distortion is caused by the different *modes* having different propagation properties, such as different *propagation constants.* The term "multimode dispersion" is often used as a synonym. However, such usage is discouraged because the mechanism is not one of dispersion. Synonymous with *intermodal distortion; modal distortion; mode distortion.*

multimode fiber. An *optical fiber* that will allow more than one *mode* to *propagate* at a given *wavelength.* The number of modes will depend on the *core diameter,* the *numerical aperture,* and the *wavelength.* Also see *mode volume.*

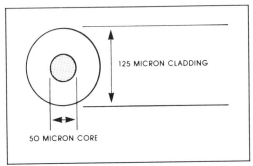

Cross Section Of Multimode Fiber

SPECIFICATIONS

Cladding diameter   *125.0 ±2.0 micron*
                    (typical variation ±0.3 nm)
Core diameter   *50.0 ±3.0 micron*
Core eccentricity   *6% maximum (typical value 1.5%)*
Core ovality   *6% maximum (typical value 4%)*
Refractive index delta   *1.3%**
Numerical aperture   *0.22*
Attenuation range   *2.6-3.5 dB/km at 850 nm*
                    0.55-1.5 dB/km at 1300 nm
Bandwidth range   *300-900 MHz-km at 850 nm*
                    300-1400 MHz-km at 1300 nm
Typical field splice loss   *0.15 dB (array);   0.20 dB (fusion)*
Coating diameter   *245 ±19 micron*
*1.0% available upon request.

M–5. This **multimode fiber** consists of a *graded-index* germanium-doped *core* and a silica *cladding.* A dual protective *coating* is applied over the cladding to cushion the fiber against *microbending losses,* provide abrasion resistance, and preserve the mechanical strength of the glass. (Courtesy AT&T.)

multimode group-delay. In an *electromagnetic wave propagating* in a *waveguide,* the variation in *group-delay time,* caused by differences in *group velocity,* among *bound propagating modes* even at a single *frequency.* The differences in arrival times of the leading and trailing edges of a *pulse* at the end of the waveguide, compared with the sending end, are caused by the different *propagation delays* of the different modes. The modes can be considered as different optical paths of different *optical path lengths.* For example, in an *optical fiber,* it is possible for *photons* or *waves* that propagate along the *optical axis* of the *fiber core* to arrive at the end sooner than those that follow a helical path through the core, thus causing the *pulse durations* at the end of the fiber to be increased. If the pulse duration is too great, *intersymbol interference,* that is, an overlapping of consecutive pulses, will occur. The received pulse duration can be reduced if the *refractive-index profile* of the core is arranged so that light rays taking a helical path along the outer edges of the core propagate through a lower *refractive-index* material, hence travel faster in the longer path than *axial rays* propagating along the *optical axis,* or in a helical path closer to the axis of the core, in the higher refractive-index material, so that they arrive at the end of the fiber at the same time. Actual propagation delay is of little consequence, as long as all rays of a given pulse arrive at the end of the fiber at the same time. So-called zero-dispersion fibers are being made. Synonymous with *differential mode delay.*

multimode laser. A *laser* that *emits radiation* containing two or more *modes.*

multimode waveguide. A *waveguide* that can support more than one *mode.* Because different *wavelengths* constitute different modes and the number of modes is also dependent on the *numerical aperture* and the *core diameter,* a given ''multimode'' waveguide might support only one mode and therefore could be called a single-mode waveguide if the operating wavelength is long enough, and conversely, a given ''single-mode waveguide'' might support several modes and therefore could be called a *multimode waveguide* if the operating wavelength is short enough.

multipath. Pertaining to *electromagnetic wave propagation* that results in the waves reaching a receiver by two or more *paths.* For *lightwaves* in dielectric waveguides, multipath may be due to many causes, such as *refractive-index variations* in *optical fibers.* For radio, video, and microwave *transmissions,* atmospheric ducting, ionospheric *reflection,* and reflection from terrestrial objects such as mountains and buildings, produce multipath effects. Multipath affects *signals* because the waves combine in a variety of ways, such as constructive reinforcement, destructive cancellation, and *phase* shifting at the receiver.

multiplexer. See *fiber-loop multiplexer; fiber optic demultiplexer (active); fiber optic demultiplexer (passive); fiber optic multiplexer (active); fiber optic multiplexer (passive); optical demultiplexer (active); optical demultiplexer (passive); optical multiplexer (active); optical multiplexer (passive); thin-film optical multiplexer.*

multiplexing. See *frequency-division multiplexing (FDM); space-division multiplexing; time-division multiplexing; wavelength-division multiplexing (WDM).*

multiplication. See *avalanche multiplication.*

multiply connected star network. See *star-mesh network.*

multiport coupler. See *fiber optic multiport coupler.*

multirefracting crystal. A *transparent* crystalline substance that is *anisotropic* with respect to the velocity of *light propagating* within it in different directions, that is, the *refractive index* is different in different directions.

N

n. In *optics,* the symbol for *refractive index,* usually implying the refractive index of a material relative to that of a vacuum.

nanometer (nm). One-billionth (U.S.) of a *meter,* that is, 10^{-9} meters or 10 Å (angstroms). Though a large part of the scientific community still use nanometers to express the *wavelength* of *light,* the trend in the *fiber optics* engineering and industrial community is toward the use of the *micron,* one-millionth of a meter or 1000 nm (nanometers), perhaps because of the convenience of the numbers and to indicate spatial relationships in *optical fibers.* The *optical wavelengths* used in fiber optics are of the order of 1 μ *(micron). Optical fiber core* and *cladding diameters; cladding, core,* and *reference tolerance areas* and *fields; longitudinal offset,* that is, fiber end-face-to-end-face distances in a *fiber optic connector; optical thin-film* thicknesses; and many other dimensional aspects of *optical waveguides,* are also expressed in microns, which simplifies comparisons between optical wavelengths and optical waveguide geometry, such as comparing *mode field diameters* and fiber *core diameters* and comparing fiber *mode volumes* using formulas that contain both geometric values and wavelengths. Many of the equations and formulas contain wavelengths and geometric values in numerators and denominators. Calculations and estimates are simplified when geometric values and wavelength values are expressed in the same units.

narrow-band. Pertaining to a group or range of *frequencies* that are within a relatively restricted part of the *spectrum* or whose limiting frequencies are relatively close together, the concept of "closeness" being dependent upon the position in the *electromagnetic frequency spectrum.* Thus, at a median frequency of 100 kHz, a band of frequencies 5 kHz wide would be considered wide. The same 5-kHz band at 100 MHz would be considered narrow. A rule of thumb is that if the band is less than 0.1% of the median or operating frequency, the band is narrow.

near-field diffraction pattern. A *diffraction pattern* for an *electromagnetic wave* observed in the *near-field region.* Synonymous with *Fresnel diffraction pattern.*

near-field pattern. See *near-field radiation pattern.*

near-field region. The region close to an *aperture* or *source* of *electromagnetic radiation* where the *radiation pattern* and *irradiance,* that is the incident power per unit area, are highly dependent upon the distance from the source, such as where the *electric field strength* varies inversely as the cube of the distance, and the *propagation times* to points within the region are negligible compared with propagation times to points in the *far-field region.*

near-field radiation pattern. The *radiation pattern* of a *source* of *electromagnetic radiation* in the *near-field region,* for example, the radiation pattern for an *optical fiber* that describes the distribution of *radiant emittance* as a function of position in the plane of the exit face of the fiber. Synonymous with *near-field pattern.*

near-field scanning technique. The method for determining the *refractive-index profile* of an optical waveguide by illuminating the entrance face with a *light source* and measuring the point-by-point *radiant emittance* as a function of position on the exit face.

near-field template. See *four-concentric-circle near-field template.*

near-infrared. Pertaining to the region of the *electromagnetic spectrum* that lies between the longer *wavelength* end of the *visible spectrum* and the shorter wavelength end of the *middle-infrared* region, that is, from about 0.8 to 3.0 μ *(microns).* Thus, the near-infrared is included in the near-*visible* region, but the middle-infrared and far-infrared regions are not included in the near-visible region. The *far-infrared* extends to the shorter wavelength end of the radio wave region, which is also the longer wavelength end of the *optical spectrum,* that is, to about 100 μ. None of the infrared regions, near, middle, or far, are included in the *visible spectrum.*

NEP. *Noise-equivalent power.*

network. **1.** An organization of stations capable of intercommunication, but not necessarily on the same *channel.* **2.** Two or more interrelated circuits. **3.** A combination of switches, terminals, and circuits in which *transmission* facilities interconnect the *user end-instruments* directly, that is, there are no *switching,* control, or *processing centers* involved. **4.** A combination of circuits and terminals serviced by a single *processing center* or *switching center.* **5.** In network topology, a group of *nodes* interconnected by *branches.* See *data-transfer network (DTN); fiber optic data-transfer network; fully connected mesh network; integrated-services digital (or data) network (ISDN); local-area network (LAN); long-haul communication network; mesh network; metropolitan-area network (MAN); ring network; short-haul communication network; single-node network; star network; star-mesh network; tree network.*

network-control center. A point in a command, control, and communication network whose main function is to control the network, but it may also serve as a

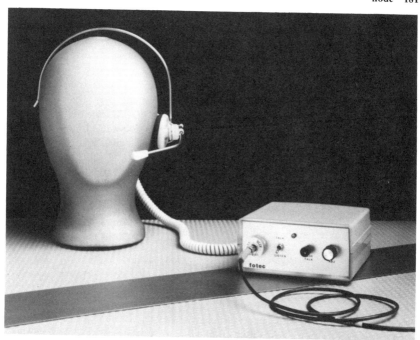

N-1.   The FOtalk® talk set allows "walkie-talkie" communication over *optical fibers* during installation and testing of *fiber optic* **networks.** Range is sufficient for *LANs* and campus networks. Other models are for military and longhaul applications. (Courtesy fotec incorporated.)

switching center, a *processing center,* or both, for local users as well as the network. A network control center is usually located at one or more *nodes* in the network.

network-interface device.   A device used to access a *bus* or a network, for example, a *modem.*

network layer.   In *open-systems architecture,* the *layer* that provides the functions, procedures, and protocol that are needed to control the transfer of *data* in a specific *transmission* facility or system. The network layer masks the routing and switching characteristics of the *data link* and *physical layers* below from the layers above. In a *fiber optic network,* the data link and physical layers consist of fiber optic and electronic components and systems. Also see *Open-Systems-Interconnection (OSI) Reference Model (RM).*

network standard.   See *synchronous optical network (Sonet) standard.*

node.   In network topology, a terminal, that is, the end, of any *branch* of a network or a terminal common to two or more branches of a network. See *end-point node.*

noise. See *Johnson noise; modal noise; quantum noise; shot noise; thermal noise.*

noise current. The electrical current caused by *noise voltage.*

noise-equivalent power (NEP). In an *optical detector,* the value of the input *radiant power* that produces an output *signal-to-noise ratio* equal to 1, for a given *wavelength, modulation frequency,* and *noise-equivalent bandwidth.* Some manufacturers and authors have defined NEP as the minimum detectable power per square root unit bandwidth. In this case, NEP has the units watts/(hertz)$^{1/2}$, and therefore NEP is a misnomer because NEP should have the units of power, such as watts. Others have defined NEP as the radiant power that produces a *signal-to-dark-current* noise ratio of unity. However, this can be misleading when dark-current noise does not dominate the noise spectrum, as is often true in *fiber optic systems.*

noise-limited operation. See *quantum-noise-limited operation.*

noise voltage. In *fiber optic systems,* a voltage that is not *coherent* with the *signal radiant power,* such as an rms component of an *optical detector* electrical output voltage. The noise voltage is usually measured with the signal power removed.

nominal central wavelength. In an *optical transmitter,* the nominal value of transmitter *wavelength* for a given application. The central wavelength is the wavelength at which the effective *optical power* may be considered to be concentrated as defined by a peak *mode* or power-weighted measurement method. The central wavelength range is defined by the minimum and maximum wavelength limits of the total range of transmitter wavelengths caused by the combined worst-case variations due to manufacturing, temperature, aging, and any other significant factors when operated under standard or extended operating conditions.

noncircularity. See *cladding noncircularity; core noncircularity; reference surface noncircularity.*

nondispersive medium. In the *propagation* of *electromagnetic waves* in material media, a *medium* in which the *phase velocity* does not vary with frequency. Thus, if an *optical pulse* consisting of more than a single *wavelength* is passed through a dispersive medium, all the different wavelengths will arrive at the end at the same time. Thus, the phase velocity and the *group velocity* are one and the same thing. All materials are dispersive to some extent. Also see *dispersive medium.*

nonequilibrium length. See *nonequilibrium-modal-power-distribution length.*

nonequilibrium modal power distribution. In an *electromagnetic wave propagating* in a *multimode waveguide,* such as an *optical fiber,* the distribution of *radiant power* among *modes* such that the fractional power in each mode changes as the wave *propagates* along the waveguide. The power distribution among the modes will continue to change until the *equilibrium length* is reached, after which the fraction of total power in each mode will remain fairly constant.

nonequilibrium-modal-power-distribution length. The distance in a *waveguide,* such as an *optical fiber,* between the entrance end face and the beginning of *equilibrium modal power distribution,* that is, the point at which the modal *radiant power* fractional distribution no longer changes with distance along the guide. Thus, the nonequilibrium length and the equilibrium length are the same, because the end of nonequilibrium and the beginning of equilibrium occur at the same point. Synonymous with *nonequilibrium length.*

nonlatching. A switch actuation method that requires continuous application of an activating force or *signal* to cause the switch to remain in the position that it is in. Also see *latching.*

nonlatching switch. In *fiber optics,* a switch that will selectively transfer *optical signals* from one *optical fiber* to another when an actuating force or *signal* is applied and will continue to transfer signals only as long as the actuating force or signal is continuously applied. When the actuating force or signal is removed, the switch will revert to its original position. Also see *latching switch.*

nonlinear device. A device whose output is not a linear function of its input. If the output is represented by *y* and the input by *x,* then the output is related to the input by a nonlinear function, such as by the relation:

$$y = ax^2 + bx + c$$

where *a, b,* and *c* are constants. For example, a device in which the output *electric field,* voltage, or current is not linearly proportional to the input electric field, voltage, or current. Such a device might have a resistance (varistor) that is a function of the current within it, such that for the relation:

$$e = ir$$

where *e* is the instantaneous output voltage, *i* is the instantaneous current, and *r* is a function of *i,* such as $r = ai + b,$ giving rise to the relation:

$$e = ai^2 + bi$$

New *wavelengths* or *modulation frequencies* would be generated by the device. The behavior of a nonlinear device can be described by a *transfer function* or an *impulse-response function.* Also see *linear device.*

nonlinear material. See *third-order nonlinear material.*

nonlinear optical (NLO) material. In *fiber optics,* a *transparent* material, such as special glasses, polymers, and organic materials, that display nonlinear properties, such as generate higher harmonic *frequencies* when energized with *optical signals.* The generation of second and higher harmonic frequencies means switching speeds of fewer than 100 femtoseconds, that is, 100×10^{-15} s. The process is also *loss*less and thus can be repeated many times. An exit wave can be frequency-mixed and output *beams* can be considerably different from input beams. For example, at the exit point of an *integrated optical circuit waveguide,* different *signals* can be switched to different *ports.* Synonymous with *NLO material.* See *polymeric nonlinear optical material.*

nonlinear scattering. 1. *Scattering* of *electromagnetic radiation* in which *wavelengths* other than the original wavelength are generated, such as occurs in *Raman scattering* and *Brillouin scattering.* **2.** Direct conversion of a *photon* from one *wavelength* to one or more other wavelengths. In an *optical waveguide,* nonlinear scattering is usually not significant below the threshold of *irradiance* for stimulated nonlinear scattering.

nonuniformly distributive coupler. A *fiber optic coupler* in which *optical power* input to each input *port* is not distributed equally among the output ports, and the optical power input to each output port is not distributed equally among the input ports. Also see *uniformly distributive coupler.*

normal incidence. A direction that makes an angle of 90° with a surface. A *light ray* that strikes a surface at an angle of 90° is a normal ray.

normalized detectivity. See *specific detectivity.*

normalized frequency. In *fiber optics,* a dimensionless quantity, usually denoted by *V,* given by the relation:

$$V = (2\pi a/\lambda_0)(n_1{}^2 - n_2{}^2)^{1/2}$$

where *a* is half the *core diameter.* λ_0 *is the wavelength* in vaccum, and n_1 and n_2 are the maximum *refractive indices* in the core and *homogeneous cladding* respectively, of an *optical fiber.* In a fiber having a *power-law refractive-index profile,* the approximate number of *bound modes,* M_b, is given by the relation:

$$M_b = (V^2/2)(g/(g+2))$$

where *g* is the *refractive-index profile parameter.* For a large number of modes, that is, $V > 5$, the number of modes, or *mode volume,* M_v, is given by the relation:

$$M_v = (1/2)V^2$$

where *V* is the normalized frequency. Synonymous with *V-number; V-value.*

NTSC. National Television Standards Code; National Television Standards Committee.

n-type semiconductor. A semiconducting material, such as silicon or germanium, that has been *doped* with minute amounts of donor-type material, that is, with material that has a valence, such as 5, that allows extra relatively free electrons to wander about the lattice structure of the semiconducting crystal, all other atomic bonds being complete. The free electrons will move readily from atom to atom under the influence of an applied *electric field,* albeit a relatively weak field, thus constituting an electric current. The *dopant,* that is, the doping material ''donates'' electrons to this current, constituting a stream of negatively charged particles, hence the name n-type semiconducting material. Also see *p-type semiconductor.*

nuclear hardness. The physical attributes of a system or component that will allow survival in an environment that includes nuclear *radiation* and *electromagnetic pulses* (EMP). Hardness may be measured in terms of susceptibility or vulnerability. Hardness is usually measured in terms of the effects of nuclear environmental conditions, such as overpressure, peak velocities, energy absorbed, and electrical stress, on the characteristics of the system or component, such as its *electrical conductivity,* molecular structure, or tensile strength. Hardness is achieved through design specifications and is verified by testing and analytical techniques. In general, *fiber optic systems* and components have demonstrated higher survivability levels than other systems and components intended to perform a similar function.

number. See *wave number.*

numerical aperture (NA). 1. The sine of one-half of the vertex angle of the largest cone of *meridional rays* that can enter or leave an *optical system* or element, multiplied by the *refractive index* of the *propagation medium* in which the cone is located. It is generally measured with respect to an image point and will vary as that point is moved. For an *optical fiber* in which the refractive index decreases monotonically from n_1 on the *optical axis* to n_2 in the *cladding,* the maximum theoretical numerical aperture is given by the relation:

$$NA = (n_1{}^2 - n_2{}^2)^{1/2}$$

where n_1 is usually taken as the maximum refractive index of the *core,* and n_2 is the refractive index of the innermost *homogeneous cladding.* However, for a *graded-index fiber,* because the NA depends on the refractive index, and the refractive index varies with distance from the center of the fiber, the NA varies with the distance from the center of the fiber, and therefore the true NA also depends on the maximum refractive index found on the fiber end face, which is at the center, and of course is progressively less as the distance from the center increases. **2.** Colloquially, the sine of the *radiation* or *acceptance angle* of an *optical fiber,* multiplied by the *refractive index* of the material in contact with the exit or entrance face. This usage is approximate and imprecise, but is often encountered. For optical fibers, the NA is a measure of the light-accepting property of a fiber inasmuch as only the light that propagates an appreciable distance, such as *propagates* beyond the *equilibrium length,* inside the core can be considered to have been accepted by the fiber. Thus, the aperture of an unaided fiber is limited to an *acceptance cone,* namely the cone within which all *rays* will undergo *total (internal) reflection* when inside the core. Typical numerical apertures for optical fibers range from 0.25 to 0.45. Loose terms, such as openness, light-gathering ability, angular acceptance, and acceptance cone have been used to describe the numerical aperture. One way to measure the numerical aperture of an optical fiber is by measuring the *launch spot* size produced by an illuminated fiber. The spot of light produced by a *light beam* from the exit end face of an optical fiber is adjusted for maximum *irradiance* on a transverse screen, that is, a surface perpendicular to the fiber axis, by aligning the fiber. The diameter of the spot is dependent upon the departure angle, that is, the *launch angle,* from the fiber end face and the distance from the fiber end face to the screen. The numerical aperture is given by the relation:

$$NA = d/(d + 4D^2)^{1/2}$$

where d is the spot diameter and D is the distance from the fiber exit end face to the screen. If $d/2D$ is less than 0.25, the numerical aperture of the element is given by the approximate relation:

$$NA = d/2D$$

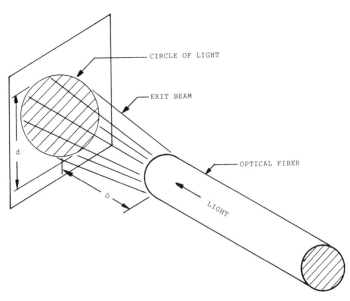

N-2. Measurement of the **numerical aperture** of an *optical fiber* as NA = d/2D, in which d is the diameter of the spot of *light* formed by the *exit beam* on a screen held perpendicular to the *optical axis* of the fiber, and D is the distance from the end of the fiber to the screen.

The NA is defined as the sine of the *acceptance angle.* However, the ratio of the spot diameter and twice the distance from the fiber end face to the screen, which is the same as the ratio between the radius of the spot and the distance to the screen, is the tangent of the acceptance angle. However, for the smaller angles the sine and tangent are nearly the same. See *launch numerical aperture (LNA).* Also see *launch spot.*

Nyquist bandwidth. The minimum *bandwidth* in the *frequency* domain needed to reproduce faithfully a sampled waveform in the time domain. If T is the time interval between adjacent equally spaced samples, then the Nyquist bandwidth is 0.5 $(1/T)$. If the time interval is in seconds, the Nyquist bandwidth will be in *hertz.* Thus, to reproduce an *analog signal* with a sufficiently high degree of fidelity, the sampling rate must be at least twice the highest significant frequency present in the signal being sampled. Also see *sampling theorem.*

Nyquist criterion. A criterion that must be satisfied to faithfully reproduce a *signal* with high fidelity using sampling techniques wherein a signal is sampled, *transmitted* in discrete *pulses,* and reconstituted at the receiving end, the sampling rate must be at least twice the highest significant frequency present in the signal being sampled.

Nyquist interval. The maximum time interval between regularly spaced samples of a *signal* that will permit the signal waveform to be completely determined. It is equal to the reciprocal of twice the *bandwidth* of the sampled signal. If the bandwidth, that is, the highest significant *frequency* in the signal being sampled, is in *hertz,* the Nyquist interval will be in seconds; if in *megahertz,* it will be in *microseconds.* Thus for 5 MHz, the interval will be 0.1 μs, that is, a sample must be taken every 0.1 μs for the *Nyquist criterion* to be satisfied.

Nyquist rate. The sampling rate required to satisfy the *Nyquist criterion,* that is, the reciprocal of the *Nyquist interval.* The Nyquist rate is equal to twice the highest significant *frequency* in a *signal.* It is usually expressed in samples per second. Also see *sampling theorem.*

O

objective. In an *optical system,* the *optical component* that receives *light* directly from the object and forms the first or primary image. In simple cameras without added lenses, such as zoom or telescopic lenses, the image formed by the objective is the final image. In telescopes, microscopes, rangefinders, binoculars, and other special instruments, the image formed by the objective is further processed by other optical components, such as magnification by an eyepiece.

OFCC. See *optical fiber cable component (OFCC).*

offline. 1. In communication systems, pertaining to the operation and use of devices and components that are not directly connected to a system and therefore are not available for immediate use on demand by the system without human intervention. Offline devices may be operated independently of the system relative to which they are considered to be offline, and they may be operated independently of each other. **2.** Pertaining to a condition wherein a device or subsystem is not connected to, is not a part of, and is not subject to the same controls as the components that are *online* to a system. Also see *online.*

offset loss. See *lateral offset loss.*

OFTF. *Optical fiber transfer function.*

old telephone service. See *plain old telephone service (POTS).*

OLTS. *Optical loss test set.*

online. 1. In communication systems, pertaining to the direct connection, operation, and control of devices and components for immediate use on demand by a system, normally without human intervention. **2.** Pertaining to a condition wherein devices or subsystems are connected to, are a part of, or are subject to the same controls as the other components of the system to which they are connected. Also see *offline.*

188

open-systems architecture. The structure, configuration, or model of a distributed-data processing system or network that enables system design, development, and operation to be described as a hierarchical structure, that is, as a layering of functions. In the concept, each layer provides a set of accessible functions that can be controlled and used by the layer above it. Thus, each layer can be implemented without affecting the implementation of other layers. The concept is useful because it permits the alteration of system performance at any layer without disturbing the huge investment in existing equipment, procedures, or protocols at that or other higher or lower levels. For example, converting from electrical wire to *optical fibers* at the *physical layer* need not affect the *data-link layer* or *network layer* except to provide more traffic capacity; the network need never be shut down for alterations; and service to the user is not interrupted during system modification or improvement.

Open-Systems-Interconnection (OSI) Reference Model (RM). An abstract description of *digital* communication among application processes running in distinct systems. This standard model is a hierarchical structure of seven layers, namely, the (1) *physical* (lowest), (2) *data link,* (3) *network,* (4) *transport,* (5) *session,* (6) *presentation,* and (7) *application* (highest) *layers.* Each layer performs value-added service at the request of the adjacent higher layer and, in turn, requests more basic services from the adjacent lower layer. See also (from lowest to highest) *physical layer; data-link layer; network layer; transport layer; session layer; presentation layer; application layer.*

open waveguide. A *waveguide* without electrically conducting walls in which *electromagnetic waves* are guided only by a *refractive-index profile,* so that the waves are confined to the guide by *refraction* and *reflection* at the outer surface of the guide or *dielectric interface* surfaces within the guide. Usually the open waveguide is constructed in such a manner that most of the *radiant power* in the electromagnetic waves *propagate,* with negligible *radiative loss,* inside an inner higher refractive-index material and along the interface between the two media. An *optical fiber* is an open waveguide of a *transparent* material, such as silica glass or plastic. Also see *closed waveguide.*

operating condition. See *extended operating condition; standard operating condition.*

operating bounce time. See *switch operating bounce time.*

operating time. See *switch operating time.*

operation. See *attenuation-limited operation; bit-rate-length-product-limited operation; distortion-limited operation; quantum-limited operation; quantum-noise-limited operation.*

optic. See *electrooptic; fiber optic; magnetooptic.*

optical. 1. Pertaining to the field of *optics.* **2.** Pertaining to eyesight. **3.** Pertaining to systems, devices, or components that generate, process, and detect *lightwaves* or light energy, such as *lasers;* lens systems; *optical fibers, bundles,* and *cables;* and *photodetectors.* **4.** Pertaining to the lightwave region of the *electromagnetic spectrum,* that is, the region in which the techniques and components used in the *visible spectrum* also apply to the region extending somewhat beyond the visible region into the *ultraviolet* and *infrared regions,* corresponding to *wavelengths* between 0.3 and 3 μ *(microns).* **5.** Pertaining to the range of *wavelengths* of *optical radiation,* which is the *electromagnetic spectrum* within the wavelength region extending from the vacuum *ultraviolet* at 0.001 μ to the *far-infrared* at 100 μ, that is, from about 1 nm (nanometer) to 0.1 mm (millimeter), which lies between the region of transition from radio waves and the transition to *x-rays.* Also see *optical spectrum.*

optical attenuator. A device used to reduce the *radiant power,* that is, the *intensity,* of *lightwaves* without otherwise altering them. Optical attenuators may be fixed, stepwise variable, and continuously variable. See *continuously variable optical attenuator; stepwise-variable optical attenuator.* Also see *fiber optic attenuator.*

optical axis. 1. In an *optical fiber,* the longitudinal geometric axis of symmetry. For example, the *optical axis* of an *optical fiber* of circular cross section is the locus of all points at the centers of cross-sectional circles, that is, the central longitudinal axis of the core. **2.** In a lens, the straight line that passes through the centers of curvature of the lens surfaces. **3.** In an *optical system,* the line formed by the coinciding principal axes of the series of *optical elements.* Also see *optic axis.*

optical blank. A piece of *optical* material having a particular shape, such as a casting molded into the desired geometry, for grinding, polishing, or reshaping into desired geometry with desired optical and mechanical properties. In *fiber optics,* the blank may be drawn or reshaped, after adding *dopants,* to make a *preform* for *drawing* into a fiber. Also see *optical fiber blank; preform.*

optical branching device. A device that has three or more *ports* and shares input *light* among its ports in a given fashion without changing the input *signal* other than distributing its energy among the ports. Types of branching devices include unidirectional, bidirectional, symmetric, and asymmetric.

optical cable. See *aerial optical cable; bidirectional optical cable; central-strength-member optical cable; fiber optic cable; peripheral-strength-member optical cable.*

optical cable assembly. See *fiber optic cable assembly.*

optical cable facility. The part of an *optical station/regenerator section* that consists of the *fiber optic cable,* including *splices,* that connects the cable splice at the *optical cable-interconnect feature* at one *optical station facility* to the splice at the cable-interconnect feature at another optical station facility, i.e., the outside plant between stations. When designing an optical station/regenerator section, the *optical losses* in the optical cable facility must be equal to or less than the *terminal/regenerator system gain.* Synonymous with *fiber optic cable facility.*

optical cable facility loss. In an *optical station/regenerator section,* the *loss (dB)* in the *optical cable facility,* i.e., in the outside plant, given by the relation:

$$L = \ell_t(U_c + U_{cT} + U_\lambda) + N_S(U_S + U_{ST})$$

where ℓ_t is the total sheath length of *spliced-fiber cable* (km), U_c is the worst case end-of-life cable *attenuation rate* (dB/km) at the transmitter *nominal central wavelength,* U_λ is the largest increase in cable attenuation rate that occurs over the transmitter central wavelength range, U_{cT} is the effect of temperature on the end-of-life cable attenuation rate at the worst-case temperature conditions over the cable operating temperature range, N_S is the number of splices in the length of the cable in the *optical cable facility,* including the splice at the *optical station facility* on each end and allowances for cable repair splices, U_S is the loss (dB/splice) for each splice, and U_{ST} is the maximum additional loss (dB/splice) caused by temperature variation. L must be equal to or less than the *fiber optic terminal/regenerator system gain, G,* for the optical station/regenerator section to operate satisfactorily. See *statistical optical cable facility loss.* Synonymous with *fiber optic cable facility loss.* Also see *fiber global attenuation-rate characteristic; terminal/regenerator system gain.*

optical cable-interconnect feature. In an *optical station facility* of an *optical station/regenerator section,* the device used to connect *optical station cables* (inside plant) to an *optical cable facility* (outside plant) via *optical connectors* and *splices.* Synonymous with *fiber optic cable-interconnect feature.*

optical cable pigtail. A short length of *fiber optic cable* permanently fixed to a component and used to couple *optical power* between it and another optical cable pigtail or fiber optic cable, such as a *transmission* cable.

optical cavity. A geometric space in vacuum or material media, bounded by two or more mirrors, in which *lightwaves* can *reflect* back and forth, thus producing *standing waves* of high *irradiance,* that is, intensity, at certain *wavelengths.* Such standing waves might be obtained in a ruby crystal *laser* with two plane or spherical mirrors, forming a *resonant cavity.* The cavity is the portion of the crystal that lies between the mirrors. The molecules in the cavity can be excited by an inert-gas lamp, which causes the cavity to generate and *emit* a narrow *beam* of *monochromatic light* of high irradiance, that is high *electric field strength* or high *radiant power,* in the direction of the crystal axis.

optical cavity mirror sensor. A *fiber optic sensor* in which operation is based on changes in the *optical reflection coefficient* caused by changes in pressure, temperature, strain, or other imposed stimuli. An arrangement of mirrors causes light to be reflected back to a *photodetector* via a *partial mirror* through which the incident *beam* has passed from the *light source*. The mirrors are positioned in such a way that one of the reflecting surfaces is at the *critical angle* of the incident light in the *fiber core*. The slightest variation in mirror position, caused by the applied stimulus to be sensed, causes the reflected light to pass into or out of *total (internal) reflection*, which changes the amount of light sent back to the photodetector.

optical cement. A permanent and *transparent* adhesive capable of withstanding extremes of temperature, such as Canada balsam. However, it is being replaced by modern synthetic adhesives, such as methacrylates, caprinates, and epoxies.

optical chopper. A device for periodically interrupting a *light beam*, for example, a rotating disk with radial slots through which a *collimated beam* must pass on to a *photodetector* will produce a *signal* with a *pulse-repetition rate* proportional to the angular rotation rate of the disk. *Optical fibers* can be used to bring the light to and from the disk, or any other rotating element.

optical circuit. See *integrated optical circuit (IOC)*.

optical combiner. **1.** In *fiber optics*, a *passive optical device* in which *radiant power* from two or more input *optical fibers* is distributed among one or more output fibers. The input fibers may each insert *light* with a different *wavelength* into the combiner. The output fiber, or fibers, will contain all the input wavelengths. **2.** A *passive optical device* in which *radiant power* from two or more input *ports* is distributed among a smaller number of output ports.

optical computer. A computer in which internal operations, such as storage, arithmetic and logic, and control functions, are performed using *optical* paths, *optical switches, integrated optical circuits (IOCs)*, and optical images. For example, a computer with an optical *modulator* tube that memorizes optical information as charge patterns while performing arithmetic functions, makes images in memory consisting of picture elements in the form of on-off state of *light sources*.

optical conduction. The *propagation* or *transmission* of *lightwaves* through a material medium, usually in which they are guided or confined. For example, in an *optical data link*, the waves are guided from a *light source* via a *fiber optic cable* to a *photodetector* with minimal *loss* of light energy by *absorption, dispersion*, deflection, *reflection, scattering*, or diffusion, such that intelligence carried by the lightwaves at the source can be recovered at the end of the cable. Care must be taken when using the term "conductor" in *fiber optics* because the terms "conductor," "conduction," and "conducting" are applied to materials that conduct electric currents, such as semiconducting materials and metals, whereas *optical* materials are *dielectric* materials which are referred to as "nonconductors." Thus, *optical fibers* are poor "conductors" of electric currents but good "conductors" of lightwaves. In fiber optics, the words *guided, transmitted*, and *propagated* are preferred over the word "conducted," thus avoiding ambiguous statements, such as "optical fibers are good conductors." Also see *optical conductor*.

optical conductor. A *transparent* material that offers a low *optical attenuation rate* to the *transmission* of *lightwaves,* such as certain glasses, plastics, and crystals. The term is obsolete and therefore deprecated. "Conductor" should be reserved for electrical and sonic systems rather than optical systems. Also see *optical conduction.*

optical connector. See *fiber optic connector; receiver optical connector; transmitter optical connector.*

optical connector variation. The maximum value *(dB)* of the difference in *insertion loss* between mating *fiber optic connectors* of the same type and model and from the same manufacturer. Synonymous with *fiber optic connector variation.*

optical contact. See *terminus.*

optical coupler. See *fiber optic coupler.*

optical data bus. A *data bus,* that is, a single *optical fiber* or *cable* to which all stations and terminals are directly connected so that they can communicate with each other via the single bus. Messages intended for specific addressees must be properly addressed depending on network protocol. However, all messages placed on the bus can be made available to any or all of the stations and terminals connected to the bus without having to pass through any specific station or terminal.

optical data processing. 1. The performance of digital data processing operations, such as execution of Boolean functions, using a combination of *optical* and electronic *elements.* **2.** The performance of analog data processing operations that model or emulate actual systems represented by mathematical functions and relationships governing the behavior of *light* passing through optical elements, such as by making use of *diffraction, interference,* and *reflection patterns* produced by variously shaped optical elements.

optical demultiplexer (active). An *active optical device* that uses *optical components* and other components, such as electrical and acoustic components, to accept *signal* streams or messages that have been previously multiplexed into a single channel and that separates the streams or messages and places them in individual independent *channels* so the messages can be used immediately or be separately routed to their ultimate destination. An active optical demultiplexer might consist of an electrically operated switching device that separates individual *wavelengths* so as to cause each color in the polychromatic light to be incident upon a different *photodetector* in order that the intelligence-bearing signal that *modulated* each *color* in the *optical transmitter* can be recovered by the *optical receiver.* Also see *fiber optic demultiplexer (active); fiber optic multiplexer (active); optical multiplexer (active).*

optical demultiplexer (passive). A *passive optical device* that uses only *optical* components to accept *signal* streams or messages that have been previously multiplexed into a single *channel* and that separates the streams or messages and places them in individual independent channels, so the messages can be used immediately or be separately routed to their ultimate destination. An optical demultiplexer might consist of a prism that can accept *polychromatic light* resulting from the multiplexing action of a *mixer* and separate the individual *wavelengths* so as to cause each color in the polychromatic light to be incident upon a different *photodetector* in order that the intelligence-bearing signal that modulated each color in the *optical transmitter* can be recovered by the optical receiver. Also see *fiber optic demultiplexer (passive); fiber optic multiplexer (passive); optical multiplexer (passive).*

optical density. An inverse logarithmic measure of *transmittance.* Transmittance is the ratio of the *transmitted radiant power* to the incident radiant power at an interface of an *optical element.* The optical density is expressed by the relation:

$$OD = \log_{10}(1/T)$$

where T is the transmittance. The analogous term, *reflectance density,* is expressed by the relation:

$$RD = \log_{10}(1/R)$$

where R is the *reflectance.* The reflectance is the ratio of the reflected power to the incident power at an interface of an optical element. The higher the optical density, the lower the transmittance. The optical density times 10 is equal to the *transmission loss* expressed in *decibels.* For example, an optical density of 0.3 corresponds to a transmission loss of 3 dB. The sum of the reflectance and the transmittance is unity, because, at an interface, the reflected radiant power plus the transmitted radiant power equals the incident radiant power. Synonymous with *transmittance density.*

optical detector. A *transducer* that generates an output *signal* when *irradiated* with *optical power,* for example, a device that converts the optical power or energy of a signal to other forms of power or energy that represent the original *optical* signal, such as a device that converts optical signals to electrical, sound, other optical, chemical-reaction, or mechanical signals. Also see *photodetector.* See *video optical detector.*

optical detector type. Characterization of an *optical detector,* i.e., a *photodetector,* by identifying the type of detector *(PIN photodiode, APD),* compositional material (Ge, Si), and possibly other features.

optical device. See *active optical device; passive optical device.*

optical disk. A flat circular disk coated with a photosensitive medium on which binary digits in the form of light and dark spots may be stored by *modulation* of *light* from a *source* such as a *laser.* Disks of the order of 30 cm in diameter can store up to 10^{13} bits, that is, up to 10 terabits, at recording rates of 10 Mbps with *bit-error ratios (BER)* less than 10^{-7}.

optical dispersion attenuation. The *attenuation* of a *signal* in an *optical waveguide* in which each *frequency* component of a *launched pulse* is attenuated in such a manner that the shorter *wavelengths* in the *spectral width* of the pulse, that is, the higher frequencies, are attenuated more than the lower frequencies, giving rise to a form of *distortion.* The dispersion attenuation factor is given by the relation:

$$DAF = e^{-df}$$

where d is an empirically determined material constant, including substance and geometry, and f is a significant frequency component of the signal being attenuated.

optical distortion. An abberation of spherical-surface *optical systems* due to the variation in magnification with distance from the *optical axis.*

optical emitter. A source of *optical radiant power,* for example, a source of *elecromagnetic radiation* in the *visible* and near-visible region of the *frequency spectrum.*

optical-energy density. In a *light beam propagating* through a unit area normal to the direction of maximum power gradient, the amount of optical energy contained in a unit volume of space within the beam, usually expressed in joules per cubic meter, which corresponds to an *irradiance* level in watts per square meter, because the *electromagnetic waves* in the beam are propagating with the speed of *light.* Thus, the time rate of energy flow across a unit area is the optical energy density multiplied by the speed, that is joules/meter^3 $\times$ meters/second, which is joules/meter^2-second. But joules/second is watts. Therefore, watts/meter^2 remain as the equivalent to joules/meter^3, which implies that the optical energy density and the irradiance, that is, the optical power density, are the same for a propagating electromagnetic wave.

optical fiber. A filament-shaped *waveguide,* made of dielectric material, such as glass or plastic, that *guides* light. It usually consists of a single discrete optically *transparent transmission* element consisting at least of a cylindrical *core* with *cladding* on the outside. Though most *optical fiber* cross sections are round, there are other cross sections, such as elliptical, rectangular, planar, and slotted, for special purposes. The *refractive index* of the core has to be higher than that of the cladding for *lightwaves* (or *photons*) to remain within and *propagate* in the fiber. If the *incidence angle* of *light rays* at the core-cladding interface exceeds a certain angle called the *critical angle,* rays headed out of the core will be *reflected* back into the core. The lightwaves can be *modulated* with an information-bearing signal. Synonymous with *optical fiber waveguide.* See *active optical fiber; compensated optical fiber; overcompensated optical fiber; polarization-maintaining optical fiber; triangular-cored optical fiber; undercompensated optical fiber.* (See Figs. O–1–O–5)

optical fiber acoustic sensor. A highly sensitive device capable of sensing sound *waves* by allowing them to impinge upon a length of *optical fiber,* the waves causing variations in the *refractive indices* of the materials in the fiber. The variation of refractive indices is used to *modulate lightwaves,* usually from a *laser,* that are *pro-*

O–1. A technician places a *"bait"* in a lathe, the first step in the *outside vapor deposition (OVD)* process in the manufacture of **optical fiber.** (Courtesy Corning Glass Works).

pagating in the fiber on their way to a *photodetector.* The spatial distribution of the fiber determines the directional capability of the sensor. For example, if the fiber is wound in the shape of a sphere, the sensor will be omnidirectional; if distributed in a plane, it will be unidirectional; if shaped in the form of a rod, it will be cylindrically directional.

optical fiber active connector. An *optical connector* with a built-in active device, such as an *LED* or *laser (transmitter),* or a *photodetector (receiver),* or both. The device is usually built into one of the mating elements of the connector, usually in the form of a semiconductor chip with an *optical fiber pigtail* for connection to a fiber. The connector is built to enable improved fiber-to-fiber or fiber-to-pigtail *coupling.* The combination provides an improved overall *coupling efficiency.* For example, one such connector provides 50 times normal coupling efficiency at 30 Mbps over a 1-km length. *Apertures* of the semiconductor element are about 0.20 mm for a 0.20-mm-diameter core with a *numerical aperture* of 0.48, adaptable for bulkhead or printed-circuit-board mounting and EMI and RFI shielding. The connector also serves as a *repeater.*

O-2. Removal of the completed **optical fiber** *preform* from a lathe. (Courtesy Corning Glass Works).

optical fiber axis. The locus of the *core* centers of an *optical fiber.*

optical fiber blank. A cylindrically shaped piece of glass of high purity, such as glass with a transition metal composition of 10 to 50 ppb (parts per billion, that is, 10^{-9}), used to make a *preform* from which *optical fibers* are *pulled.* In the *doped-deposited-silica (DDS)* process for making optical fibers, the glass and *dopants* are deposited on a *mandrel* mounted on a lathe using an *inside vapor-phase-oxidation (IVPO) process* or an *outside vapor-phase-oxidation (OVPO) process* to control the *refractive-index profile* of the fiber. After sufficient glass has been deposited on the mandrel to form the *optical fiber blank,* it is removed from the lathe. The mandrel is then removed and the blank is consolidated into a dense glass preform. The fiber is subsequently pulled from the preform. Also see *optical fiber mandrel; optical fiber preform.* Also see *optical blank.*

O-3. The *spot preform* is lowered into the **optical fiber** consolidation furnace. (Courtesy Corning Glass Works.)

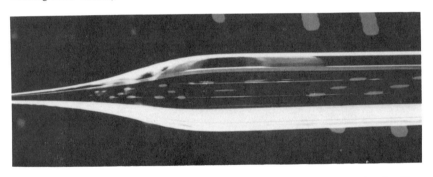

O-4. The consolidated glass *blank* will be drawn into **optical fiber.** (Courtesy Corning Glass Works.)

optical fiber bundle. 1. An assembly of un*buffered optical fibers.* Usually an optical fiber bundle is used as a single *transmission channel,* as opposed to multifiber cables in which each fiber provides a separate channel. Bundles used only to *transmit light,* as in optical communications and illumination systems, are flexible and are typically unaligned. *Aligned bundles,* that is, coherent bundles, are used to transmit and display images. **2.** Two or more *optical fibers* in a single protective sheath or *jacket.* The number of fibers might range from a few to several hundred, depending on the application and the characteristics of the fibers. Synonymous with *fiber optic bundle.*

optical fiber cable. See *fiber optic cable.*

O-5. Precise measurments are required for quality control in the manufacture of **optical fiber.**
(Courtesy of Corning Glass Works).

optical fiber cable component (OFCC). **1.** A component or part of a *fiber optic cable.* **2.** A *waveguide* portion of a *fiber optic cable,* such as a 12-fiber *ribbon* complete with fibers, a *buffered fiber* complete with *strength members* and *jacket* suitable for cabling with other similar components, or a buffered fiber augmented with a concentric layer of strength members and an overall jacket. **3.** A *buffered fiber* augmented with a concentric layer of strength members and an overall jacket.

optical fiber coating. **1.** A protective material, often called a *buffer,* bonded to an *optical fiber* over the *cladding* for various purposes, such as preserving fiber strength, inhibiting cabling *losses,* protecting against mechanical damage *(microbending),* protecting against moisture and other debilitating environments, providing compatibility with fiber and *cable* manufacturing *processes,* and providing compatibility with the *jacketing* process. Coating application methods include dip-coating, extrusion, spray-coating, and electrostatic coating. Coating materials include fluoro-polymers, Teflon,

Kynar, polyurethane, and many others. **2.** Special materials bonded to *optical fibers* to make them sensitive to a specific physical variable that is to be sensed, such as magnetostrictive materials to measure magnetic fields, metals to measure electric currents, and thermally sensitive materials for measuring temperature.

optical fiber communications. The use of physical paths, such as lines, circuits, and links, that are made of *optical fibers* between terminals, from terminals to *user end-instruments,* between exchanges and *switching centers,* or between other components of a communication system.

optical fiber concentrator. A communication system that *multiplexes* a number of separate incoming optical fiber communication *channels.* For example, a *distribution frame* that accepts 3000 voice *optical fiber channels* from home *optical* transceivers and *time-division multiplexes* them using ten 10-to-1 multiplexers. The output of each goes into 30 more 10-to-1 multiplexers, the final output *modulating* a *lightwave carrier* in an optical fiber. Another example is a unit that multiplexes 32 full-duplex *digital data* channels on a single multiplexed *fiber optic data link* or *cable.*

optical fiber concentricity error. In an *optical fiber,* the distance between the center of the two concentric circles that specify the *cladding diameter* and the center of the two concentric circles that specify the *core diameter.* The measurement is used in conjuction with *tolerance fields* to specify optical fiber core and cladding geometry. Synonymous with *core eccentricity.*

optical fiber connector set. The total of the *fiber optic connector* parts required to provide demountable *coupling* between two or more *fiber optic cables.*

optical fiber coupler. See *fiber optic coupler.*

optical fiber delay line. An *optical fiber* of precise length used to introduce a delay in a *lightwave pulse* equal to the time required for the pulse to *propagate* from beginning to end. The delay line may be used for such operations as *phase* adjustment, *pulse positioning,* pulse interval coding, or *data* storage.

optical fiber demultiplexer (active). See *fiber optic demultiplexer (active).*

optical fiber demultiplexer (passive). See *fiber optic demultiplexer (passive).*

optical fiber hazard. A feature of a *fiber optic system* that has the potential of causing injury. Examples of *optical fiber* hazards include *fiber optic cables* that have been treated with a special flame-retardant material that is also caustic and can cause skin lesions, the output of high-*radiation lasers* that can cause eye injury, fragments of fibers that can cause eye damage, and grinding and polishing residues suspended in air that can cause lung damage if inhaled. Precautions include the use of gloves, goggles, masks, ventilators, *filters,* and frequent washing with warm water and soap during handling and processing operations. Particular precaution must be taken in regard to viewing *radiation* from the end of an energized fiber because the radiation is not in the *visible spectrum* of the *electromagnetic spectrum,* thus giving the impression the fiber is not energized or the appearance that the radiation is at a lower *irradiance* level than it actually is.

optical fiber interconnection box. See *fiber optic interconnection box.*

optical fiber jacket. The material that is placed over or wrapped around the *buffered* or unbuffered *optical fiber, strength members,* added buffers, *fillers,* and any other elements used in the construction of a *fiber optic cable.* Except perhaps for *overarmor,* the jacket forms the outside of the cable. Markings are placed on it. Synonymous with *sheath.*

optical fiber junction. An *interface* formed by butting the end faces of two *optical fibers* to allow direct fiber-to-fiber *optical transmission.*

optical fiber link. See *optical link.*

optical fiber mandrel. A cylindrically shaped piece of glass of high purity, such as glass with a transition metal composition of 10 to 50 ppb (parts per billion, that is, 10^{-9}), used to make an *optical fiber blank,* which is used to make an *optical fiber preform.* In the *doped-deposited-silica (DDS) process* for making optical fibers, the glass and *dopants* are deposited on the mandrel mounted on a lathe using an *inside vapor-phase-oxidation (IVPO) process* or an *outside vapor-phase-oxidation (OVPO) process* to control the *refractive-index profile* of the fiber. After sufficient glass has been deposited, the blank is removed from the lathe. The mandrel is then removed, and the blank is consolidated into a dense glass preform. The fiber is subsequently *pulled* from the preform under tight melt, tension, flow, and diameter control conditions. Also see *optical fiber blank; optical fiber preform.*

optical fiber merit figure. A figure that indicates the ability of an *optical fiber* to handle high-*frequency signals* over given distances with specified *distortion* limits and *bit-error ratios (BERs).* For example, the *pulse-repetition rate* times the fiber length when *pulse dispersion* is the specified limiting factor in signal distortion rather than *attenuation.* Another such figure of merit (FM) for *multimode fibers* is given by the relation:

$$FM = PL^s$$

where s varies between 0.5 and 1, L is the length of the fiber, and P is the pulse repetition rate. Empirically determined, s depends on materials and geometry.

optical fiber multiplexer (active). See *fiber optic multiplexer (active).*

optical fiber multiplexer (passive). See *fiber optic multiplexer (passive).*

optical fiber organizer. See *fiber-and-splice organizer.*

optical fiber pigtail. A short length of *optical fiber* extending from and permanently attached to a component, such as a *fiber optic coupler, connector, repeater, tap,* or *mixer,* to which an optical fiber can be *spliced.* The pigtail is used to facilitate connecting the component to another optical fiber or component, such as for *coupling optical power* or *signals* between the component and a *transmission* line. It performs a function similar to that of an electrical pigtail, which is a length of wire extending from an electrical circuit element, such as a resistor, transistor, capacitor, diode, or inductor, used for connecting the element to another circuit element or terminal. An optical fiber pigtail on a *light source* is also called a *launching fiber.*

optical fiber polarizer. An *optical fiber,* usually with a specially shaped cross section, that produces a single well-defined *polarized mode* by selecting one of the two polarized modes from a *lightwave propagating* in a *single-mode fiber.* The unwanted polarization is extinguished while the other is *transmitted.* When the optical fiber polarizer is *spliced* into a conventional single-mode system, the need for *collimination* is eliminated and the sensitivity to vibration and other disturbances is reduced. Synonymous with *fiber optic polarizer.*

optical fiber preform. A specially shaped piece of *optical* material from which *optical fibers* may be *pulled* or rolled. For example, a solid rod made with a higher *refractive-index* glass than the tube into which it is slipped, to be heated and pulled or rolled into a *cladded* optical fiber; or four lower refractive-index rods surrounding a higher refractive-index rod heated and drawn into a cladded fiber. The *drawing process* results in fiber many times longer than the preforms. For round solid preforms, the length of the drawn fiber is equal to the square of the ratio of the preform and fiber radii times the length of the preform, resulting in a fiber tens of kilometers long from a single preform. In the *doped-deposited-silica (DDS) process* for drawing optical fiber, a cylindrically shaped piece of glass of high purity, such as glass with a transition metal composition of 10 to 50 ppb (parts per billion, or 10^{-9}), used to make a *fiber optic blank,* which is used to make the preform. The glass and *dopants* are deposited on a *mandrel* mounted on a lathe. The *inside vapor-phase-oxidation (IVPO) process* or an *outside vapor-phase-oxidation (OVPO) process* is used to control the *refractive-index profile* of the fiber. After sufficient glass has been deposited, the blank is removed from the lathe. The mandrel is then removed, and the blank is consolidated into a dense glass preform. The fiber is subsequently *pulled* from the preform. Synonymous with *fiber optic preform.* Also see *optical fiber blank; optical fiber preform.*

optical fiber pulse compression. In the *transmission* of *pulses* in *optical fibers,* the compression or shortening of certain *frequency-modulated pulses* arriving at the end of the fiber when longer *wavelengths emitted* later catch up with shorter wavelengths emitted earlier.

optical fiber radiation damage. The increased *attenuation* of *lightwaves propagating* in an *optical fiber* caused by increased *losses* from *absortion,* diffusion, *scatter-*

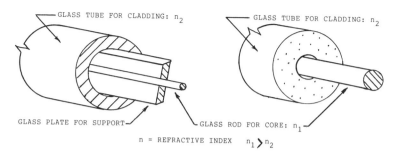

GLASS TUBE FOR CLADDING: n_2

GLASS TUBE FOR CLADDING: n_2

GLASS PLATE FOR SUPPORT

GLASS ROD FOR CORE: n_1

n = REFRACTIVE INDEX $n_1 > n_2$

O-6. For the **optical fiber preform** on the left, a central rod, supporting plate, and tubing will all be drawn into a single *optical fiber*. On the right, the preform consists only of a glass tube surrounding a glass rod.

ing, or deflection at specific sites created by exposure to high-energy bombardment, such as by *gamma rays,* cosmic rays, or high-energy neutrons. Degradation can be catastrophic or graceful. Effects of *radiation* an optical fibers are cumulative and for the most part irreversible, though there have been indications of considerable recovery after exposure to short bursts of high-energy radiation.

optical fiber ribbon. A cable consisting of *optical fibers* laminated within a flat plastic strip.

optical fiber ringer. A signaling system for soliciting attention in which a *photodetector* and *transmitter* convert enough *lightwave* energy from information-bearing *modulated-carrier signals* in *optical fibers* into an audio tone for sound signaling, that is, for ringing. The ringer power may also be used for other applications, such as providing power for electronic circuits.

optical fiber source. An electronically powered *light source,* such as an *LED* or *laser,* that can insert *radiant power* into an *optical fiber,* usually via a *pigtail* for *splicing* or *coupling* to the fiber. Light *emitted* by some sources can be directly *modulated* by electronic *signals* before the light enters the fiber. In other cases, the radiant power from a constant-output source can be modulated after the light is coupled into the fiber. Typical power output of such sources is of the order of several milliwatts at *optical wavelengths.*

optical fiber splice. A connection between two *optical fibers* made by joining an end of one fiber to an end of another fiber. It consists of the optical fiber joint and a fiber *splice* housing. It is usually contained within an *interconnection box* or a *cable splice closure.* Types of splices include *fusion, ultraviolet* cured or bonded, *mechanical, rotary mechanical,* and *ribbon.* Also see *fiber optic splice.*

optical fiber tensile strength. The maximum load that a short piece of *optical fiber* can sustain without breaking, normally calculated by the relation:

$$T_S = Er/r_{\min}$$

where T_s is the safe tension stress; E is Young's modulus, that is, the modulus of elasticity, which normally for glass is about 10^7 lb/in.^{2} (7×10^9 kg/m^2 or 70×10^9 N/m^2); r is the radius of the fiber; and r_{min} is the *minimum bend radius* before breaking. The T_s thus calculated is approximate and must be adjusted for probabilistic increases in occurrences of fractures and *microcracks* in longer fibers.

optical fiber transfer function (OFTF). The transformation that an *optical fiber* performs on an *electromagnetic wave* that enters it, such that if the input *signal* composition is known, and the *transfer function* is known for the fiber, the output signal can be determined, for example by the relation:

$$\text{OFTF} = e^{-af}$$

where a is a constant for the fiber, and f is a *frequency* component of the input signal, that is, the input electromagnetic wave.

optical fiber trap. An *optical fiber* that breaks easily when stressed; can be placed on approaches to areas to be secured, such as fences and fields; can signal the location of a break and so cannot be cut without detection; and thus can be used to detect the presence of trespassers. Also, the fiber itself is almost invisible, particularly in dimly lit areas.

optical fiber waveguide. See *optical fiber.*

optical filter. An *optical element* capable of altering the *spectral* composition of *lightwaves* that are incident upon it. The *transmitted light* emanating from the filter might have the *radiant power* of certain wavelengths in the incident light reduced or wholly removed. The transmitted light will not contain wavelengths that are not in the incident wave, and the radiant power input will always be greater than the *optical power output.*

optical glass. See *bulk optical glass.*

optical harness. A number of *multifiber cables* or *jacketed bundles* placed together in an array that contains branches. A harness is usually installed within other equipment and mechanically secured to that equipment.

optical harness assembly. An *optical harness* that is terminated and ready for installation.

optical impedance discontinuity. An abrupt spatial variation in the *refractive index, absorption coefficient, scattering coefficient,* geometric configuration, or other *parameters* of an *optical waveguide,* such as an *optical fiber* or *slab-dielectric waveguide.* The *electric permittivity, magnetic permeability,* and *electrical conductivity* can also contribute to optical *impedance* discontinuities. Optical impedance discontinuities are not only a function of the intrinsic physical properties of the material from which the guide is constructed, but they are also a function of other extrinsic parameters, such as the *frequency* or *wavelength* of an *electromagnetic wave propagating* in the waveguide, the temperature, and the applied pressure.

optical isolator. 1. A device that prevents *light* energy from *propagating* in a given direction at a point in a *transmission* path. For example, it prevents *reflections* from proceeding toward the *source* of the *incident* light. **2.** A two-*port optical device* that *attenuates optical radiation* more in one direction than in the opposite direction. Synonymous with *optoisolator*. Also see *fiber optic isolator*.

optical junction. Any *optical interface,* that is, a physical interface, in a *fiber optic system*. *Source* to *optical fiber,* fiber to fiber, fiber to *detector, beam* in air to prism or lens, fiber to lens, and lens to fiber are examples of optical junctions.

optical length. See *optical path length*.

optical lever. A means of *detecting* and measuring small angular displacements of an object by *reflecting* a *beam* of *light* from the object, such as from a smooth reflecting surface of the object or from a mirror or prism attached to the object. The reflected beam is allowed to form a spot of light on a scale to measure the angular displacement.

optical line rate. The *data signaling rate (DSR),* that is, the *interface rate,* in bits/second at a given *wavelength,* that appears at the *interface,* that is, the boundary, between an *optical fiber* and *transmission* equipment.

optical link. An *optical transmission channel* designed to connect two end terminals or to connect other channels in series, that is, in tandem. The link usually consists of a *transmitter,* that is, a *light-emitting* unit, a *fiber optic cable,* a *receiver,* that is, a *photodetector,* and necessary connecting elements, such as *fiber optic connectors, penetrators,* and *repeaters*. Sometimes terminal hardware, such as transmitter/receiver modules, *modulators,* and *modems* are also considered as parts of the optical link to which they are connected. Synonymous with *fiber optic link; optical fiber link*. See *repeatered optical link; repeaterless optical link; satellite optical link*.

optical-loss test set (OLTS). A device consisting of a stabilized *light source,* such as a *laser* or an *incoherent light-emitting diode (LED),* and a calibrated optical power meter. The set measures the amount of *optical power loss* between two points by *launching* a known amount of power at one of the points and comparing it with the amount of power received at the other point. The set can also be used to measure *insertion loss* and *attenuation rate*. Also see *optical time-domain reflectometry (OTDR)*.

optically active material. A material that can rotate the *polarization* of a *lightwave* that *propagates* within it. For example, an optically active material exhibits different *refractive indices* for *left-hand* and *right-hand circular polarizations,* that is, circular *birefringence.* See *nonlinear optical (NLO) material; polymeric nonlinear optical material.*

optical maser. A source of *monochromatic* and *coherent electromagnetic radiation* produced by the synchronous and cooperative *emission* of optically pumped ions introduced into a crystal host lattice, a gas, or a liquid whose atoms are excited in a discharge tube. The *radiation* is a *lightwave* with a sharply defined *wavelength,* that is, with a narrow *spectral width,* that *propagates* in a high-power, highly directional *beam.*

optical mixing. The production of a *lightwave* that contains all the wavelengths that are in all the input lightwaves, for example, the production of *dichromatic* or *polychromatic radiation* from two or more *monochromatic* lightwaves through the use of an *optical mixing box, mixing rod,* or other means, or the production of a pulsating electrical current proportional to the difference in wavelength between two *interfering* monochromatic lightwaves *incident* upon the same *photodetector.*

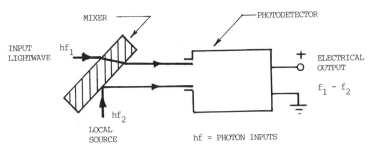

O–7. **Optical mixing** by means of a partially *reflecting* mirror and partially *transmitting mirror (mixer)* and a *photodetector.*

optical mixing box. A *fiber optic coupler* consisting of a piece of *transparent* material that receives several *optical wavelengths,* or *frequencies,* and that mixes them to produce *dichromatic* or *polychromatic lightwaves* for dispatch via one or more outputs for *transmission* elsewhere and perhaps subsequent separation into the constituent wavelengths to produce the original intelligence introduced by the *modulation* of each of the constituent wavelengths. The optical mixing box usually has *reflective* inner surfaces, except at the *ports.* The lightwaves entering the box are usually a group of monochromatic waves each of a different wavelength and each having been modulated separately.

optical mixing rod. An *optical mixing box* that has the general shape of a right circular cylinder, usually with *pigtails* for entrance and exit *ports.*

optical modulator. See *thin-film optical modulator.*

optical multimode dispersion. *Dispersion* of the various *frequencies* comprising the *pulses* in an *optical waveguide* caused by *mode scrambling* that occurs when two or more *transmission* modes are supported by the same guide. Optical multimode dispersion is greatly reduced in *graded-index (GI) fibers* and somewhat reduced by using a *monochromatic light source,* such as a *laser.*

optical multiplexer (active). A device that uses *optical components* and other types of components, such as electrical and acoustic components, to create *data links,* circuits, or channels to increase the *transmission capacity* of a system, such as by using *space-division, wavelength-division (optical), frequency-division (baseband),* and *time-division multiplexing* schemes, singly or in combination. Examples of active optical multiplexers include *optoelectronic multiplexers* and *electrooptic multiplexers.* Additional energy is required to operate the active optical multiplexer over and above that contained in the *lightwaves* being multiplexed. See *thin-film optical multiplexer.* Also see *fiber optic demultiplexer (active); fiber optic multiplexer (active); optical demultiplexer (active); optical multiplexer (passive).*

optical multiplexer (passive). A device that uses only *optical components* to create *data links* or *channels* to increase the *transmission capacity* of a system, such as by using *space-division* or *wavelength-division multiplexing* schemes, for example, an *optical mixer.* No energy is required to operate the passive optical multiplexer other than that contained in the *lightwaves* being multiplexed. See *thin-film optical multiplexer.* Also see *fiber optic demultiplexer (passive); optical demultiplexer (passive); fiber optic multiplexer (active); fiber optic multiplexer (passive); optical multiplexer (active).*

optical network standard. See *synchronous optical network (Sonet) standard.*

optical path component. A component whose only function is to convey an *optical signal* through a point or from one point to another, that is , simply to provide a path for the signal. Examples of optical path components are a *dielectric waveguide, optical fiber, fiber optic cable, splice, connector, terminus, penetrator, rotary joint, distribution frame, splice tray,* or *raceway.* The ideal optical path component does not introduce *propagation delay, distortion, attenuation, noise,* or in any way change the *optical signals* that are passing through. Also see *passive optical device.*

optical path length. In a *propagation medium* of constant *refractive index, n,* the *geometric distance* times the refractive index. If n varies with position the optical path length is the line integral of *nds,* that is, the optical path length is given by the relation:

$$L = \int nds$$

where *n,* the refractive index, is a function of position, and *ds* is the elemental path length in the integration path. Optical path length is proportional to the *phase* shift of a *lightwave propagating* along the path. Synonymous with *optical length.*

optical phase modulator. An *optical device* that can control or vary the *phase* of a *lightwave* relative to a fixed reference or relative to another lightwave, such as a phase modulator that uses the same *Sagnac interferometric* principle to make fiber optic gyroscopes. *Signals* are kept amplified, *coherent,* and in phase prior to entering the Sagnac fiber loop. Synonymous with *fiber optic phase modulator.*

optical power. In *fiber optics,* the *radiant power.* See *transmitter optical power.*

optical power budget. In an *optical transmission system,* the distribution of the available *radiant power* required for *transmission* within specified distortion limits or *bit-error ratios (BERs).* The distribution is usually made in *decibels* of *attenuation* for each component of the system from source to sink. Usually an allowance is made for a *power safety margin.* In a *optical link,* the *power level* change from source radiant power output to *photodetector* input power *threshold* of *sensitivity* is typically about 30 dB. The power is distributed among the intervening components, such as *fiber optic cables, connectors, splices, couplers, multiplexers, demultiplexers, attenuators,* and *mixers.*

O-8. The OPM-6 Handheld Power Meter and OPS-5 Handheld Power Source assembled as a kit with accessories and range of *connector* adapters. The kit is used to measure **optical power.** (Courtesy of Philips Telecom Equipment Company, a division of North American Philips Corporation).

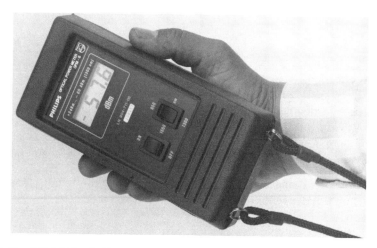

O-9. The OPM-6 Handheld Power Meter for measuring **optical power.** (Courtesy Philips Telecom Equipment Company, a division of North American Philips Corporation).

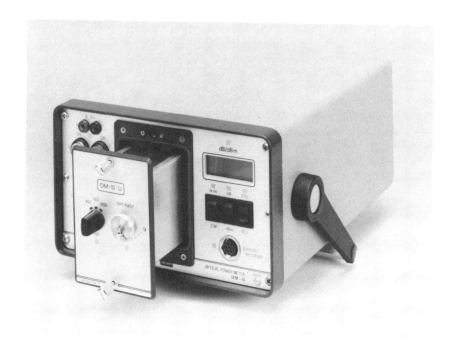

O-10. The OPM-10 Precision Power/Attenuation Meter, showing ruggedized case and interchangeable *receiver* module. The meter is used to measure **optical power** and *optical attenuation.* (Courtesy Philips Telecom Equipment Company, a division of North American Philips Corporation).

optical power density. See *irradiance.*

optical power efficiency. The ratio of *electromagnetic output power* from an *optical source* to the electrical input power to the source.

optical power output. The power in the *optical spectrum,* which is within the *electromagnetic spectrum, emitted* by a device. It may be expressed in total watts, total lumens per second, or watts per unit solid angle per unit of projected area of the source in a given direction, that is, the *radiance.*

optical protective coating. Films that are applied to a *coated* or uncoated *optical* surface primarily for protecting the surface from mechanical abrasion, chemical corrosion, or both. An important class of protective coatings consists of evaporated thin films of titanium dioxide, silicon monoxide, or magnesium fluoride. A thin layer of silicon monoxide may be added to protect an aluminized surface to prevent corrosion.

optical pulse broadening. See *optical pulse stretching.*

optical pulse distortion. An unintentional change in a property that characterizes an *optical pulse,* such as its shape, amplitude, *phase, wavelength, position,* or *duration.*

optical pulse lengthening. See *optical pulse stretching.*

optical pulse stretching. The increase in *optical pulse duration,* that is, the length or width, caused by *transmission* through one or more *optical elements,* such as a length of *optical fiber,* or a series of fibers connected in tandem. The amount of stretching, $T_2 - T_1$, can be calculated from the empirical relationship:

$$T_2 = T_1 (L_2/L_1)^k$$

from which the relation:

$$T_2 - T_1 = T_1 [(L_2/L_1)^k - 1]$$

is derived, where T_2 is the pulse duration at the end of a fiber of length L_2, T_1 is the pulse duration at the end of fiber length L_1, L_2 is the overall system fiber length including L_1, and k is a factor that is dependent upon the system length, L_2, and the *refractive index profile.* Typically k varies between 0.5 and 0.9, but more usually varies between 0.7 and 0.9. Pulse stretching is caused primarily by *dispersion.*

optical pulse suppressor. A device that *attenuates* an *optical pulse,* such as 1 km (kilometer) of *optical fiber.* Optical pulse suppressors were originally conceived to compensate for the *dead zones* of *optical time-domain reflectometers (OTDRs)* so that measurements could be made beginning at the entrance face of an *optical fiber* or in a *fiber optic connector* and *pigtail.*

optical pulse widening. See *optical pulse stretching.*

optical radiation. *Electromagnetic radiation* at *wavelengths* between the region of transition to x-rays and the region of transition to radio *waves,* that is, the *electromagnetic spectrum* between about 1 nm (nanometer) and about 0.1 mm (millimeter), that is, between 0.001 μ *(microns)* and 100 μ.

optical ray. A representation of the direction of *propagation* of *electromagnetic radiation* as defined by directional parameters. A *ray* is perpendicular to a *wavefront* and in the direction of the *Poynting vector* at each point in a *propagation medium.*

optical receiver. A *detector* that is capable of *demodulating* a *lightwave* and that can be *coupled* to a *propagation medium,* such as an *optical fiber.*

optical receiver maximum input power. The maximum input *optical power (dBm)* coupled to an *optical receiver,* as measured using specified standard procedures, from the line side of the receiver *fiber optic connctor* when operated under standard or extended operating conditions, that the receiver will accept and still maintain the manufacturer-specified *bit-error ratio (BER).*

optical reflection. See *maximum optical reflection.*

optical regenerative repeater. See *optical repeater.*

optical repeater. In an *optical transmission system,* an *optoelectronic device* or module that receives *optical signals,* amplifies them, or, in the case of a regenerative optical repeater for *digital signals,* reshapes, retimes, or otherwise reconstructs the signals, and *transmits* the result as an optical signal representing the same information as the input signals. The repeater is inserted at a point in a *propagation medium,* such as a long *fiber optic cable,* to overcome the effects of *attenuation* and *dispersion.*

optical repeater power. In *optical systems,* the power required to operate an optical repeater. The power may be delivered to the repeater by such means as (1) an electrical conductor in the cable that also contains the *optical waveguides,* (2) a separate electrical wire or cable, (3) a solar cell, (4) a local battery, primary or secondary, such as a seawater battery, (5) an electrical power system local power outlet, (6) an optical power cable separate from the optical signal cable or a *fiber bundle* in the same cable with the signal waveguides, or (7) by tapping the signal power contained in the optical signal fibers without disturbing the information content of the signals.

optical rotation. **1.** The angular displacement of the *polarization plane* of an *electromagnetic wave,* such as *lightwave propagating* through a *propagation medium.* **2.** The azimuthal displacement of the field of view achieved through the use of a rotating prism.

optical rotor. A *fiber optic device* that permits *optical signals propagating* in one optical element to pass across the axial *interface* into another element while they are axially rotating at different speeds, such as between an *optical fiber* rotating about its central longitudinal axis and a stationary fiber, which might occur when a *fiber optic system* fixed to a platform that is rotating with respect to the ground is connected to a system that is fixed to the ground. Also see *fiber optic rotary joint.*

optical sensitivity. See *receiver optical sensitivity.*

optical serrodyne. A *modulator* in which the *phase* of a *coherent lightwave* is modulated by controlling the transit time.

optical signal. See *two-level optical signal.*

optical source. A device that generates and *emits radiant power.* See *standard optical source; video optical source.*

optical source type. Characterization of a *light source* by identifying the type of source *(laser, LED, SLED),* compositional material (InGaAs), generic device structure (edge-emitting), and possibly other features.

optical spectrum. The range of *wavelengths* of *optical radiation.* Generally, it is the *electromagnetic spectrum* within the wavelength region extending from the vacuum *ultraviolet* at 0.001 μ *(micron)* to the *far-infrared* at 100 μ, that is, from about 1 nm *(nanometer)* to 0.1 mm (millimeter). This region lies between the region of transition from radio *waves* and the transition to x-rays. Thus, the *lightwave* region lies within the *optical spectrum,* and the *visible spectrum* lies within the lightwave region. The lightwave region has meaning in the sense that techniques and components designed for the human visible region also apply to regions somewhat beyond the human visible region, that is, the lightwave region is "visible" to the components though not in its entirety to humans. Thus, the optical spectrum is divided into several regions. See table: OPTICAL SPECTRUM on facing page. Also see *electromagnetic spectrum; light; optical.*

optical speed. In lens systems, the reciprocal of the square of the *f*-number, which is inversely proportional to twice the *numerical aperture.* Thus, the optical speed of an *optical fiber* is directly proportional to the numerical aperture squared, which is equal to the difference of the squares of the *refractive indices* of the *core* and *cladding.*

optical splice. See *fiber optic splice.*

optical station cable. In an *optical station/regenerator section,* the *fiber optic cable* from the *optical transmitter* or *receiver fiber optic pigtail* to the *splice* on the *optical cable facility* side (line or outside-plant side) of the *optical station facility.* It includes the cable to and within the *cable-interconnect feature.* The cable is used within a

Optical Spectrum

ELECTROMAGNETIC SPECTRUM REGION FOR OPTICAL DEVICES*	WAVELENGTH LIMITS *(microns)*	
	Lower	Upper
Radio*	100	–
Optical	0.001	100
Ultraviolet	0.001	0.4
Lightwave	0.3	3
Visible	0.4	0.8
Infrared	0.8	100
Near-infrared	0.8	3
Middle-infrared	3	30
Far-infrared	30	100
X-ray*	–	0.001

*The optical spectrum lies between the lower wavelength limit of radio waves, that is, 100 μ (micron) or 0.1 mm (millimeter), and the upper wavelength limit of x-rays, that is, 0.001 μ, 1 nm *(nanometer),* or 10 Å (angstrom).

building environment to connect the outside plant *fiber optic cable* to the *fiber optic system* terminal equipment. The station cable may provide the optical path in the form of *fiber optic patch panels* that allow rearrangement of paths to the outside plant fibers. Synonymous with *fiber optic station cable; station optical cable.*

optical station facility. The part of an *optical station/regenerator section* that lies within a station, that is the inside plant, including the *fiber optic transmitter, station cable, connectors,* and *cable interconnect features.* Synonymous with *fiber optic station facility.*

optical station/regenerator section. A *fiber optic transmission system* consisting of (1) an *optical station facility (fiber optic transmitter, connectors, station cables,* and a *cable interconnect feature,* (2) an *optical cable facility* (cable sections joined by splices between station facilities, and (3) another station facility (fiber optic cable interconnect feature, station cables, connectors, and receiver). Synonymous with *fiber optic station/regenerator section.*

optical surface. In an *optical system,* a *reflecting* or *refracting* surface of an optical element, or any other identifiable geometric surface in the system. Normally, optical surfaces occur at boundaries of discontinuity, that is, at abrupt changes of *refractive indices, absorptive* qualities, *transmittance,* vitrification, or other optical quality or characteristic of material.

optical switch. A circuit that enables *signals* in *optical fibers, integrated optical circuits (IOC),* or other *optical waveguides* to be selectively switched from one circuit or path to another or to perform logic operations with these signals by using *electrooptic effects, magnetooptic effects,* or other effects or methods that operate on *transmission modes,* such as *transverse-electric* versus *transverse-magnetic* *modes,* type and direction of *polarization,* or other characteristics of *electromagnetic waves.* For example, a *multimode* achromatic electrooptic waveguide switch or an optical polarization switch. Synonymous with *photonic switch.* See *thin-film optical switch.*

optical system. A group of interrelated parts and equipment designed to accept, process, and utilize *lightwaves transmitted* or guided for the most part in material media, such as lenses, prisms, *optical fibers, filters, multiplexers, couplers, connectors, mixers,* and various *waveguides,* to accomplish such functions as communication, illumination, inspection (endoscopy), sensing, and display. Examples of optical systems include optical fiber communication *links,* telemetry systems, optical gyrocompasses, telescopes, microscopes, and *aligned bundles* of fibers.

optical taper. An *optical fiber* with a diameter that is a linear function of its length, thus having a conical shape with either increasing or decreasing diameter in a longitudinal direction. The taper can be used in *aligned bundles* to increase or decrease the size of an image or as a *transition fiber* for *splicing* fibers with different diameters.

optical tapoff. See *optical tapping.*

optical tapping. The removal or extraction of some of the *radiant power,* that is, *optical power,* from *electromagnetic waves,* such as *lightwaves,* that are *propagating* in an *optical waveguide,* such as an *optical fiber, bundle, cable, slab-dielectric waveguide,* or a waveguide in an *integrated optical circuit (IOC).* Optical tapping is usually accomplished by using *connectors, couplers, taps,* and branches. Synonymous with *optical tapoff.*

optical thickness. **1.** The physical thickness times the *refractive index* of an optical element consisting of an *isotropic propagation medium.* **2.** The total *optical path length* through an optical element or a series of elements.

optical thin film. A thin layer of an *optical propagation medium,* usually deposited or formed on a substrate, with a geometric shape and *refractive* quality so as to enable control, *transmission,* and guidance of *lightwaves* for specific purposes, usually to accomplish logical functions by forming *optical switches* and gates. For example, a thin layer of *transparent* material in an *integrated optical circuit (IOC).*

optical time-domain reflectometer (OTDR). A device used to characterize an *optical waveguide,* such as *optical fiber,* by means of reflectometry. See figure on facing page. (See Fig. O–11.)

optical time-domain reflectometry (OTDR). A method for characterizing an *optical fiber* wherein an *optical pulse* is *coupled* into the fiber, and the *light* that is *backscattered* or *reflected* back to the input is measured as a function of time. The method is useful in estimating the *attenuation rate* of the fiber, that is, the *optical attenuation coefficient,* as a function of distance; identifying the nature and location of defects;

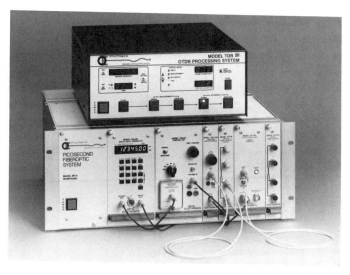

O-11. A millimeter resolution **optical time-domain reflectometer (ODTR)** system that uses 100 ps *(picosecond) light pulses,* measures *Fresnel reflections* along an *optical fiber* with a *time resolution* less than 10 ps and a corresponding *distance resolution* less than one mm. *Multimode* or *single-mode* operation at *wavelengths* from 0.820 to 1.550 μ *(microns)* are provided by a variety of *transmitter/receiver* modules as plugin units. (Courtesy Opto-Electronics, Inc.)

determining other localized *losses,* such as *insertion losses* caused by *fiber optic connectors, couplers,* and *splices;* and measuring other parameters of optical fibers and other components. The distance to a reflection *interface surface* or other discontinuity is determined by measuring the time it takes for a *lightwave pulse* to travel to the discontinuity and back. Reflection surfaces include the ends of *fiber optic cables,* breaks in fibers, splices, connector interfaces, cracks, fractures, or other *anisotropic* features and discontinuities of the *propagation medium.* The measurement equipment is called an *optical time-domain reflectometer.* It displays the reflected waves on a time axis for precise reading by showing, for example, the leading edge of a *transmitted optical pulse* and the various reflections that occur, usually before the next pulse is *launched.*

optical transimpedance. In an *optical transmission system,* the ratio of the output voltage at the *photodetector* to the input current at the *light source.*

optical transmitter. A *light source* capable of being *modulated* and *coupled* to a *propagation medium,* such as an *optical fiber* or an *integrated optical circuit (IOC).*

optical video disk (OVD). A disk on the surface of which *digital data* may be recorded at high packing densities in concentric circles or in a spiral, using a *laser beam* to record spots that are read by means of a *reflected* laser beam of lower *irradiance,* that is, intensity, than the recording intensity. Up to 10^{13} bits are being recorded on a single disk at rates of 1 Mbps, making optical video disks suitable for hours of television programming playback.

optical waveguide. Any structure having the ability to guide the flow of *optical energy* along a path parallel to its axis and, at the same time, to contain the energy within, or adjacent to, its surface(s). See *diffused optical waveguide; step-index optical waveguide; thin-film optical waveguide.*

optical waveguide connector. A device whose purpose is to transfer *optical power* between two *optical waveguides* or *optical fiber bundles,* and that is designed to be connected and disconnected repeatedly.

optical waveguide coupler. 1. A device whose purpose is to distribute *optical power* among two or more ports. **2.** A device whose purpose is to *couple optical power* between a *waveguide* and either a *light source* or a *photodetector.*

optical waveguide splice. A permanent joint whose purpose is to *couple optical power* between two *waveguides.*

optical waveguide termination. The point in an *optical waveguide* at which *optical power propagating* along the waveguide leaves the waveguide and continues in a nonwaveguide mode of propagation. For most applications, care should be taken to prevent *leakage, losses,* and *reflection.*

optic attenuator. See *fiber optic attenuator.*

optic axis. In an *anisotropic medium,* a direction of *lightwave propagation* in which orthogonal *polarizations* have the same *phase velocity.* Also see *optical axis.*

optic borescope. See *fiber optic borescope.*

optic branching device. See *fiber optic branching device.*

optic bundle. See *optical fiber bundle.*

optic cable. See *fiber optic cable; Tempest-proofed fiber optic cable.*

optic cable assembly. See *fiber optic cable assembly.*

optic cable core. See *fiber optic cable core.*

optic cable feed-through. See *fiber optic cable feed-through.*

optic connector. See *fiber optic connector.*

optic coupler. See *fiber optic coupler.*

optic cross-connection. See *fiber optic cross-connection.*

optic data link. See *fiber optic data link.*

optic data-transfer network. See *fiber optic data-transfer network.*

optic demultiplexer (active). See *fiber optic demultiplexer (active).*

optic demultiplexer (passive). See *fiber optic demultiplexer (passive).*

optic display device. See *fiber optic display device.*

optic distribution frame. See *fiber optic distribution frame.*

optic dosimeter. See *fiber optic dosimeter.*

optic drops. See *fiber optic drop.*

optic endoscope. See *fiber optic endoscope.*

optic filter. See *fiber optic filter.*

optic flood illumination. See *fiber optic flood illumination.*

optic gyroscope. See *fiber optic gyroscope (FOG).*

optic illumination detection. See *fiber optic illumination detection.*

optic illuminator. See *fiber optic illuminator.*

optic interconnection. See *fiber optic interconnection.*

optic interconnection box. See *fiber optic interconnection box.*

optic isolator. See *fiber optic isolator.*

optic jumper. See *fiber optic jumper.*

optic light source. See *fiber optic light source.*

optic link. See *fiber optic link; TV fiber optic link.*

optic mixer. See *fiber optic mixer.*

optic mode stripper. See *fiber optic mode stripper.*

optic modulator. See *fiber optic modulator.*

optic multiplexer (active). See *fiber optic multiplexer (active).*

optic multiplexer (passive). See *fiber optic multiplexer (passive).*

optic multiport coupler. See *fiber optic multiport coupler.*

optic patch panel. See *fiber optic patch panel.*

optic penetrator. See *fiber optic penetrator.*

optic photodetector. See *fiber optic photodetector.*

optic receiver. See *fiber optic receiver.*

optic ribbon. See *fiber optic ribbon.*

optic rotary joint. See *fiber optic rotary joint.*

optics. The branch of science and technology devoted to the study of (1) the nature and properties of *electromagnetic radiation* in the *optical spectrum,* with most of the emphasis placed on the *visible* and near-visible spectra, (2) the application of the radiation in useful systems, and (3) the phenomena of vision. See *acoustooptics; active optics; coated optics; electrooptics; fiber optics; fixed optics; geometric optics; integrated optics; physical optics; ultraviolet fiber optics; woven fiber optics.*

optic scrambler. See *fiber optic scrambler.*

optic sensor. See *fiber optic sensor.*

optic splice. See *fiber optic splice.*

optic splice tray. See *fiber optic splice tray.*

optic splitter. See *fiber optic splitter.*

optic spot illumination. See *fiber optic spot illumination.*

optic supporting structures. See *fiber optic supporting structures.*

optic switch. See *fiber optic switch.*

optic telemetry system. See *fiber optic telemetry system.*

optic terminus. See *fiber optic terminus.*

optic test method. See *fiber optic test method (FOTM).*

optic test procedure. See *fiber optic test procedure (FOTP).*

optic transmitter. See *fiber optic transmitter.*

optoacoustic transducer. A device that converts an audio-frequency-*modulated lightwave* into a sound tone by causing a crystal to vibrate at the audio modulation *frequency* of the lightwave.

optoelectrical cable. A cable that contains *optical waveguides,* such as *optical fibers,* and electrical conductors.

optoelectronic. Pertaining to devices containing both *optical* and electronic components, usually involving the conversion of *optical power* or energy into electrical power or energy, such as the conversion of an *optical signal* into an electrical signal, or vice versa, representing the same information. Optoelectronic devices must have at least one electrical *port.* They respond to, *emit,* or modify *optical radiation,* or use optical radiation for internal operation. An optoelectronic device may function as an electrical-to-optical or optical-to-electrical *transducer. Photodiodes, light-emitting diodes (LEDs), injection lasers, repeaters,* and *integrated optical circuits (IOCs)* are optoelectronic devices commonly used in *fiber optic systems.* Synonymous with *optronic.* Not synonymous with *electrooptic.* Also see *electrooptic; electrooptic effect.*

optoelectronic detector. **1.** A device that *detects radiation* by utilizing the influence of *light* in forming an electrical *signal.* **2.** A device in which a *modulated optical signal* is converted into an electrical signal having the original modulation information content.

optoelectronic directional coupler. A directional *coupler* in which the coupling function is electronically controllable, usually wth a *photodetector* that permits an electronic circuit to be driven by the coupler by *tapping* some of the passing optical power.

optoelectronic receiver. A receiver that accepts an *optical signal* from a *fiber optic cable* and converts it to an electrical signal for further processing or use.

optoelectronics. The combination of pure and applied electronics, *optics,* and *electromagnetics.* Synonymous with *optronics.*

optoelectronic transmitter. A *transmitter* that accepts an electrical *signal* and converts it to an *optical signal* that represents the same information as the electrical signal.

optoisolator. See *optical isolator.*

optooptics. The branch of science and technology devoted to the use of *optical power* to control the generation, *transmission,* reception, and processing of *optical signals* and power. Optooptics is being applied in *integrated optical circuits (IOCs)* using *nonlinear* optical materials to perform *modulation, multiplexing,* and *switching* functions. Also see *acoustooptics; electrooptics; magnetooptics; photonics.*

optronic. See *optoelectronic.*

optronics. See *optoelectronics.*

orderwire. The *channel* or *path* in a communication *network* used for *signals* to control or direct the control of network operations. For example, an *analog* voice orderwire may operate at 300 to 400 *Hz; a digital* orderwire, or datawire, at 16 kbps; and a telemetering orderwire at 2kbps. In a *fiber optic cable,* the orderwire may be a metallic conductor or an *optical fiber,* either one used for regular communication or a separate conductor or fiber.

O–12. A rack-mounted version of FOtalk®, a *fiber optic* **orderwire** talkset for *network* use. (Courtesy fotec incorporated).

ordinary ray. A *light ray* that has *isotropic* velocity in a *doubly refracting* crystal. An ordinary ray obeys *Snell's law* of *refraction* at the crystal surface. Also see *extraordinary ray.*

organizer. See *fiber-and-splice organizer.*

OSI. *Open-Systems Interconnection.*

OSIRM. *Open-Systems-Interconnection Reference Model.*

OTDR. **1.** *Optical time-domain reflectometry.* **2.** *Optical time-domain reflectometer.*

outages. See *mean time between outages (MTBO).*

output. See *optical power output.*

output angle. See *radiation angle.*

output aperture. The *aperture* at the point of exit of an *optical system,* usually equal to the aperture of the final optical component, such as an eyepiece lens or the end of an *optical fiber,* from which the image or exit *beam* is projected for viewing or for input to another device. The output aperture of an optical fiber can be measured by projecting the exiting beam on a screen and taking the ratio of half the beam diameter at the screen and the distance to the screen.

outside vapor deposition (OVD). Pertaining to a process for the production of *optical fibers* in which materials are deposited in vapor or soot form on the outside of *optical fiber blanks,* which are then sintered to make *optical fiber preforms.* Fibers with desired *refractive-index profiles* are then *pulled* from the preforms, *buffered,* and wound on spools or reels for subsequent *cabling.*

outside vapor-phase-oxidation (OVPO) process. A *chemical vapor-phase-oxidation (CVPO) process* for the production of *optical fibers,* in which the soot stream and heating flame are deposited on the outside surface of a rotating glass rod.

OVD. **1.** *Optical video disk.* **2.** *Outside vapor deposition.*

overall length. See *borescope overall length.*

overarmor. An additional cover, *jacket,* or *sheath* placed over a *cable,* such as a *fiber optic cable,* to provide more strength and added protection against harsh environments and for uses in addition to *transmission* of *signals,* such as use as a tether cable from ship to a buoy or ship to ship, as a shore-to-ocean interface, or as a ground cable where abrasive abuse is expected. The overarmor is usually braided and made of metal or high-strength polymers. Smooth metal tubing that serves as the jacket over optical fibers is not considered overarmor.

overcompensated optical fiber. An *optical fiber* whose *refractive-index profile* is adjusted so that the *high-order propagating modes* arrive ahead of the *low-order modes.* Higher order modes have higher eigenvalues in the solution of the *wave equations,* higher *frequencies,* and shorter *wavelengths* than lower order modes. Also see *compensated optical fiber; undercompensated optical fiber.*

overfill. When *coupling lightwaves* from a *light source* to an *optical fiber,* the situation that occurs when the *core diameter* is smaller than the diameter of the cone of light emanating from the source at the end face of the fiber, that is, the *launch spot* is larger than the core diameter, and the *launch angle* of the source exceeds the *acceptance angle* of the fiber.

overhead rate. In a *transmission system,* the bit rate at an interface corresponding to the bits required to operate the system, usually consisting of bits for error detection and correction, synchronization, buffer stuffing, control, and other system purposes. The interface may be between a *user end-instrument* and the system or between functional units within the system. Overhead bits representing information generated by the user are not considered system overhead bits, whereas overhead bits generated within the system and not delivered to the user are considered system overhead bits. Thus, user throughput is reduced by both overheads while system throughput is reduced only by system overhead. The interface rate is the payload rate plus the system overhead rate. Also see *interface rate; payload rate.*

overtone absorption. In the *propagation* of an *electromagnetic wave* in material media, *absorption* of *frequency* components above the fundamental frequency in the waves. For example, absorption caused by the presence of the hydroxyl ion, which gives rise to fundamental resonance vibrations and hence an absorption centered at 2.8 μ *(microns),* with higher harmonics at *wavelengths* of 1.4, 0.97, and 0.75 μ. The hydroxyl ion (OH) originates primarily from water as a contaminant in optical glass.

OVPO. *Outside vapor-phase oxidation.*

OVPO process. See *outside vapor-phase-oxidation process.*

oxidation process. See *axial vapor-phase-oxidation (AVPO) process; chemical vapor-phase-oxidation (CVPO) process; inside vapor-phase-oxidation (IVPO) process; outside vapor-phase-oxidation (OVPO) process.*

P

packing density. In an *optical fiber bundle,* the end cross-sectional total *core area* per unit of total cross-sectional area, usually within the ferrule, including *jacket, strength-member, buffer, core, cladding,* and interstitial areas, of the assembly or bundle of all the fibers whose cross-sectional areas are being counted, that is, per unit of the total area within the perimeter of the bundle, such as within the jacket. Packing density varies with the size of fiber, core areas relative to total fiber area, pressure applied to the bundle, and other factors.

packing fraction. In an *optical fiber bundle,* the ratio of the aggregate cross-sectional *core area* to the total cross-sectional area of the bundle of fibers. It may be necessary to specify the conditions under which the packing fraction is to be measured. For example, the end area of an optical fiber bundle or *fiber optic cable* might be the inside area of a *coupler, connector,* or *termination.* A packing fraction might be given by the relation:

$$PF = N(a/d)^2$$

where N is the number of optical fibers, a is the diameter of the core in each fiber, and d is the diameter of the whole assembly or bundle.

packing fraction loss. The *irradiation,* that is, the *optical power,* lost because the *packing fraction* is always less than unity, except, of course, when there is only one *optical fiber* in the bundle. The loss is usually expressed in *decibels.*

PACVD. *Plasma-activated chemical vapor deposition.*

PACVD process. See *plasma-activated chemical-vapor-deposition (PACVD) process.*

pad. See *attenuate; attenuator.*

PAM. *Pulse-amplitude modulation.*

panel. See *fiber optic patch panel.*

parabolic refractive-index profile. In an *optical fiber,* a *power-law refractive index profile* in which the *profile parameter, g,* is equal to 2. Synonymous with *quadratic refractive-index profile.* See *linear refractive-index profile; power-law refractive-index profile; radial refractive-index profile.*

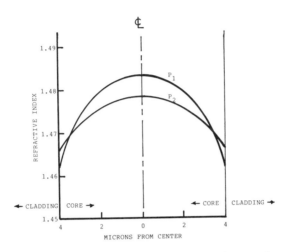

P-1. **Parabolic refractive-index profiles** of two *optical fibers* with *core diameters* of 8 μ (*microns*).

parameter. See *application parameter; global fiber parameter; material-dispersion parameter; profile-dispersion parameter; refractive-index profile parameter; signal parameter; wave parameter.*

paraxial ray. In *optical systems,* a *ray* that is close to and nearly parallel with the *optical axis.* For computation, the angle θ, between the ray and the optical axis is small enough for $\sin \theta$ to be practically equal to $\tan \theta$, and either or both can be replaced by θ in radians.

Parseval's theorem. An extension of the superposition theorem applied to nonperiodic *wave* shapes for *spectral irradiance,* that is, *spectral power density.* In the periodic case, the various *signals* are considered orthogonal over the period, but in the nonperiodic case, the interval of orthogonality extends over the entire time axis from minus to plus infinity.

partial coherence. In an *electromagnetic wave* or *waves, coherence* such that the *electromagnetic fields* at two points in space or two instants of time have a low statistical level of correlation.

partial mirror. A surface that simultaneously *transmits* and *reflects* a significant portion of incident *electromagnetic radiation,* such as a *beam* of *lightwaves.* Usually the surface consists of a substrate, such as glass, with a layer of reflective *coating,* such as a metal, so thin that it reflects approximately half of the incident *radiation* and transmits approximately half, with negligible *absorption*. If two persons, *A* and *B,* are on opposite sides of such a mirror, *A* might see *B* but *B* can't see *A*. If the mirror is constructed with precise dimensions, such as a half- or quarter-wavelength thickness, with a thin reflective and transmitting coating on both sides of the glass, it can be used as a *beam splitter, combiner,* or *interferometer* to mix two *coherent* lightwaves from different directions that are incident on opposite sides of the mirror. Depending on the nature of the reflective and transmitting coatings on both sides of the *transparent* substrate, the output of the mirror would then consist of either (1) two coherently mixed beams, each component of the mixture down 3 dB from its original *power level,* or (2) one coherently mixed beam. Also see *beam splitter.*

passive optical device. A device that operates with or performs specific operations on *electromagnetic waves* having wavelengths that are in the *optical spectrum,* that is, in the optical region of the *electromagnetic spectrum,* usually within the *visible* and near-visible spectra. The operations are performed by the *propagation media* through which the waves pass without the use of input energy other than that contained in the waves themselves. *Optical couplers, splitters, mixers, filters, attenuators, gratings, lenses, prisms,* and all-*optical multiplexers* and *demultiplexers* are passive optical devices. *Fiber optic cables, splices,* and *connectors* are also passive optical devices. However, these three perform no specific operation on *lightwaves* except transport them between and through points. Also see *optical path component.*

patch panel. See *fiber optic patch panel.*

path. **1.** In network topology, a route between any two *nodes* of a *network.* The route may consist of several *branches* and several nodes that lie between the two nodes connected by the path. The path has a start point and an end point. **2.** In communication systems, a route between any two points in a *network.* The points may be two telephones, a control panel and a motor controller of a ship propulsion system, a remote *data* input terminal and a computer, or two *switching centers.*

path attenuation. In communications, the *power losses* that occur in a *wave* or *signal propagating* between a *transmitter* and a *receiver.* Usually measured in *decibels,* it is caused by many effects, such as *free-space loss, refraction, reflection, aperture-propagation-medium coupling* loss, and *absorption.*

path component. See *optical path component.*

path length. See *optical path length.*

pattern. See *acceptance pattern; equilibrium radiation pattern; far-field diffraction pattern; far-field radiation pattern; fiber pattern; Fraunhofer diffraction pattern; Fresnel diffraction pattern; near-field diffraction pattern; near-field radiation pattern; radiation pattern; speckle pattern.*

payload rate. In a *transmission system,* the *interface rate* minus the *overhead rate.* Synonymous with *information payload capacity.* Also see *interface rate; overhead rate.*

PCS. *Plastic-clad silica.*

PCS fiber. See *plastic-clad silica (PCS) fiber.*

PD. *Photodetector.*

peak-emission wavelength. In *fiber optics,* the *wavelength* at which a *light source emits* its maximum value of *optical power.*

peak-radiance wavelength. The *wavelength* at which the *radiance* of a *light source* in a given direction is a maximum. Synonymous with *peak wavelength.* Also see *radiance.*

peak spectral power. The *maximum radiance,* that is, *optical power, emitted by* a *light source,* usually occurring at a specific *wavelength* or within a specific range of wavelengths. For example, a gallium arsenide *light-emitting diode (LED)* emits its *peak spectral power* at a wavelength of 0.910 μ *(microns).*

peak spectral wavelength. The *wavelength* at which a *light source emits* its maximum total *optical power.* For example, a gallium arsenide *light-emitting diode (LED)* emits its *peak spectral wavelength* of 0.910 μ *(micron)* when operating at a peak diode current of 50 mA at 1 mW of total optical power output with about 0.30 mW of the optical power output *coupled* to an *optical fiber pigtail.*

peak wavelength. See *peak radiance wavelength.*

penalty. See *dispersion power penalty; reflection power penalty.*

penetration. In *fiber optic systems,* the passage through a partition, wall, hull, tank, bulkhead, or other obstacle, usually by means of a *fiber optic penetrator* or continuous optical cable.

penetrator. See *fiber optic penetrator.*

periodically distributed thin-film waveguide. A *slab-dielectric waveguide* formed by evaporating, growing, etching, cementing, depositing, masking, or otherwise placing thin films of *transparent dielectric* material on substrates. For example, an *integrated optical circuit (IOC)* chip having a deposited *waveguide* with a periodically varying width, in order that only selected *modes* are *propagated* while other modes are eliminated because of the geometric shape. See figures on facing page. (See Figs. P–2 and P–3.)

peripheral-strength-member optical cable. An *optical cable* containing *optical waveguides,* such as *optical fibers,* that are surrounded by a group of high-tensile-

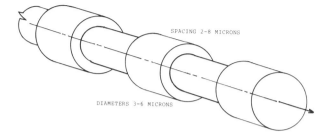

P-2. A circular, *dielectric,* **periodically distributed thin-film waveguide**. Arrow indicates direction of *propagation* of *electromagnetic waves.*

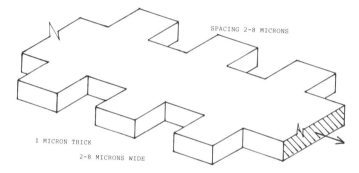

P-3. A *planar, dielectric,* **periodically distributed thin-film waveguide**. Arrow indicates direction of *propagation* of *electromagnetic waves.*

strength materials, such as stranded or solid contrahelical or longitudinal steel wires, Nylon or Kevlar strands, steel or copper tubes, or other material, with a crush-resistant *jacket,* sheath, or *overarmor* on the outside of the cable. Also see *central-strength-member optical cable.*

permeability. See *magnetic permeability; relative magnetic permeability.*

permittivity. See *electric permittivity; relative electric permittivity.*

Perot fiber optic sensor. See *Fabry-Perot fiber optic sensor.*

perturbation. See *phase perturbation.*

petahertz (PHz). A unit of *frequency* equal to 1 thousand trillion *hertz,* or 10^{15} Hz.

phase. In a periodic function or phenomenon, such as the *electric field vector* in an *electromagnetic wave,* the instant, event, or position at which a specified significant *parameter* of the function occurs relative to a given time reference or relative to a *significant instant,* event, or position in another function or phenomenon. An example of a *phase* relationship is the angular displacement between the peak values of two sinusoidal *waves* of the same *frequency.* Thus, when *lightwaves propagate* in an *optical fiber,* waves of different *wavelength* are shifted in phase over given lengths of the fiber by different amounts due to their different propagation velocities, giving rise to *distortion* of *signals,* because they invariably consist of more than one optical frequency, that is, more than one wavelength.

phase constant. See *phase term.*

phase departure. In a *wave propagating* in a *transmission* line or *waveguide,* an unintentional deviation from the nominal *phase* value of a point in the wave, that is, an unintentional phase deviation not caused by *modulation.*

phase detector. A circuit or device that *detects* the difference in *phase* between corresponding points on two synchronous *signals.*

phase-front velocity. In an *electromagnetic wave,* such as a *lightwave,* the velocity of a *wavefront,* the wavefront being a surface described by the locus of all points at which the same *significant instant* occurs in the wave, such as the surface that describes the locus of all points at which the peak value of the *electric field vector* is a constant.

phase jitter. A rapid or repeated *phase* perturbation.

phase modulator. See *optical phase modulator.*

phase perturbation. The cause of a rapid shifting of the relative *phase* of a *signal,* or the momentary phase shift itself. The phase shift may be random, cyclic, or a single shift. Other signal *parameter* perturbations may also occur along with phase perturbations, caused by superimposed *noise,* spikes, *propagation media* anomalies and changes, and *interference.* The amount of phase perturbation may be expressed in electrical degrees or radians. A cyclic (recurring) perturbation of any sort may be expressed in *hertz.* Phase perturbation may be negligible in a given *transmission* system. Also see *jitter.*

phase term. In *transmission* lines and waveguides, the imaginary part of the axial *propagation constant* for a particular *mode,* usually expressed in radians per unit length. Synonymous with *phase constant.*

phase velocity. The velocity of *propagation* of a *uniform plane-polarized electromagnetic wave,* given bv the *light source* or *free space wavelength* times the *frequency* divided by the *refractive index* of the *propagation medium* in which the wave is propagating. In International System (SI) units, the refractive index of free space may be considered to be unity because the refractive indices of materials are normalized with respect to that of a vacuum. Strictly speaking, this concept can be applied only to a single-*frequency* wave, such as an *unmodulated carrier wave* from a single-frequency or *monochromatic source.* The phase velocity is the velocity an observer would have to move to make the wave characteristic appear to remain constant in phase in a given propagation medium. For a given frequency of an electromagnetic wave, which remains constant at interfaces, the wavelength and speed are less in material media than in free space. In *nondispersive media,* that is, media that do not cause *dispersion,* the phase velocity and the *group velocity* are equal.

phonon. An acoustic energy packet similar in concept to a *photon,* because its energy is a function of the *frequency* of vibration of the sound source that produced it. Phonon streams form a sound *wave,* depending on whether one speaks in terms of the *quantum* theory or the wave theory of sound, just as is true for *light.* The energy of a phonon is usually less than 0.1 eV. It is an order of magnitude less than the energy of a photon. However, when dealing with *band-gap energies* in semiconductors, the energy of a phonon is not negligible and, with appropriate *coupling,* can contribute appreciably to the movement of electrons into higher energy levels on a statistical basis. When these excited electrons move to lower energy levels, photon *emission* occurs. Thus, the existence of phonons can affect the *spectral* composition of *radiation* emitted by a light source.

phosphorescence. The *emission* of *electromagnetic radiation* by a material after stimulation from an outside source of energy ceases. However, phosphorescence may also occur during such stimulation. Also see *fluorescence; luminescence.*

photocell. A device that produces an output electric current, voltage, or power proportional to incident *irradiance,* that is, *optical power,* integrated over the illuminated area of the surface that is sensitive because of *photovoltaic, photoconductive,* or *photoemissive effects.*

photoconductive effect. The phenomenon in which some nonmetallic materials exhibit a marked increase in *electrical conductivity* upon *absorption* of *photon energy.* Photoconductive materials include (1) gasses that have been ionized by the energy in photons and (2) certain crystals. They are used in conjunction with semiconductor materials that are ordinarily poor electrical conductors, but become distinctly conducting when subjected to photon absorption. The incident photons excite electrons into the conduction band where they move more freely, resulting in good electrical conductivity. The increase in conductivity is due to the additional free carriers generated when photon energies are absorbed in energy transitions. The rate at which free carriers are generated and the length of time they remain in conducting states, that is, their lifetime, determines the amount of change in conductivity produced by the incident stream of photons. Synonymous with *photoconductivity; internal photoelectric effect.*

photoconductive film. A film of material whose electrical current-carrying capacity is enhanced when illuminated by *electromagnetic radiation,* particularly in the *visible* and near-visible spectra, that is, the *lightwave* region, of the *electromagnetic frequency spectrum.*

photoconductivity. The *photoelectric effect* in which electrons are *emitted* from the surface of a material *irradiated* by *optical electromagnetic radiation.*

photocurrent. The electrical current, in a material medium, resulting from the *absorption* of *electromagnetic radiation,* such as *light* energy, by the material. In a strict sense, photocurrents result from the *photovoltaic* or *photoemissive* effects, rather than from an increase in conductivity, because then the actual source of current is a power source external to the *photodetector.* However, because the electric current is a function of incident *electromagnetic power,* the term may be applied in any case. Photocurrents occur in *photodiodes* and other photodetectors. Photocurrents may be calculated from the relation:

$$I_{ph} = \Gamma e P / hf$$

where Γ is the *carrier* collection *quantum efficiency, e* is the electron charge, P is the incident *optical power, h* is *Planck's constant,* and f is the *frequency* of the incident *radiation.* Synonymous with *light current.*

photodarlington. A combination of a *photodetector* and a Darlington-pair transistor circuit. The photodarlington is capable of *detecting* an *optical signal* and producing an amplified electrical version of the signal.

photodetector (PD). A *transducer* capable of accepting an *optical signal* and producing an electrical signal containing the same information as in the optical signal. See *avalanche photodetector; fiber optic photodetector; photoelectromagnetic photodetector; photoemissive photodetector; photovoltaic photodetector.* Also see *optical detector.* (See Figs. P–4, P–5, and P–6.)

photodetector responsivity. The ratio of the RMS value of the output current or voltage of a *photodetector* to the RMS value of the incident *optical power.* In most cases, photodetectors are *linear* in that their *responsivity* is independent of the *irradiance,* that is, the power density or intensity, of the incident *radiation.* Thus, the photodetector output in amperes or volts is proportional to the incident optical power in watts. Differential responsivity applies to small variations in input optical power. Photodetectors are square-law detectors in that they respond to the irradiance of the incident *electromagnetic wave,* that is, to the square of the *electric field strength* associated with the incident wave, which is proportional to the optical power.

photodetector signal-to-noise ratio. The *signal-to-noise ratio* for a *photodetector* is given by the relation:

$$SNR = (N_p / B)(1 - e^{-hf/kT})$$

where N_p/B is the number of incident *photons* per unit of *bandwidth, h* is *Planck's constant, f* is the *frequency* of the incident photons, k is *Boltzmann's constant,* and T is the absolute temperature.

P-4. The PD20 Ultra Fast **Photodetector** with power supply. The PD20 is designed for monitoring and analyzing ultra fast *optical pulses* or ultra high *frequency optical modulation*. Many models are available in the PD series, each with a different set of options. (Courtesy Opto-Electronics, Inc.).

P-5. The Picosecond Fiberoptic System performs many *fiber optic* measurements. It includes picosecond **photodetectors** (Si, Ge, InGaAsP, non-avalance, *avalanche,* and amplified), picosecond *pulsed diode lasers* (0.800 to 1.500 μ (*micron*), externally *modulatable* diode lasers (GHz and Gbps regimes), *optical fiber couplers* (4 *port, single-mode, multimode*), and many other components. (Courtesy Opto-Electronics, Inc.).

231

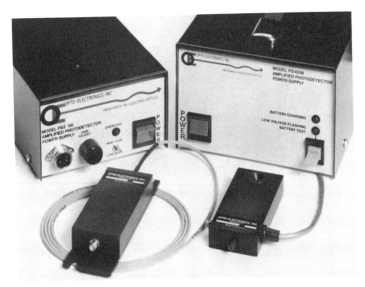

P-6. Amplified **photodetectors** and power supplies for monitoring fast *optical pulses* or high-*frequency modulation*. High-speed *photodiodes,* including *PIN,* PN, and *APD* diodes, are coupled to microwave amplifiers. Line and rechargeable battery operation are available. (Courtesy Opto-Electronics, Inc.).

photodiode. A *photodetector* consisting of a *p-n junction,* that is, an *interface* between two semiconductors of different composition or between a semiconductor and a metal, that either (1) absorbs *radiation* in the area of the junction, producing a *photocurrent,* or (2) increases its conductivity with increased *power level* in incident *electromagnetic* radiation, that is, in *irradiance,* in which case a source of voltage is required because *photoemissive* or *photovoltaic effects* are not being used. See *avalanche photodiode (APD); PIN photodiode.* Synonymous with *diode photodetector.*

photoeffect detector. See *internal photoeffect detector.*

photoelastic effect. That property of a material in which Young's modulus, that is, the modulus of elasticity, mechanical elasticity, or the coefficient of elasticity (stress/strain), at a point in the material is a function of the instantaneous *electromagnetic irradiance* at the point; and conversely, the effect that changes in elasticity have on *electromagnetic waves propagating* in a material medium. When an electromagnetic wave is propagating in a material or *free space,* the irradiance of the wave is (1) the instantaneous *electromagnetic power* passing through a cross-sectional area, that is, power density in watts per square meter, or (2) the electromagnetic energy contained in a unit volume of the radiation, that is, power density in joules per cubic meter, at the point in the material. These are identical units of irradiance.

photoelastic sensor. A *fiber optic sensor* in which the *photoelastic effect* is used for sensing based on the change in *refractive index* or the *double refraction* that is produced when stress is applied to a *transparent* material, especially transparent plastics. When the several *lightwaves* caused by the changes in refractive indices reinforce or cancel one another, *light* patterns indicate points of stress. A *photodetector* can be used to produce a *signal* indicating the resultant *modulation* of light from specific points being subjected to stress in a mechanical system.

photoelectric effect. 1. The changes in the electrical characteristics of a material caused by *photon energy absorption*. **2.** The phenomenon of (1) the liberation of electrons and other electrical charge carriers in a material, (2) their increased flow as a result of increased conductivity, (3) their aggregation at material *interfaces* to produce a voltage, or any combination of these, as a result of *photon energy absorption* exciting the material. The energy of the incident individual *photon* is given by the relation:

$$E_p = hf$$

where *h* is *Planck's constant* and *f* is the *frequency* of the photon, that is, the frequency of the incident *radiation*. If the incident radiation consists of a mixture of frequencies within a given range, the energy level of each photon will be determined by one of the frequencies in the range. The electrons may be released into a vacuum or into another material. The excited material may be a solid, liquid, or gas. Thus, *photoconductive, photoelectromagnetic, photoemissive,* and *photovoltaic effects* are all photoelectric effects. See *internal photoelectric effect.*

photoelectric equation. The equation that relates *photon energy,* the *work function* of a material, such as certain metals and oxides, and the emitted-electron energy, that is, the kinetic energy with which an electron will be emitted by the material when the photon strikes it, given by the relation:

$$hf = W + (1/2)mv^2$$

where *h* is *Planck's constant, f* is the *frequency* associated with the incident photon, *W* is the work function of the material the photon strikes, *m* is the mass of the electron, and *v* is the velocity with which the electron is ejected. If the emitted-electron energy is zero, that is, the electron has negligible velocity, the photon has just sufficient energy to overcome the work function. The frequency corresponding to this condition is the *threshold frequency.* If the energy of every photon in a stream of incident photons is less than that of the work function, the photons cannot cause an electron to be emitted no matter how many photons strike the material during a given time or at what rate they strike the material. Also see *work function.*

photoelectromagnetic effect. The production of an electric potential difference as a result of the interaction of a magnetic field with a *photoconductive* material when subjected to incident *lightwaves.* The incident *radiation* creates hole-electron pairs that diffuse into the material. The magnetic field causes the paired components to separate as they move in the same direction, but they constitute oppositely directed electric currents, resulting in the production of an electric potential difference across the material. In most applications, the light is made to fall on a flat surface of an intermetallic semiconductor located in a magnetic field that is parallel to the surface. Excess hole-electron pairs are created by the magnetic field to produce a current flow through the semiconductor that is at right angles to both the *light rays* and the magnetic field. This is caused by transverse forces acting on electrons and holes that are diffusing into the semiconductor from the surface. Synonymous with *photomagnetoelectric effect.*

photoelectromagnetic photodetector. A *photodetector* that makes use of the *photoelectromagnetic effect,* that is, a photodetector that needs an applied magnetic field, accepts incident *optical electromagnetic radiation,* and produces an electric current proportional to the *electric field strength* of the incident radiation.

photoemissive effect. The ejection of electrons, usually into a vacuum, from a material surface *irradiated* by *optical radiation.* The *emission* is a result of *photon absorption.* Synonymous with *external photoelectric effect.*

photoemissive photodetector. A *photodetector* that makes use of the *photoemissive effect.* An applied electric field is necessary to attract or collect the emitted electrons, thus producing an electric current. Were it not for the applied field, the density of emitted electrons would build up and create an electric field, or space-charge cloud that would inhibit further emission.

photomagnetoelectric effect. See *photoelectromagnetic effect.*

photon. A *quantum* of electromagnetic energy. The energy of a photon is given by the relation:

$$E_p = hf$$

where *h* is *Planck's constant* and *f* is *frequency.* The energy of a photon with a *wavelength* of 1 μ *(micron)* is approximately 1.2 eV (electron-volts). A value for the energy of a typical *phonon* is approximately 0.1 of that of the typical photon. The photon has some particlelike characteristics; for example, they can be counted and their individual energy measured. See *gamma photon.*

photon detector. A device that responds to incident *photons,* that is, it *signals,* with some reasonable probability of being correct, the *absorption* of a photon, that is, a *quantum* of *electromagnetic* energy. The photon detector contains a sensitive material that exhibits a change in property when it absorbs a photon, that is, the material becomes *photoemissive, photoconductive, photovoltaic,* or *photoelectromagnetic.*

photon energy. The energy associated with a *photon,* equal to the product of the photon *frequency* and *Planck's constant.*

photonics. The branch of science and technology devoted to the study, control, and use of *photons,* or *lightwaves,* including the generation, *transmission,* reception, and processing of *optical signals* and *power.* The control of lightwaves may be effected by electrical, optical, acoustic, or other means. Also see *acoustooptics; electrooptics; magnetooptics; optooptics.*

photonic switch. See *optical switch.*

photon noise. See *quantum noise.*

phototransistor. A transistor that produces electrical output *signals* corresponding to input incident *optical* signals. Phototransistors may be used as *photodetectors* and in other circuits, such as in *photodarlington circuits.*

photovoltaic. Pertaining to the capability of generating a voltage as a result of exposure to *electromagnetic radiation* in the *optical spectrum.*

photovoltaic effect. The production of an electromotive force, that is, a voltage, usually across a semiconductor *p-n junction* because of the *absorption* of *photon energy.* The electrical potential is caused by the diffusion of hole-electron pairs across the junction potential barrier, which the incident photons cause to shift or increase, leading to direct conversion of a part of the absorbed energy into usable electromotive force (voltage). The effect is used to produce voltages in nonhomogeneous semiconductors, such as silicon, or to produce voltages at a junction between two different types of material.

photovoltaic photodetector. A *photodetector* that makes use of the *photovoltaic effect.* A source of voltage is not needed for the photovoltaic photodetector, because it is a source of voltage.

physical layer. In *open-systems architecture,* the layer that provides the functions that are used or needed to establish, maintain, and release physical connections and to conduct *signals* between data terminal equipment, data circuit-terminating equipment, and switching equipment, that is, between *user end-instruments* and *end-point nodes,* between *switching nodes* and end-point nodes, or between switching nodes and control nodes. For example, *fiber optic* and electronic components can be used to implement the physical layer. Also see *Open-Systems-Interconnection (OSI) Reference Model (RM).*

physical optics. The branch of *optics* in which *light* is considered as a form of *wave* motion in which energy is propagated by *electromagnetic radiation* in the form of waves described by *wavefronts,* that is, as surfaces, rather than as rays as in *geomet-*

ric optics. The wavefront is perpendicular to the direction of *propagation* at every point, that is, it is perpendicular to the *Poynting vector.* The plane of the wavefront contains, and is defined by, the *electric* and *magnetic field vectors* at any given point. Synonymous with *wave optics.*

PHz. *Petahertz.*

picosecond. One-millionth of one-millionth of a second, that is, 10^{-12} s.

picowatt. One-millionth of one-millionth of a watt, that is, 10^{-12} W.

pigtail. See *optical fiber pigtail.*

PIN. *Positive-intrinsic-negative.*

PIN diode. A *junction* diode whose junction consists of three semiconducting materials joined in sequence in the forward-current conducting direction; the first of the three materials being *p-type semiconductor,* thereby creating holes; that is, acceptor sites; the second being undoped, that is, it is made of intrinsic material; and the third being *n-type* material, thereby creating free electrons, that is, donor sites. PIN diodes are used extensively as *photodetectors* in fiber optic systems and integrated optical circuits (IOCs).

PIN-FET integrated receiver. An *optical receiver* formed by combining a *PIN photodiode* and a field effect transistor (FET) in a single housing. The packaging is performed in such a way that the performance of the combination is better than that of the individual components employed as discrete components.

PIN photodiode. A *PIN diode* that changes its *electrical conductivity* in accordance with the *irradiance,* that is, the *optical power density* or intensity, and with the *wavelength* of incident *light.* The PIN photodiode has an intrinsic region between *p-type* and *n-type semiconductor* regions. *Photons absorbed* in this region create electron-hole pairs that are separated by an *electric field,* thus generating an electric current, that is, a *photocurrent.*

pipe. See *light pipe.*

plain old telephone service (POTS). Telephone service that provides voice and low-speed *digital data* services only. Wire pairs are used in *local loops.* Therefore, video, high-speed digital, and *broadband analog signals* cannot be carried. If POTS wires to homes and offices were replaced with *fiber optic cables,* all services could be provided with one *optical fiber.*

planar diffused waveguide. A *thin-film slab-dielectric waveguide,* usually 1 to 10 μ *(microns)* thick and usually with a *graded refractive-index profile* that is made by controlling the diffusion of *dopants* during its construction.

Planck's constant. A physical constant equal to 6.626×10^{-34} joule-seconds. It is usually designated by h. The energy of a *photon* is given by the relation:

$$E_p = hf$$

where f is the *frequency* of the *radiation* associated with the photon. The energy of a photon of 1 μ *(micron) wavelength* is about 1.2 eV (electron-volts). Electrons can exist only in certain energy levels. Therefore, only certain transitions between these levels can occur, giving rise to *absorption* and *emission* of only certain *quanta* of energy, as expressed by hf, where h is Planck's constant and f is the frequency of the absorbed or emitted photon. Therefore, only certain discrete wavelengths of *optical* radiation exist in a given *spectral line.* For example, photon energy is given by the relations:

$$PE = hf = hkc$$

where h is Planck's constant, k is the *wave number,* and c is the velocity of *light.* The wave number turns out to be the number of wavelengths per unit distance in the direction of *propagation.* The *work function* of a material is given by the relation:

$$W = hf_\circ$$

where h is Planck's constant and f_0 is the *threshold frequency* required to release an electron from the material. *Photocurrents* may be calculated from the relation:

$$I_{ph} = \Gamma eP/hf$$

where Γ is the carrier collection *quantum efficiency,* e is the electron charge, P is the incident *optical power,* h is Planck's constant, and f is the frequency of the incident radiation. The signal-to-noise ratio for a photodetector is given by the relation:

$$SNR = (N_p/B)(1 - e^{-hf/kT})$$

where N_p/B is the number of incident photons per unit of *bandwidth,* h is Planck's constant, f is the frequency of the incident photons, k is *Boltzmann's constant,* and T is the absolute temperature. The equation that relates photon energy, the work function of a material, such as certain metals and oxides, and the emitted-electron energy, that is, the kinetic energy with which an electron will be emitted by the material when the photon strikes it, is given by the relation:

$$hf = W + (1/2)mv^2$$

where h is Planck's constant, f is the frequency associated with the incident photon, W is the work function of the material the photon strikes, m is the mass of the electron, and v is the velocity with which the electron is ejected. In a *photoemissive detector,* the remaining energy of an electron that escapes from the emissive material due to the energy imparted to it by an incident photon, given by the relation:

$$E_e = hf - qw$$

where E_e is the remaining energy of the electron, h is Planck's constant, f is the photon frequency, q is the charge of an electron, and w is the work function of the emissive material. In quantum mechanics, the product of the total energy in a stream of photons and the time during which the flow occurs, expressed by the relation:

$$A = h \sum_{i=1}^{m} f_i n_i t_i$$

where f_i is the ith frequency; n_i is the number of photons of the ith frequency; t_i is the time duration of the ith frequency summed over all the frequencies, photons, and time durations of each in a given *light beam* or beam *pulse;* and h is Planck's constant.

Planck's law. The *quantum* of energy associated with an *electromagnetic field* is given by the relation:

$$E_p = hf$$

where E_p is the energy of each *photon, h* is *Planck's constant,* and f is the *frequency* of the *radiation* associated with the photon. The product of the energy and the time is sometimes referred to as the action. Hence, h is sometimes referred to as the elementary quantum of action. Planck's law is the fundamental law of quantum theory. It has direct application in *optical fiber communication systems.* It describes the concept of the particle, granular, or corpuscular theory of electromagnetic radiation, particularly of *light.*

plane. See *polarization plane.*

plane polarization. 1. In an *electromagnetic wave,* such as a *lightwave* or radio wave, *polarization* of the *electric* and *magnetic field vectors* in such a manner that they describe a plane that remains perpendicular to a constant direction of *propagation.* At a given instant, if successive planes were to be identified they would be parallel. Thus, the orientation of the plane is constant with respect to distance coordinates in the direction of propagation. The orientation of the plane at a point in space or in a *propagation medium* does not change with time except perhaps slowly with respect to *frequency.* A slowly rotating source may be considered to be capable of producing *plane-polarized electromagnetic waves* at each instant of time. **2.** The *polarization* of a *plane-polarized electromagnetic wave.*

plane-polarized electromagnetic wave. An *electromagnetic wave* whose *electric field vector* is contained in a plane that is perpendicular to a fixed direction of *propagation* so that successive planes in space remain parallel. Thus, *plane-polarized wave-*

fronts are planar rather than spherical. See *uniform plane-polarized electromagnetic wave.*

plane (electromagnetic) wave. An *electromagnetic wave* that predominates in the *far-field region* of an antenna, that has a *wavefront* that is essentially in a flat plane, and in which all the wavefront planes are parallel. Thus, a wave whose surface of constant phase are infinite parallel planes normal to the direction of *propagation.* A *uniform plane-polarized electromagnetic wave* propagating in *free space* or in *dielectric* media has a *characteristic impedance,* that is, the ratio between the *electric* and *magnetic field strengths,* given by the relation:

$$Z = E/H$$

in which E and H are orthogonal and are in a direction perpendicular to the direction of propagation, namely perpendicular to the *Poynting vector,* which is the vector obtained from the cross (vector) product of the *electric* and *magnetic field vectors,* with the direction of a right-hand screw obtained when rotating the electric vector into the magnetic vector over the smaller angle. In free space, the characteristic impedance of a plane wave is 377 ohms. From *Maxwell's equations,* the characteristic impedance of a plane wave in a *linear, homogeneous, isotropic, dielectric,* and electric-charge-free *propagation medium* is given by the relation:

$$Z = (\mu/\epsilon)^{1/2}$$

where μ is the *magnetic permeability* and ϵ is the *electric permittivity.* For free space, $\mu = 4\pi \times 10^{-7}\ H/m$, and $\epsilon = (1/36\pi) \times 10^{-9}\ F/m$, from which 120π, or 377, ohms is obtained. For dielectric media, the electric permittivity is the dielectric constant, which for glass, ranges from about 2 to 7. However, the *refractive index* is the square root of the dielectric constant. Thus, the characteristic impedance of dielectric materials is $377/n$, where n is the refractive index.

plasma-activated chemical-vapor-deposition (PACVD) process. A *chemical-vapor-deposition (CVD) process* for making *graded-index (GI) optical fibers* by depositing a series of thin layers of materials of different *refractive indices* on the inner wall of a glass tube as chemical vapors flow through the tube. A microwave *resonant cavity* is used to stimulate the formation of oxides by means of a nonisothermal plasma generated by the microwave cavity.

plastic-clad silica (PCS) fiber. An *optical fiber* consisting of a pure silica glass *core* with a plastic *cladding,* thus being a *stepped-refractive-index* fiber. *Optical power loss* and *dispersion* are generally higher in plastic-clad silica fibers than in all-glass fibers. Usually the plastic is a soft silicone material.

plastic fiber. See *all-plastic fiber.*

plotter. See *acceptance angle plotter.*

p-n junction. In a semiconductor device, such as a transistor or diode, the *interface* between *p-type* and *n-type semiconductor material,* thus creating holes, that is, acceptor sites, on one side of the interface, and relatively free electrons, that is, donor sites, on the other side of the interface. The junction is usually assumed to be abruptly or linearly graded in its transition region across the interface from the p-type to the n-type material, that is, from the positively *doped* to the negatively doped region. The holes are less mobile than the electrons. Both contribute to the total electric current.

Pockels cell. A material, usually a crystal, whose *refractive index* changes linearly with an applied *electric field,* the material being configured so as to be part of an *optical system.* Thus, the cell provides a means of *modulating* the *light* in the optical path in which it is placed. The modulation depends on the rotation of the *polarization plane* of the light in a *beam,* the rotation being caused by an applied electric field. The beam is passed through a *polarizer* that *transmits* only certain parts of the beam. The parts of the beam that are transmitted depend on the orientation of the polarization plane of the part. Lithium niobate can be used to make a Pockels cell.

Pockels effect. In *birefringent media,* an increase in their normal or natural *birefringence* caused by an applied *electric field.* The change is linearly proportional to the applied *electric field strength,* which causes a rotation of the *polarization plane* of a *plane-polarized electromagnetic wave.* When combined with *polarizers,* the effect can be used in *modulators* and other *active optical devices,* such as *optical switches.* Also see *Kerr effect.*

Pockels-effect sensor. A *fiber optic birefringent sensor* in which an applied *electric field* (1) produces *birefringence* in a material that is not ordinarily a *birefringent medium* without the applied field or (2) enhances birefringence in a material that is inherently birefringent. The voltage applied across the material causes *plane-polarized light propagating* within it to be resolved into two orthogonal vectors. The change in *phase* retardation between the two vectors (ellipticity), is directly proportional to the applied *electric field strength.* An *optical fiber polarizer* analyzes the output *beam,* causing *irradiance,* that is, intensity, *modulation* of *light* at a *photodetector.* The effect is found in particular crystals that are capable of advancing or retarding the *phase* of the induced *ordinary ray,* relative to the phase of the *extraordinary ray,* when the electric field is applied. Thus, the applied electric field strength controls the output of the photodetector.

polariscope. A combination of a *polarizer* and an analyzer used to *detect birefringence* in materials placed between them or to detect *polarization plane* rotation caused by materials placed between them.

polarization. The property of an *electromagnetic wave* describing the time-varying direction and amplitude of the *electric field vector;* specifically, the figure traced as a function of time by the extremity of the vector at a fixed location in space. In general, the figure is elliptical and is traced in a clockwise or counterclockwise sense. The commonly referenced *circular* and *linear polarizations* are obtained when the ellipse becomes a circle or a straight line, respectively. When observed along the direction of *propagation,* clockwise rotation of the electric vector is designated right-hand polarization, and counterclockwise rotation is designated left-hand polarization. See *circular polarization; elliptical polarization; left-hand circular polarization; left-hand helical polarization; plane polarization; right-hand circular polarization; right-hand helical polarization; rotating polarization.*

polarization diversity. 1. Pertaining to the ability to change the *polarization* of an *electromagnetic wave,* usually at the source of *radiation,* by changing the direction of the *polarization plane* or by changing the *linear, circular, elliptical,* or *helical polarization* of the wave. **2.** Pertaining to two or more types of *polarization* mixed in the same *light beam* or in the same *transmission.* **3.** Pertaining to any method of *transmission* and reception of *electromagnetic waves* in which the same information *signal* is transmitted and received simultaneously using orthogonally *polarized* waves with independent *propagation* characteristics. **4.** Pertaining to the ability to change the direction of *polarization* of an *electromagnetic wave,* usually at the source, by changing the horizontal or vertical polarization of the *transmitted* wave.

polarization-maintaining optical fiber. An *optical fiber* constructed in such a manner that the *polarization planes* of *lightwaves launched* into the fiber are maintained during *propagation* of the wave with little or no cross-coupling of *optical power* between the *polarization modes.* The fiber is designed to operate at specific *wavelengths,* such as 0.85, 1.30, or 1.55 μ *(microns).* Cross sections of polarization-maintaining fibers range from elliptical to rectangular. Synonymous with *polarization-preserving optical fiber.*

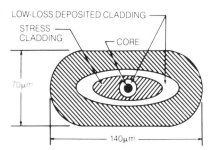

P-7. This **polarization-maintaining optical fiber** has a *depressed cladding* and a rectangular cross section and operates in *single-mode.* The stress cladding allows two orthogonal *polarizations* to *propagate* with minimal cross coupling. A protective *coating* is applied over the cladding to cushion the fiber against *microbending losses.* Applications include *fiber optic sensors,* gyroscopes, *coherent* communications and switching. (Courtesy AT&T.)

polarization modulation. The *modulation* or an *electromagnetic wave* in such a manner that the *polarization* of the electromagnetic wave, such as the direction of polarization of the *electric* and *magnetic field vectors,* or their relative *phasing,* to produce changes in polarization angle in linear, circular, or elliptical polarization, is varied according to a characteristic of an intelligence-bearing *signal,* such as a *pulse-*or*-no-pulse digital signal.* Polarization modulation can be accomplished in *waves propagating* in *waveguides.* For example, in *dielectric waveguides,* such as *optical fibers,* polarization shifts that are made in accordance with an input signal are a practical means of modulation.

polarization plane. In a typical *transverse electromagnetic (TEM) wave,* the plane defined by the *electric* and *magnetic field vectors* of the wave, that is, both field vectors at a point in space or in a material medium. They lie in and therefore define the polarization plane at any instant and any position. The direction of *propagation,* or *electromagnetic power* flow, of the wave is perpendicular to both the electric and magnetic field vectors at a point and therefore is perpendicular to the surface of the instantaneous polarization plane. In *geometric optics,* the unit vector that would represent the direction of the polarization plane is always and everywhere in the same direction as the *Poynting vector,* that is, in the direction of the *ray.* In *physical optics,* the *wavefront* at a point lies in the polarization plane.

polarization-preserving optical fiber. See *polarization-maintaining optical fiber.*

polarization sensitivity. In an *optical device,* the change in a performance *parameter,* such as the *transmittance* of an *attenuator* of the power *splitting* ratio of a *coupler,* per unit change in the angle of the *polarization plane.*

polarization sensor. In *fiber optics,* a *sensor* in which the *polarization plane* of an *electromagnetic wave,* such as a *lightwave* in an *optical fiber,* is rotated by a longitudinal *magnetic field* applied by a coil wrapped around the fiber. The total number of times the polarization plane is rotated when the fiber passes through the magnetic field is a function of the *magnetic field strength* and the distance that the fiber is in the field. A change in the number of polarization plane rotations produced by a change in the electric current level in the coil can be *detected* by a *photodetector* placed after a polarization analyzer, the output of which is a maximum when the analyzer grating is parallel to the polarization plane and is zero when perpendicular. A finite number of output *signal* peaks occurs for a finite change of current in the coil. The instantaneous *pulse-repetition rate* or the *frequency* of the output pulses is proportional to the rate of change of the current in the coil. Thus, the time derivative of an *amplitude-modulated* input electrical signal to the coil results in a *frequency-modulated* output signal from the photodetector.

polarized electromagnetic wave. See *left-hand polarized electromagnetic wave; plane-polarized electromagnetic wave; right-hand polarized electromagnetic wave; uniform plane-polarized electromagnetic wave.*

polarized mode. See *linearly polarized (LP) mode.*

polarized wave. See *clockwise-polarized wave; counterclockwise-polarized wave.*

polarizer. An *optical device* capable of transforming unpolarized *light,* that is, *diffused* or *scattered* light, into *polarized* light, or capable of altering the type of polarization of polarized light. See *optical fiber polarizer.*

polychromatic radiation. *Electromagnetic radiation* consisting of two or more *frequencies* or *wavelengths.*

polymeric nonlinear optical material. A *transparent* polymer, such as a polymer that functions as a third-order nonlinear material, that displays *nonlinear optical* properties when energized by *lightwaves,* e.g., a polymer that produces second and higher order harmonic frequencies when energized with single-frequency lightwave *signals.* The generation of higher harmonic frequencies means switching speeds of fewer than 100 *femtoseconds* and less. The process is also *loss*less and thus can be repeated many times. An exit wave can be *frequency*-mixed, and output *beams* can be considerably different from input beams. For example, at the exit point of an *integrated optical circuit waveguide* made of polymeric nonlinear optical material, different signals can be switched to different ports. Polymeric materials are being used to manufacture *optical modulators, multiplexers,* and *switches.*

population inversion. A redistribution of electron *energy levels* in a population of atoms in chemical elements and molecules in chemical compounds, such that instead of having more atoms with lower-energy-level electrons, there are fewer atoms with higher-energy-level electrons. Thus, an increase in the total number of electrons in the higher excited states occurs at the expense of the energy in the electrons in the lower and ground states and at the expense of the resonant energy source, that is, the pump. Population inversion is not an equilibrium condition. Population inversion is brought about by, and must be maintained by, the pumping action of an energy source. When population inversion occurs, the probability of downward energy transitions, giving rise to *radiation,* is greater than the probability of upward energy transitions, giving rise to *photon absorption.* This results in a net output radiation level, thus obtaining stimulated emission, that is, the *laser action* that occurs in a laser. Also see *pump frequency.*

port. 1. In a *communication network,* a point at which *signals* can enter or leave the network enroute to or from another network. For example, the point in a shipboard *data transfer network (DTN)* at which a ship-to-shore communication *link* can be connected is considered as a port in the DTN. **2.** In an *optical fiber, cable,* or *bundle,* a point at which *signals* can enter or leave. **3.** A place of access to a device or *network* where energy may be supplied or withdrawn or where the device or network variables may be observed or measured. (See Fig. P–8, p. 244)

positive-intrinsic-negative (PIN) diode. See *PIN diode.*

positive-intrinsic-negative (PIN) photodiode. See *PIN photodiode.*

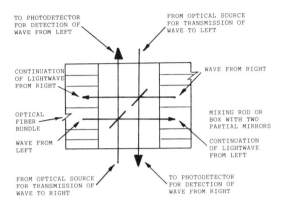

P-8. *Bundles* of *optical fibers butt-coupled* to an *optical mixing rod,* showing entry and exit **ports.**

POTS. *Plain old telephone service.*

powder. See *glass powder.*

power. See *chromatic resolving power; noise-equivalent power (NEP); optical power; optical receiver maximum input power; optical repeater power; peak spectral power; radiant power; resolving power; source-to-fiber coupled power; theoretical resolving power; transmitted power; transmitter optical power.*

power budget. In a system, the allocation of available power, such as *optical* or electrical *power,* among the various functions that need to be performed and the various *losses* that have to be sustained. See *optical power budget.*

power density. See *irradiance.*

power distribution. See *core-cladding power distribution; nonequilibrium modal power distribution.*

power distribution length. See *nonequilibrium-modal-power-distribution length.*

power efficiency. See *optical power efficiency; source power efficiency.*

power-law refractive-index profile. For a round *optical fiber,* the variation of *refractive index* of the *core* as an exponential function of the distance from the *optical axis,* that is, the variation is such that the refractive index within the core at any distance from the axis, r, is given by the relation:

$$n_r = n_1(1 - br^g)^{1/2}$$

where n_1 is the refractive index at $r = 0$, that is, at the axis; r is the radial distance from the axis, g is the *refractive-index profile parameter* that determines the shape of the *refractive-index profile,* and b is a constant given by the relation:

$$b = 2\Delta/a^g$$

where a is the value of r for which the *refractive index* becomes uniform, that is, the radius of the *core,* and Δ is given by the relation:

$$\Delta = (n_1{}^2 - n_2{}^2)/2n_1{}^2$$

where n_1 again is the refractive index at $r = 0$, that is, at the axis, and n_2 is the refractive index at the outer edge, that is, at $r = a$. For most optical fibers, the indices are nearly equal. Therefore Δ is also given by the relation:

$$\Delta = 1 - (n_2/n_1)$$

where the *parameters* are as defined above. Because $n_r = n_2$ at $r = a$, this relation: is also given by the equivalent relation:

$$n_r = n_1[1 - 2\Delta \ (r/a)^g]^{1/2}$$

Another form of the equation, approximately the same as those above, is given by the relation:

$$n_r^2 = n_1{}^2[1 - 2\Delta \ (r/a)^g]$$

where the parameters are as defined above. The profile parameter, g, defines the shape of the profile. The Δ is the *refractive-index contrast* when the refractive index of the *cladding* is constant. Nevertheless, in all these expressions, if the g is about 2.25, *intermodal dispersion* will be minimized or nearly eliminated for most fibers. If $g = 2$, a *parabolic index profile* results. If $g = 1$, a straight line, that is, a triangular variation of refractive index as a function of radial distance from the axis, occurs. If g is infinite, the core refractive index is constant and equal to n_1, which is the case for the *step-index fiber.* Also see *g; linear refractive-index profile; parabolic refractive-index profile; radial refractive-index profile.*

power level. At any point in a *transmission* system, the ratio of the power at that point to some arbitrary amount of power chosen as a reference. This ratio is usually expressed either in *decibels* referred to 1 milliwatt, dBm, or in decibels referred to 1 watt, dBW.

power margin. See *safety margin.*

power output. See *optical power output.*

power penalty. See *dispersion power penalty; reflection power penalty.*

power safety margin. See *safety margin.*

Poynting vector. In an *electromagnetic wave,* the resulting vector obtained from the cross-product, that is, the vector product, when the *electric field vector* is rotated into the *magnetic field vector* through the smaller angle, the resulting vector *propagating* in the direction of the end face of a right-hand screw. The Poynting vector, with *propagation media parameters* and physical constants, defines the *irradiance,* that is, the power density, and the direction of propagation of the wave.

precision-sleeve splicer. A round tube that has a round hole with a diameter equal to the outer diameter of two *optical fibers* to be *spliced,* that contains an *index-matching material,* such as an epoxy, into which the two fibers may be inserted from opposite ends. If necessary, the ends of the sleeve may be crimped to hold the fibers tightly, at least until the material cures. The material may be an *optical cement.*

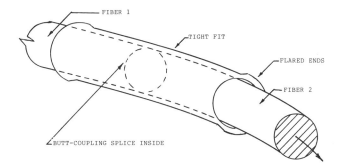

P–9. A **precision-sleeve splicer** for *butt-coupling* two *optical fibers.*

preform. See *optical fiber preform.*

prelasing condition. The operating condition of a *laser* in which its *radiation* is primarily spontaneous and not *coherent.*

presentation layer. In *open-systems architecture,* the layer that provides the functions, procedures, services, and protocol that are selected by the *application layer.* For example, the functions may include *data* definition and control of data entry, data exchange, and data display. It is directly below the application layer in the *Open-Systems-Interconnection (OSI) Reference Model (RM).*

primary coating. In the manufacturing of *optical fibers,* a *coating* that is placed in contact with the fiber *cladding.* It is applied to preserve and protect the cladding outer surface.

principle. See *Fermat's principle.*

procedure. See *fiber optic test procedure (FOTP).*

process. See *axial vapor-phase-oxidation (AVPO) process; cabling process; chemical-vapor-deposition (CVD) process; chemical vapor-phase-oxidation (CVPO) process; doped-deposited silica (DDS) process; double-crucible (DC) process; inside vapor-phase-oxidation (IVPO) process; ion-exchange process; modified chemical-vapor-deposition (MCVD) process; molecular stuffing process; plasma-activated chemical-vapor-deposition (PACVD) process; vapor-phase axial-deposition (VAD) process.*

processing. See *optical data processing.*

processing center. In a switched communication *network,* a point at which *signal* and *data* processing is accomplished for the network or for users of the network. A processing center is usually located at a *node* in the network.

product. See *bit-rate-length product.*

profile. See *equivalent step-index (ESI) profile; graded-index profile; linear refractive-index profile; parabolic refractive-index profile; power-law refractive-index profile; radial refractive-index profile; refractive-index profile; step-index profile; uniform refractive-index profile.*

profile dispersion. In a *waveguide,* the *dispersion* caused by variation of the *refractive-index profile* with *wavelength* of the waves *propagating* in the guide. In an *optical fiber,* the profile variation also can be caused by variation in the *refractive-index contrast* and variation in the *profile parameter.*

profile-dispersion parameter. In an *optical fiber,* the *parameter* that characterizes the part of the *refractive-index profile dispersion* caused by a variation of the *refractive-index contrast* with *wavelength,* given by the relation:

$$P(\lambda) = (n_1/N_1)(\lambda/\Delta)(d\Delta/d\lambda)$$

where P is the profile dispersion parameter, λ is the *source,* that is, *free space,* wavelength of the *light,* n_1 is the maximum *refractive index* of the *fiber core,* N_1 is the *group index* corresponding to n_1, where N_1 is given by the relation:

$$N_1 = n_1 - \lambda(dn_1/d\lambda)$$

and Δ is the refractive-index contrast, given by the relations:

$$\Delta = (n_1^2 - n_2^2)/2n_1^2 \text{ or } (n_1 - n_2)/n_1$$

for optical fibers in which the refractive indices are not different from each other by more than 1%.

profile parameter. See *refractive-index profile parameter.*

proofed fiber optic cable. See *Tempest-proofed fiber optic cable.*

propagation. The motion or passage of waves within or through a *propagation medium.*

propagation constant. See *transverse propagation constant.*

propagation delay. In an *optical device,* such as a *fiber optic transmitter, receiver, cable,* or *coupler,* the delay between the leading edge of an input *signal,* such as an *optical pulse,* and the leading edge of the corresponding output signal.

propagation constant. In an *electromagnetic wave propagating* in a *waveguide,* such as an *optical fiber* or a metal pipe, the factor in the expression for the exponentially varying characteristics of the wave given by the relation:

$$e^{-pz} = e^{-ihz-az}$$

in which the term $p = ih + a$ includes both the *phase term, h,* and the *attenuation term, a,* governing the propagation characteristics of the wave in the guide, and z is the longitudinal distance. *Dispersion* occurs because the propagation constant is a function of *frequency,* as well as a function of the materials of construction of the guide. Synonymous with *axial propagation constant.*

propagation medium. Any material substance that can be used for the *propagation* of *signals,* usually in the form of *modulated* radio, *light,* or acoustic waves, from one point to another, such as metals for electrical current signals, glass and other *dielectric* materials for *lightwave* signals, and air for sound wave signals, with the possible exception that vacuum is considered as a propagation medium for *electromagnetic wave* signals, such as light, radio, video, and microwave signals. Except for electromagnetic *transmission* in vacuum, the propagation medium is usually shaped, or available in specific forms in order to guide the energy in a signal from a point of dispatch to a point of reception. Shaped media include *optical fibers, cables,* or *bundles;* wires, including coaxial cables; *slab-dielectric waveguides;* and the sea-atmosphere interface. Synonymous with *transmission medium.* See *anisotropic propagation medium; isotropic propagation medium.*

propagation mode. 1. One of the possible electric and magnetic field configurations in which *electromagnetic* energy *propagates* in a *waveguide* or along a *transmission* line. The mode is an allowable electromagnetic field condition, distribution, or configuration that can exist in a waveguide relative to the direction of propagation of the wave in the guide. Each mode has a factor, that is, an eigenvalue, that defines the *propagation constant* but not the *attenuation term* in the *wave equation* for the discrete mode. The entire field can be described in terms of these modes. *Bends* and discontinuities may lead to mode conversion, that is, the transfer of energy from mode to mode, but not necessarily to energy that would *radiate* away from the waveguide. The number of modes a waveguide can support is dependent upon the dimensions of the guide, the *wavelength,* and the *refractive indices* of the *propagation medium.* With proper choice of waveguide size, *refractive index,* and operating wavelength, single-mode transmission can be achieved. In an *open waveguide,* that is, a *dielectric waveguide* such as an *optical fiber, evanescent fields* are established in a transverse plane. These modes are guided by the gradient of the refractive index. **2.** In radio *transmission,* the manner in which radio *signals propagate* from a transmitting antenna to a receiving antenna, such as by ground wave, sky wave, direct wave, ground reflection, or *scatter.*

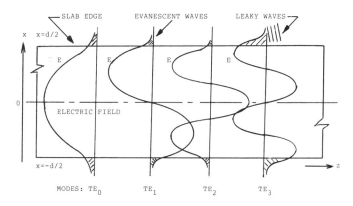

P-10. *Lower order* **propagation modes** in a *slab-dielectric (planar) waveguide,* showing *evanescent waves* outside the guide, but *coupled* to waves inside the guide, and *leaky waves* outside the guide, but becoming decoupled from the waves inside the guide. The leaky waves will *radiate* into space. They are usually associated with the *higher order modes.*

protective coating. See *optical protective coating.*

protective covering. In *fiber optics,* a covering placed over a component, such as a *fiber optic cable, splice,* or *connector,* usually in the form of a wrapped tape or bonded *coating* designed to protect the device, against the environment, such as humidity, abrasion, and *bending.* Caps, dust covers, and similar devices are also considered protective covering.

proximity coupling. In *fiber optics,* the transfer of *radiant energy,* that is, *optical power,* from one *optical fiber* to another by stripping their *cladding* for a short distance and placing their *cores* close together, that is, adjacent to each other for the stripped length. The amount of power that is transferred can be controlled by the stripped length and the proximity of the fibers.

PRR. *Pulse-repetition rate.*

p-type semiconductor. A semiconductor material, such as silicon or germanium, that has been *doped* with minute amounts of acceptor-type material, that is, material with a chemical valence that creates molecular centers lacking an electron to complete their energy shells. These centers will attract an electron from their neighbors, leaving a hole among the neighbors. These holes migrate when electrons moving under the influence of an *electric field,* albeit a relatively weak field, fill the holes, thus constituting an electric current that is oppositely directed to the flow of electrons, that is, the flow constitutes a positive electric current. Thus, the *dopant* creates holes, or positive centers, that trap electrons; hence the material is called p-type material. The positive current constituted by the migrating holes must be added to the oppositely directed negative current of electrons when calculating the total electric current. Also see *n-type semiconductor.*

pulling machine. See *fiber-pulling machine.*

pulse. A temporal or spatial variation, usually characterized as a rise and decay from a baseline or reference, of the magnitude, such as *amplitude, phase, frequency,* or other *parameter* of a physical quantity, such as the *electric field strength* of an electromagnetic wave, the variation being short relative to the time or space schedule of interest. In *fiber optics,* the physical quantity might be a *lightwave propagating* in an *optical fiber* and the pulse is a sudden increase in *irradiance,* that is, *optical power density* or electric and *magnetic field strength,* a sudden shift in phase, a sudden change in frequency, or a change in *polarization,* and a return to the original state that existed before the change occurred. The pulse magnitude is the extent to which a parameter deviates from a reference. Thus, a pulse may have a spatial width measured in microns; a temporal width measured in *nanoseconds;* a *bandwidth* measured in *hertz;* or a phase shift measured in radians. Workers in the field tend to use pulse length or width for spatial dimensions and *pulse duration,* length, or width for temporal dimensions. *Pulse modulation* is the creation of pulse patterns or sequences to represent *digital data* or discrete samples of *analog signals.* Pulses are usually coded or *modulated* and *transmitted* in streams at high *data signaling rates (DSRs),* that is, high *pulse-repetition rates,* to represent and transport information. See *Gaussian pulse.*

pulse amplitude. A measure of the extent to which a physical quantity, such as *optical power,* electric current, or voltage used to represent a *pulse* changes from a zero or other baseline value for the *pulse duration.* For example, the momentary change in optical power output of a *laser* to represent a single pulse or the amount of the change in *photocurrent* of a *photodetector* from the *dark current* value to the *light* current value. Often it is necessary to use modifiers, such as average, instantaneous, peak, and root-mean-square, to indicate the significance of the units of measure used to define the pulse amplitude. Synonymous with *pulse height.* Also see *pulse magnitude.*

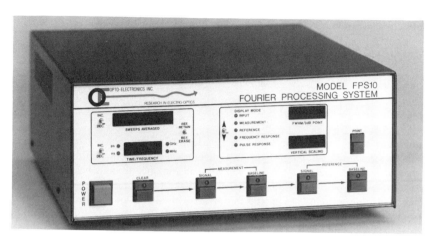

P-11. The FPS10 Fourier Processing System, an instrument that performs measurements on an *optical fiber* or any other passive or active device and converts these measurements into a *frequency* response curve and a **pulse** response curve. The conversions are carried out using highly accurate fast Fourier transform and inverse transform algorithms. The FPS10 is also an efficient *signal* averager, capable of improving the signal-to-noise ratio of detected signals up to 220X. (Courtesy Opto-Electronics, Inc.).

pulse-amplitude modulation (PAM). A form of *modulation* in which the *amplitude* of the individual *pulses* of a constant-*pulse-repetition-rate carrier,* such as the amplitude of the individual pulses in a stream of *light* pulses, is varied in accordance with some characteristic of the modulating *signal.*

pulse broadening. Increasing the *pulse width* (spatial) or *pulse duration* (time) of a pulse as a result of *dispersion.* The *distortion* is caused by the *spreading* in time and space of an electromagnetic pulse, such as an optical pulse, *propagating* along a *propagation medium* or path, such as an electrical *transmission* line or *optical fiber.* The dispersion is caused by the variation of the *propagation constant* for each *wavelength* in the signal pulse, that is, the different wavelengths propagate at different speeds. This phenomenon limits the useful transmission *bandwidth* of the path or, for a given bandwidth, limits the *data signaling rate (DSR),* that is, limits the *pulse-repetition rate* that may be transmitted because of the *intersymbol interference* that occurs when the *pulse duration* at the receiving end increases to a point where pulses begin to overlap each other upon arrival. For an optical fiber, the amount of spreading depends on the *spectral width* of the *light,* the *refractive-index profile* of the *fiber,* the length of the fiber, and the duration of the *launched* pulse. Because of the dependence on length, a significant measure of optical fiber performance is the *bandwidth-length product.* Pulse broadening may be specified by the impulse *response* or by the *full-duration-half-maximum* pulse broadening. Synonymous with *pulse dispersion; pulse spreading.*

pulse compression. See *optical fiber pulse compression.*

pulse-decay time. See *fall time.*

pulse dispersion. See *pulse broadening.*

pulse distortion. See *optical pulse distortion.*

pulse duration. The time interval that a *pulse* lasts to represent a binary digit. For example in *amplitude modulation,* the time interval between the points on the leading and trailing edges of a pulse at which the instantaneous value bears a specified relation to the peak pulse amplitude, such as the time interval between the *full-wave-half-power* points in rise and fall or between the 10% values in rise and fall; or in *frequency* or *phase*-shift *modulation,* the period during which the frequency or the phase remains changed to represent a binary digit. Because pulses have a duration at a point, that is, they have a temporal width, and they are *propagating* at a specific speed, they also have a spatial width, that is, they occupy a space in the *propagation medium.* Synonymous with *pulse length; pulse width.* See *root-mean-square (rms) pulse duration.*

pulse half-duration. In a *pulse,* half of the full duration of the pulse, the full duration being defined as the time interval between specified points on the pulse waveform, such as between the 90%, half, or $1/e$ maximum value points. Synonymous with *pulse half-width.*

pulse half-width. See *pulse half-duration.*

pulse height. See *pulse amplitude.*

pulse jitter. The *jitter* of a *pulse parameter,* such as the jitter of the *pulse width,* pulse time or space position, or *pulse amplitude.*

pulse length. See *pulse duration.*

pulse magnitude. A measure of the extent to which the physical *parameter,* quantity, or phenomenon used to represent a *pulse* changes from a baseline or reference for a short time, such as the amount of change in *optical power* output of a *laser,* the extent of the *phase* shift of a *monochromatic wave,* or the extent of a *frequency* change in a *frequency-modulated* system. Often it is necessary to use modifiers, such as average, instantaneous, peak, and root-mean-square, to indicate the significance of the units or measure used to define the magnitude of the pulse. Also see *pulse amplitude.*

pulse modulation. The temporal changing of one or more characteristics or *parameters,* such as *pulse duration,* pulse position, or *pulse amplitude* of a *carrier wave* in accordance with an intelligence-bearing *signal.* For example, in *fiber optics, light-*

wave carrier pulse modulation can be accomplished by an intelligence-bearing electrical modulating signal sent to the *light source,* or by modulating the constant lightwave carrier output of the source after *emission* when the lightwave is in an *optical fiber* and the fiber is subjected to a continuously varying or pulsating physical variable, such as pressure, sound, temperature, or interferometric variations.

pulse rate. See *maximum pulse rate.*

pulse-repetition rate (PRR). The number of *pulses* that occur per unit of time at a particular point in a *propagation medium,* usually expressed as pulses per second. In *fiber optic systems,* desired *bit-error ratios, dispersion, attenuation,* available power, *detector sensitivity, noise* levels, and other factors limit the pulse-repetition rate that a given transmission system can handle. Also see *data signaling rate (DSR).*

pulse rise time. See *rise time.*

pulse spreading. See *pulse broadening.*

pulse stretching. See *optical pulse stretching.*

pulse suppressor. See *optical pulse suppressor.*

pulse width. 1. The spatial length or the duration of a *pulse.* 2. See *pulse duration.*

pulse-width distortion. 1. In an *optical fiber,* the difference between the *pulse width* of an *optical input pulse* and the width of the corresponding optical output pulse at the distal end of the fiber. 2. In a *fiber optic transmitter,* the difference between the width of an *optical input pulse* and the width of the corresponding electrical output pulse. 3. In an *optical receiver,* the difference between the *pulse width* of an *optical input pulse* and the width of the corresponding optical output pulse. 4. In an *optical link,* the difference between an electrical input *pulse* at the transmitter and the electrical output pulse at the receiver.

pump frequency. The *frequency* of an oscillator used to provide the sustaining power to certain specially designed devices, such as parametric amplifiers or *lasers.* Pumps usually are designed to provide resonant frequencies to raise *power levels* in the devices they pump energy into. Also see *population inversion.*

Q

quadratic refractive-index profile. See *parabolic refractive-index profile.*

quadruply clad fiber. An *optical fiber* construction that has four *claddings* surrounding the *fiber core.* The core has a relatively very high *refractive index* compared with the four fiber claddings, which are of very low, high, low, and medium refractive-index materials, respectively, from the core radially outward. The quadruply clad fiber is usually designed so as to achieve *single-mode* operation.

qualification testing. Formal testing designed to demonstrate that the software and hardware of a system meet specified requirements. Qualification testing may be accomplished at any time during the life of a system, such as during prototype development, manufacturing, shipment, storage, installation, and operation. Most often the qualification testing is conducted to determine the extent to which a system passes a specified set of performance criteria.

quality factor. See *intrinsic quality factor (IQF).*

quantization. A process in which the continuous range of values of a *signal* is divided into nonoverlapping, but not necessarily equal, subranges; and, to each subrange, a discrete value of the output is uniquely assigned. Whenever the signal value falls within a given subrange, the output has the corresponding discrete value.

quantum. A unit of *electromagnetic* energy equal in magnitude to *hf,* where h is *Planck's constant* and f is the *frequency* of the *radiation.* A quantum of electromagnetic energy is released when an electron in an excited or radioactive chemical element moves from a higher to a lower *energy level.* A *photon* is a quantum of electromagnetic energy in the *optical spectrum.* The lowest energy photon would have the lowest frequency, that is, the longest *wavelength,* at the extreme end of the *far-infrared* region of what is considered to be the optical spectrum, 100 μ *(microns),* corresponding to a frequency of 3 THz, or 3×10^{12} Hz. The highest energy photon wave length is 0.001 μ, which corresponds to 0.3 EHz *(exahertz)* or 0.3×10^{18} Hz. The energy of a 1-μ photon is about 1.2 eV.

quantum efficiency. In a quantum device, the ratio of the countable elementary events at the output to the countable elementary events at the input. The ratio defines the device *transfer function* for the countable events. For example, in an *optical* semiconductor *source,* such as *photodiode,* it is the ratio of the number of photons emitted to the number of electrons applied by an input electrical *pulse;* in a *photodetector,* it is the ratio of the number of electrons generated in the *photocurrent* to the number of photons applied by an input *optical pulse.* See *differential quantum efficiency; response quantum efficiency.*

quantum-limited operation. In the operation of a *photodetector,* the inability of the detector to measure *incident radiation* levels below a *threshold level* because of fluctuations in the output current that are not due to the incident *photons,* such as *dark currents* and *noise.*

quantum noise. *Noise* attributed to the discrete or particle nature of *light,* that is, light in the form of streams of *photons.* The *absorption* of each photon produces a discrete contribution to the total noise. Synonymous with *photon noise.*

quantum-noise-limited operation. The condition that prevails in a device or system when *quantum noise* at a point in the device or system limits its performance or the performance of the device or system to which it is connected. For example, the condition that prevails in an *optical link* when *quantum noise* is the predominant mechanism that limits link performance.

quartz. See *fused quartz.*

R

rad. The basic unit of *radiation* absorbed dose (rad) that produces ionization of the material upon which it is incident. A dose of 1 rad is equivalent to the *absorption* of 100 ergs of *radiant* energy per gram of absorbing material. An erg is a dyne-centimeter. A dyne is 1/980th of a gram of force, or about 10^{-5} newtons. A joule is 10^7 ergs. Also see *radiation hardness*.

radial refractive-index profile. In an *optical fiber* with a circular cross section, the *refractive index* described as a function of the radial distance from the center and the index at the center, that is, a function whose general form may be described by the relation:

$$n_r = n_0 f(r)$$

where n_r is the refractive index at a radial distance r from the center, n_0 is the refractive index at the center, and $f(r)$ is the function of r that expresses the index at the distance r from the center. Zero-point symmetry, that is, axial symmetry, of the *refractive-index profile* is assumed. Also see *linear refractive-index profile; parabolic refractive-index profile; power-law refractive-index profile*.

radiance. *Radiant power,* in a given direction, per unit solid angle per unit of projected area of the source, as viewed from the given direction. Specifically, radiance at a given point is (1) the radiant power in a given direction at the given point on a real or imaginary surface transmitted by an elementary *beam* passing through the given point and *propagating* in the solid angle containing the given direction; divided by (2) the product of the value of this solid angle, the cross-sectional area of that beam containing the given point, and the cosine of the angle between the normal to that section and the direction of the beam. The surface may be any surface that emits, intersects, or receives a light beam. Radiance is usually expressed in watts per steradian per square meter. Synonymous with *brightness*. See *spectral radiance*. Also see *irradiance; peak-radiance wavelength*.

radiance conservation law. A *passive optical device* or *optical system* cannot increase the quantity given by the relation:

$$Q = L/n^2$$

where L is the *radiance* of a *beam* and n is the local *refractive index*. Q would be constant if losses, such as losses caused by *absorption* and *scattering*, were zero. Synonymous with *brightness conservation; brightness theorem*.

radiant efficiency. In an *optical system*, the ratio of (1) the forward useful *radiant power* from a *source*, that is, the *radiance* integrated over the total forward solid angle from a source to (2) the total power input to the source. Thus, the radiant efficiency is a measure of the percentage of the input power that is converted into useful *optical power output*. Also see *radiation efficiency*.

radiant emittance. 1. *Radiant power emitted* into a full sphere, that is, 4π steradians, per unit area of a *source*, expressed in watts per square meter. **2.** The *radiant power emitted* per unit area of a source. For example, radiant emittance at a point is (1) the radiant power leaving an element of the surface containing the point divided by (2) the cross-sectional area of that element. Synonymous with *radiant exitance*.

radiant energy. Energy *emitted*, transferred, or received in the form of *electromagnetic waves*, that is, the time integral of *radiant power*, usually expressed in joules. Radiant energy is not considered to involve the motion or action of material matter to achieve its *propagation*, in contrast to elastic, kinetic, and some forms of potential energy.

radiant exitance. See *radiant emittance*.

radiant flux. Deprecated term. See *irradiance; radiance; radiant power*.

radiant intensity. Deprecated term. See *radiant emittance*.

radiant power. The time rate of flow of *radiant energy*, usually expressed in watts. If the radiant power is incident upon a surface, the amount of radiant power distributed per unit area of the surface is the *irradiance*, that is, the power density. Thus, the radiant power in a *beam* is the total power obtained by integrating the irradiance over the cross-sectional area of the beam. If the radiant power in a beam is distributed uniformly over the cross-sectional area of the beam, the irradiance is the radiant power divided by the cross-sectional area, which is also the average irradiance. For an *electromagnetic wave* propagating in a vacuum or a *propagation medium*, the radiant power propagating per unit area in the direction of maximum power gradient is also the irradiance, which is usually expressed in watts per square meter. Because the wave is propagating with a velocity given by the relation:

$$v = c/n$$

where c is the speed of light in a vacuum, and n is the *refractive index* of the material at the point at which the speed is being considered, the irradiance can also be expressed in terms of the energy per unit volume of vacuum or propagation medium, expressed as joules per cubic meter, hence giving rise to the term "electromagnetic energy density." A convex lens a few centimeters in diameter and a centimeter thick can increase the irradiance of solar *radiation,* that is, sunlight, when focused on the surface of a combustible material, to a level high enough for combustion, but the total radiant power, that is, the radiance, at the lens and at the material is the same. Flux is a deprecated synonym. Also see *irradiance.*

radiation. Energy in the form of *electromagnetic waves* or *photons.* See *coherent radiation; electromagnetic radiation (EMR); gamma radiation; incoherent radiation; light amplification by stimulated emission of radiation (laser); optical radiation; polychromatic radiation; secondary radiation; visible radiation.*

radiation angle. In *fiber optics,* half the vertex angle of the cone of *light emitted* at the exit face of an *optical fiber.* The cone is usually defined by the angle at which the *far-field irradiance* has decreased to a specified fraction of its maximum value or as the cone within which can be found a specified fraction of the total *radiant power* at any point in the far field. Synonymous with *output angle.*

radiation damage. See *optical fiber radiation damage.*

radiation efficiency. The ratio of the *radiant emittance,* that is, the total *radiant power emitted* by a *source* or radiator, to the total power supplied to the radiator. The radiation efficiency of a source or radiator is usually quoted at a given *frequency* or *wavelength.* Also see *radiant efficiency.*

radiation hardness. A measure of the ability of a system, such as a *fiber optic* system for communications; an *optical fiber sensor;* or a computer, control, or data processing system, to function satisfactorily after exposure to a given *irradiance* level, that is, *radiant power* distributed over a given area of the system, for a specified time or after absorbing a given number of *rads,* that is, radiation dose. Also see *rad.*

radiation mode. In an *optical waveguide,* a *mode* that transfers *radiant power* in a direction transverse to the direction in which power is intended to be transferred or guided, whose *electric* and *magnetic fields* are transversely oscillatory everywhere external to the waveguide, and whose fields exist even at the limit of zero *wavelength.* Radiation modes correspond to *refracted rays.* A radiation mode satisfies the relation:

$$\beta \leq [n_a^2 k^2 - (\ell/a)^2]^{1/2}$$

where β is the imaginary part (*phase term*) of the axial *propagation constant,* n_a is the *refractive index* at the radial distance a from the *optical axis* at which β is taken,

ℓ (an integer) is the azimuthal index of the mode at a, and k is the *free-space wave parameter*, equal to $2\pi/\lambda$, where λ is the *wavelength*. The symbol k is also used to represent *wave number*, equal to $1/\lambda$.

radiation pattern. 1. In *fiber optics*, the relative *radiant power* distribution, as a function of position or angle, *emitted* from a surface, such as a *light source* or the exit face of an *optical fiber*. The *near-field radiation pattern* is a description of the *radiant emittance* as a function of position in the plane of the exit face of an optical fiber. The *far-field radiation pattern* describes the *irradiance* as a function of angle in the *far-field region* of the exit face of an optical fiber. The radiation pattern may be a function of the length of the *waveguide*, the manner in which it is excited, and the *wavelength*. **2.** The variation of the *radiant emittance* of an antenna as a function of direction. The pattern is normally represented graphically for the *far-field* conditions in either horizontal or vertical planes. See *equilibrium radiation pattern; far-field diffraction pattern; near-field radiation pattern*.

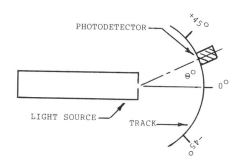

R–1. A **radiation pattern** plotter.

radiation scattering. The *scattering* of *radiant power*, such as *radiation* from thermal, *electromagnetic*, or nuclear *sources*, from its original path as a result of interactions or collisions with atoms, molecules, or large particles in the atmosphere, *waveguides*, or other *propagation media* between the source of radiation, such as a *light-emitting diode (LED)*, *laser*, or nuclear explosion and a point some distance away. As a result of *scattering, radiation*, that is, *radiant power*, expecially neutron and *gamma radiation*, will be received at such a point from many directions instead of only from the direction of the source. In an *optical fiber*, scattering contributes to *signal attenuation* and *pulse broadening*, that is, pulse spreading.

radiator. See *Lambertian radiator*.

radiometry. The branch of science and technology devoted to *radiation* measurement. See TABLE OF RADIOMETRIC TERMS on page 260.

radius. See *critical radius*.

Table of Radiometric Terms

TERM	SYMBOL	QUANTITY	UNIT
Radiant energy	Q	Energy	joule (J)
Radiant power (optical power)	P_o	Power	watt (W)
Irradiance	E	Power incident per unit area irrespective of angle	$W \cdot m^{-2}$
Spectral irradiance	E_λ	Irradiance per unit wavelength interval at a given wavelength	$W \cdot m^{-2} \cdot nm^{-1}$
Radiant emittance (radiant exitance)	W	Power emitted (into a full sphere) per unit area	$W \cdot m^{-2}$
Radiant intensity	I	Power per unit solid angle	$W \cdot sr^{-1}$
Radiance	L	Power per unit solid angle per unit projected area	$W \cdot sr^{-1} \cdot m^{-2}$
Spectral radiance	L_λ	Radiance per unit wavelength interval at a given wavelength	$W \cdot sr^{-1} \cdot m^{-2} \cdot nm^{-1}$

Raman scattering. The generation of many different *wavelengths* of *light* from a single-*wavelength source* by means of *laser action* and interaction with molecules, thereby creating many different excited molecular energy levels that will produce *photons* of various energies, which correspond to various wavelengths when transitions to lower excited states occur. In addition, when two *frequencies* beat together, they induce dipole moments in molecules at the difference frequency. This causes *modulation* of laser-molecule interaction, which produces light at sideband frequencies, that is, additional wavelengths.

range. See *dynamic range; measurement range; transmitter central-wavelength range.*

range designation of data signaling rates (DSRs). A method of referring to a range, or *band*, of *data signaling rates*. The designator is a two- or three-letter abbreviation for the name. Each range is a decade wide. The range limits for each decade have coefficients of 3. Thus, the logarithmic midrange value of each range (decade) is close to 1 times a power of 10. Each range is designated by a number that is the power of 10 of the midrange value. For example, the logarithmic midrange value of the 30–300 *Gbps* EHR (Extremely High Data Signaling Rate) is 10^{11}. Therefore the numerical designator for this range is 11. The data signaling rate (DSR) describes the rate at which *data* pass through a point, for example, in bits, characters, words, or other data units per unit time. It is not the speed at which *lightwaves propagate* in an *optical fiber*, such as in *meters* per second. Also see *data signaling rate (DSR); spectrum designation of frequency*. The data-signaling-rate ranges and their designators are given in the TABLE OF DATA-SIGNALING-RATE (DSR) RANGES on the facing page.

rate. See *data signaling rate; interface rate; maximum pulse rate; Nyquist rate; optical line rate; payload rate; pulse-repetition rate; system information rate.*

Table of Data-Signaling-Rate (DSR) Ranges

DATA SIGNALING RATE RANGE* (Lower Limit Exclusive, Upper Limit Inclusive)	LETTER GROUP AND WORD DESIGNATOR	NUMERICAL DESIGNATOR
Below 300 bps	ELR (Extremely Low Data Signaling Rate)	2
300–3000 bps	ULR (Ultra Low Data Signaling Rate)	3
3–30 kbps	VLR (Very Low Data Signaling Rate)	4
30–300 kbps	LR (Low Data Signaling Rate)	5
300–3000 kbps	MR (Medium Data Signaling Rate)	6
3–30 Mbps	HR (High Data Signaling Rate)	7
30–300 Mbps	HR (Very High Data Signaling Rate)	8
300–3000 Mbps	UHR (Ultra High Data Signaling Rate)	9
3–30 Gbps	SHR (Super High Data Signaling Rate)	10
30–300 Gbps	EHR (Extremely High Data Signaling Rate)	11
300–3000 Gbps	THR (Tremendously High Data Signaling Rate)	12
3–30 Tbps	AHR (Awesomely High Data Signaling Rate)	13
30–300 Tbps	PHR (Phenomenally High Data Signaling Rate)	14
300–3000 Tbps	(To be determined)	15
3–30 Pbps	(TBD)	16
30–300 Pbps	(TBD)	17
300–3000 Pbps	(TBD)	18
3–30 Ebps	(TBD)	19
30–300 Ebps	(TBD)	20
300–3000 Ebps	(TBD)	21

(Note: The theoretical upper limit for the numerical designator is about 30)
*Prefixes:

k = kilo = 10^3	M = mega = 10^6	G = giga = 10^9
T = tera = 10^{12}	P = peta = 10^{15}	E = exa = 10^{18}

rate characteristic. See *wavelength-dependent attenuation-rate characteristic.*

rate-length product. See *bit-rate-length product.*

ratio. See *axial ratio; bit-error ratio (BER); photodetector signal-to-noise ratio.*

ray. An infinitesimally narrow *beam* of *electromagnetic radiation,* such as a *light-wave,* that is represented by a straight line drawn in the direction of propagation at a point in a *propagation medium* or *free space.* The line representing the ray is drawn tangent to and in the direction of the line representing the path taken by the wave. Rays are used for analyses of *lightwave* behavior in the terminology of *geo-*

metric optics. Rays are in the direction of the *Poynting vector* and therefore perpendicular to *wavefronts* in the terminology of *physical optics.* See *axial ray; direct ray; extraordinary ray; guided ray; leaky ray; light ray; meridional ray; optical ray; ordinary ray; paraxial ray; reflected ray; refracted ray; skew ray; x-ray.* Also see *light beam.*

ray bundle. **1.** A group of *ordinary rays* in which the ordinary rays differ from one another in some specific respect, such as *wavelength (color)* or *radiant intensity,* yet have one or more aspects in common, such as direction or velocity. For example, the rays of a given bundle may be separated from each other by *dispersion* caused by a prism, in which case a multiwavelength bundle of rays incident on the prism are spatially separated into distinct ordinary rays of different *color;* in an *optical fiber,* dispersion will cause different wavelengths to arrive at the end at different times, thus temporally separating the rays of an incident bundle. **2.** A group of *light rays* considered as a unit for some purpose or discussion. Synonymous with bundle of rays.

Rayleigh distribution. A mathematical statement of the *frequency* distribution of random variables, for the case where two orthogonal variables are independent and normally distributed with the same variance. The Rayleigh distribution occurs because of the intrinsic molecular structural pattern of the *propagation* media, such as glass. Also see *Rayleigh scattering.*

Rayleigh fading. *Phase* interference fading that is approximated by the *Rayleigh* distribution and is caused by recombination after *multipath.*

Rayleigh scattering. *Scattering* caused by *refractive-index* fluctuations that are small, with respect to *wavelength.* The refractive-index variations are caused by inhomogeneities in material *optical density,* composition, and molecular structure. The scattered *radiant power* is inversely proportional to the fourth power of the wavelength. Also see *Rayleigh distribution.*

ray method. See *refracted-ray method.*

ray optics. See *geometric optics.*

ray trajectory. The path or course taken by a *light ray* in a *propagation medium* or a vacuum. The trajectory is perpendicular to the *wavefront.*

real time. **1.** The absence of delay, except for the time required for the *transmission* of *elecromagnetic energy,* between the occurrence of an event or the transmission of *data* and the knowledge of the event or reception of the data at some other location. **2.** The actual time during which a physical process occurs. **3.** Pertaining to the performance of a computation during the actual time that the related physical process occurs, in order that results of the computation can be used in guiding the physical process.

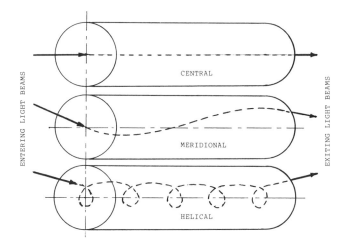

R–2. Some of the possible **ray trajectories** for *graded-index optical fibers.*

receiver. See *fiber optic receiver; optical receiver; optoelectronic receiver; PIN-FET integrated receiver.*

receiver-information descriptor. A unique descriptor from which information about a receiver can be determined, such as the manufacturer, terminal equipment association, system design application *(single-mode, multimode),* performance specifications, *detector* type *(APD, PIN photodiode),* temperature controller, and manufacturer product change designation (issue, revision).

receiver input. See *maximum receiver input.*

receiver maximum input power. See *optical maximum input power.*

receiver optical connector. Th *optical connector* provided at the input of an *optical receiver* that is attached to the receiver *pigtail.* The receiver optical connecter description should include the manufacturer, type (biconic, FC, etc.) model number, classification *(multimode, single-mode),* and mating connector model number.

receiver optical sensitivity. The worst-case value of *input optical power (dBm) coupled* into a receiver, as measured on the line side of the receiver connector under specified *standard* or *extended operating conditions,* that is necessary to achieve the manufacturer-specified *bit-error ratio (BER)* as measured under standard procedures. The worst-case value combines manufacturing, temperature, aging, *extinction ratio,* and rise-fall time variations in a worst-case fashion. The receiver sensitivity should not include power penalties associated with *dispersion* or *reflection.*

recording. See *fiberscopic recording.*

REED. *Restricted edge-emitting diode.*

reference model. See *Open-Systems-Interconnection (OSI) Reference Model (RM).*

reference surface. In a round *optical fiber,* the cylindrical surface used for reference purposes when joining optical fibers. The reference surface usually is the outer surface of the outermost *cladding.* The *core-cladding interface* surface, that is, the inner surface of the innermost cladding, has also been used as the reference surface because this surface identifies the core.

reference-surface center. 1. In the cross section of an *optical fiber,* the center of the circle that best fits the outer limit of the *reference surface.* The reference-surface center may not be the same as the *core* and *cladding* centers. The method of best fit must be specified. **2.** The center of the smallest circle into which the *reference surface* can be fitted.

reference-surface concentricity. See *core reference-surface concentricity.*

reference-surface noncircularity. The difference between the diameters of the two circles used to define the *reference-surface tolerance field* divided by the *reference-surface* diameter.

reference-surface tolerance field. In the cross section of an *optical fiber,* the region between the smallest circle concentric with the center of the *reference surface* circumscribed about the *core area* and the largest circle, concentric with the first one, that fits inside the reference surface.

reference test method (RTM). In *fiber optics,* a test method in which a given characteristic of a specified class of *fiber optic devices,* such as *optical fibers, fiber optic cables, connectors, photodetectors,* and *light sources,* is measured strictly according to the definition of this characteristic and that gives accurate and reproducible results relatable to practical use. Also see *alternative test method.*

reflectance. In *optics,* the ratio of the reflected to the incident *irradiance,* that is, *optical power,* at an *interface* point. The conditions under which the reflectance occurs should be stated, such as the spectral (*wavelength*) composition and *polarization* of the incident *wave,* the geometrical shape of the reflecting surface, and the composition of the *propagation media* on both sides of the interface. In *optics,* the reflectance is often expressed as a percentage or as *reflectance density,* that is, as the logarithm to the base 10 of the reciprocal of the reflectance. In communications, it is usually expressed in decibels.

reflectance density. The logarithm to the base 10 of the reciprocal of the *reflectance.* Also see *optical density.*

reflected ray. When an *electromagnetic wave* is *incident* upon an *interface* surface between two different *propagation media,* such as a boundary between *dielectric* materials with different *refractive indices,* the *ray* that is turned back into the medium containing the incident ray. If the refractive indices at the interface vary as a step function across the interface, *spectral reflection* takes place. The angle that the reflected ray makes with the normal to the interface surface, namely the *reflection angle,* has the same value as the incident ray has with the normal, that is, the *incidence angle* is equal to the reflection angle. The *power* of the reflected ray is equal to the power in the incident ray times the *reflection coefficient,* as determined by the refractive indices and the incidence angle, assuming no power is *absorbed* at the interface, that is, at the reflection surface. For example, in an *optical fiber,* the reflected ray is the portion of a ray that is in the *core* and incident to the core-*cladding* interface that is returned to the core. For core-cladding interface incidence angles greater than the *critical angle,* all the *optical power* in the incident ray is contained in the reflected ray. For incidence angles equal to the critical angle, a refracted ray will propagate along the core-cladding interface surface. If there is a smooth transition of refractive index at the core-cladding interface of an optical fiber, such as in a *graded-index fiber,* there will be some penetration of an incident ray a short distance into the cladding, and the incident ray will be returned to the core by successive or continuous *reflection.* Also see *Goos-Haenchen shift; refracted ray.*

reflecting loss. See *reflection loss.*

reflection. The changing of direction of an incident *wave* at an *interface* between two dissimilar *propagation media* so that it is directed partially or totally back into the medium from which it originated. See *diffuse reflection; Fresnel reflection; internal reflection; maximum optical reflection; specular reflection; total reflection.*

reflection coefficient. 1. The ratio between the amplitude of the *reflected wave* and the amplititude of the incident wave. For large smooth surfaces, the reflection coefficient may be near unity. At near grazing incidence, even rough surfaces may reflect relatively well. 2. At any specified point in a *transmission* line between a *source* of power and a sink, that is, an absorber of power, the complex ratio of the *electric field strength* associated with the reflected wave to that associated with the incident wave. The reflection coefficient magnitude is given by the relation:

$$RC = |(Z_2 - Z_1)/(Z_2 + Z_1)| = (SWR - 1)/(SWR + 1)$$

where Z_1 is the impedance toward the source, Z_2 is the impedance toward the load, the vertical bars designate absolute magnitude, and *SWR* is the *standing wave* ratio. 3. The ratio of the *reflected electric field strength* to the incident *electric field strength* when an *electromagnetic wave* is incident upon an *interface* surface between *dielectric propagation media* of different *refractive indices.* For example, if, at oblique incidence, the *electric field vector* of the incident *plane-polarized electro-*

magnetic wave is parallel to the interface, the reflection coefficient is given by the relation:

$$R = (m_2 \cos A - m_1 \cos B)/(m_2 \cos A + m_1 \cos B)$$

where m_1 and m_2 are the reciprocals of the refractive indices of the incident and transmitted media, respectively, and A and B are the *incidence* and *refraction angles,* respectively. This is one of the *Fresnel equations.* The sum of the reflection coefficient and the *transmission coefficient* is not necessarily unity. Also see *transmission coefficient.*

reflection density. See *optical density.*

reflection law. When a *ray* of *electromagnetic radiation* is *reflected* in whole or in part, the *reflection angle* is equal to the *incidence angle,* the incident *ray,* reflected ray, and normal to the surface all being in the same plane.

reflection loss. The ratio, usually expressed in *decibels,* between the incident and reflected *wave radiant powers* at any discontinuity or *impedance* mismatch. When the two waves have opposite *phases* and appropriate magnitudes, a reflection gain may be obtained. The reflection loss for a given frequency at the junction of a source of power and a load is given by the relation:

$$\text{Reflection loss} = -20 \log_{10} |(Z_1 + Z_2)/(4Z_1 Z_2)^{1/2}|$$

where the reflection loss is in *decibels,* the vertical bars designate absolute magnitude, and Z_1 and Z_2 are the impedances of the *source* of power and the load. Synonymous with *reflecting loss.*

reflection method. See *Fresnel reflection method.*

reflection power penalty. In an *optical receiver,* the additional power (dB) required by a receiver, when the manufacturer-specified value of maximum *optical reflection* is introduced at the line side of the associated transmitter connector to achieve the same *bit-error ratio (BER)* that is obtained without the introduced reflection using standard measurement procedures.

reflection sensor. See *frustrated total (internal) reflection sensor.*

reflective star coupler. An *optical fiber coupler* in which *signals* in one or more fibers are *transmitted* to one or more other fibers by entering the signals into one end of an optical fiber, cylinder, or other piece of *transparent* material with a *reflecting* back surface in order to reflect the diffused signals into output *ports* for further transmission in one or more fibers.

reflectivity. The *reflectance* of an opaque object, that is, an object of such thickness that further increases in thickness do not alter the reflectance. The term is no longer in common use. See *spectral reflectivity.*

reflectometer. See *optical time-domain reflectometer (OTDR).*

reflectometry. See *optical time-domain reflectometry (OTDR).*

reflector. 1. In *optics,* a surface with a high *reflection coefficient* at all *incidence angles,* such as a *mirror.* **2.** In radio, one or more electrical conductors or conducting surfaces for *reflecting radiant energy,* such as the parabolic dish of a receiving antenna. See *Lambertian reflector.*

refracted ray. When a *ray* of an *electromagnetic wave* is incident upon an *interface* surface between two different *propagation media,* such as a boundary between *dielectric* materials with different *refractive indices,* the ray that emerges on the side of the boundary opposite the side of incidence. When the incident ray is in a higher refractive-index material, such as in the *core* of an *optical fiber,* the refracted ray is bent away from the normal and toward the *interface* surface; if in a lower refractive-index material, the refracted ray is bent toward the normal. For example, in an optical fiber, the portion of a ray incident to the core-*cladding* interface that enters the cladding and does not return to the core at the point of incidence is a refracted ray. For *incidence angles* greater than the *critical angle,* there is no refracted ray. For incidence angles equal to the critical angle, the refracted ray will *propagate* along the core-cladding interface surface. The amplitude of the refracted ray relative to the incident ray is given by the *refraction coefficient.* In an optical fiber, a ray at radial position r is refracted into the cladding from the core when it has such a direction that it satisfies the relation:

$$[n_r^2 - n_a^2]/[1 - (r/a)^2 \cos^2 \phi_r] \leq \sin^2 \theta_r$$

where n_r is the refractive index at the radial distance r from the *optical axis,* that is, the *core center,* n_a is the refractive index at radial distance a from the *optical axis,* but a is the core radius, that is, half the *core diameter,* therefore it is the refractive index on the core side of the *core-cladding interface,* a is the radius of the core, ϕ_r is the azimuthal angle of projection of the ray on the transverse plane of the fiber, that is, a plane normal to the optical axis, and θ_r is the angle the ray makes with the optical axis of the fiber. A refracted ray corresponds to a *radiation mode* in the terminology used for mode description. Also see *reflected ray.*

refracted near-field scanning method. See *refracted-ray method.*

refracted-ray method. In *fiber optics,* a method for measuring the *refractive index profile* of an *optical fiber* by scanning the entrance face of the fiber with the vertex of *light* from a high-*numerical-aperture* cone and measuring the change in *power* of the *refracted rays*. The refracted rays leave the *core* and therefore are un*guided*. Synonymous with *refracted near-field scanning method*.

refracting crystal. See *multirefracting crystal*.

refraction. The bending of a sound wave, radio wave, or *lightwave* as it passes obliquely across the *interface* from one *propagation medium* to another with a different *refractive index* or the bending that occurs in a propagation medium in which the refractive index is a continuous function of spatial position, for example, in a medium with a *graded-index profile*. See *double refraction*.

refraction angle. When an *electromagnetic wave* strikes a surface of another *propagation medium* and is wholly or partially *transmitted* into the new medium, the acute angle between the normal to the surface and the *refracted ray*.

refraction coefficient. See *transmission coefficient*.

refractive index. At a point in a *propagation medium* and in a given direction, the ratio of the velocity of *light* in vacuum to the magnitude of the *phase velocity* of a sinusoidal *electromagnetic plane wave propagating* in that direction. It is also the ratio of the sines of the *incidence* and *refraction angles* as a light *ray* passes across an *interface* from one medium to another. If one of the media is a vacuum, the measured refractive index will be the refractive index or that medium relative to a vacuum. The refractive index of a vacuum is defined as 1.000000, of air it is 1.000292, of water it is about 1.333, and of ordinary crown glass it is 1.156. The refractive index of a *dielectric* substance, that is, whose *electrical conductivity* is zero, is also given by the relation:

$$n = (\mu\epsilon/\mu_0\epsilon_0)^{1/2}$$

where μ is the *magnetic permeability* and ϵ is the *electric permittivity* of the substance, and the other variables are the corresponding values for a vacuum, which are nearly the same as for air, namely unity. In absolute units for *free space,* the electric permittivity is $\epsilon = 8.854 \times 10^{-12}$ F/m (farads per *meter*), and the magnetic permeability is $\mu = 4\pi \times 10^{-7}$ H/m (henries per meter). The reciprocal of the square root of the product of these values yields approximately 2.998×10^8 m/s (meters per second), the velocity of light in a vacuum. Because the refractive index for a propagation medium is defined as the ratio between the velocity of light in a vacuum and the velocity of light in the medium, the ratio becomes unity when the propagation medium is a vacuum. Thus, the *relative refractive index* is given by the relation:

$$n_r = (\mu\epsilon)^{1/2}$$

where the n_r is the refractive index relative to vacuum (or air). The refractive indices for the materials used in *optical fibers* are always given as relative to a vacuum and are therefore dimensionless. Also, except for ferrous materials, cobalt, and nickel, and some diamagnetic materials, the relative magnetic permeability of materials, especially *dielectric* materials, is approximately 1. For a vacuum it is 1. See *relative refractive index*. Also see *electric permittivity; magnetic permeability; relative electric permittivity; relative magnetic permeability; wave impedance.*

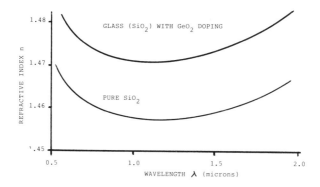

R-3. The measured **refractive index** of materials, such as glasses, varies with the *wavelength* of the *propagating lightwave*. The refractive index can be increased by adding certain *dopants,* such as germanium, to the intrinsic glass. Many *optical waveguides* have *cladding* made of pure silicon dioxide.

refractive-index contrast. A measure of the relative difference in *refractive index* across the *interface* surface between *propagation media* with two different refractive indices. For an *optical fiber,* the refractive index contrast, Δ, is given by the relation:

$$\Delta = (n_1^2 - n_2^2)/2n_1^2$$

or, for interfaces where the indices are not different by more than 1% of each other, by the relation:

$$\Delta = (n_1 - n_2)/n_1$$

where n_1 is the maximum refractive index in the *core,* and n_2 is the refractive index of the *homogeneous cladding.*

refractive-index difference. See *ESI refractive-index difference.*

refractive-index dip. In an *optical fiber,* a reduced *refractive index* at the central region of the *core.* The dip is an imperfection that occurs only when certain fiber fabrication and *drawing* techniques are used.

refractive-index profile. In a *dielectric waveguide,* the variation of *refractive index* in the direction transverse to the direction of *wave propagation,* that is, in a cross section of the guide. In a round *optical fiber,* the refractive-index profile is a description of the value of the refractive index as a function of distance from the *optical axis* along any *fiber diameter*. The refractive-index variation is symmetrical about the *optical axis* of the fiber, that is, there is zero-axis symmetry. See *linear refractive-index profile; parabolic refractive-index profile; power-law refractive-index profile; radial refractive-index profile.*

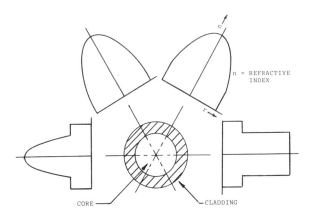

R-4. Four *graded* and *stepped* **refractive-index profiles** for *optical fibers.* The *refractive index* is a function of the radial distance from the *optical axes* of the round fibers. (The coordinate axes for all the profiles are the same.)

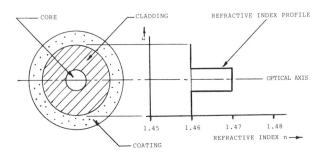

R-5. The **refractive-index profile** for a *step-index optical fiber* is a step-function of the radial distance from the *optical axis*. The abrupt change in *refractive index* occurs across the *core-cladding interface*. The core index must be higher than the cladding index for *lightwaves* to remain confined to the core. The number of *modes* the fiber is capable of supporting is a function of the *normalized frequency,* which is given by $V = (2\lambda a(NA))/\lambda$, where NA is the *numerical aperture, a* is half of the *core diameter,* and λ is the *wavelength* of the *light source*. A mode may also be considered as a possible path for a *light ray*. When the normalized frequency (*V*-value or *V*-number) is ≥ 2.4 for a step-index fiber, the fiber will support many modes, approximately $V^2/4$ modes.

refractive-index-profile parameter. The exponent, *g*, in the several relations for the *power-law refractive-index profile*. Also see *power-law refractive-index profile*.

refractive-index template. See *four-concentric-circle refractive-index template*.

regenerator section. See *optical station/regenerator section*.

regenerator system gain. See *statistical terminal/regenerator system gain; terminal/regenerator system gain*.

region. See *far-field region; intermediate-field region; near-field region*.

relations. See *constitutive relations*.

relative electric permittivity. The incremental or absolute *electric permittivity* of a material, ϵ_r, normally represented as ϵ and implied to be relative to *free space*, that is, compared with that of a vacuum, ϵ_0. The actual electric permittivity of free space in SI (International System) units, absolute or incremental, is 8.854×10^{-12} (coulombs per square *meter*)/(volts per meter), which is the same as farads per meter. The refractive index for the *silica glass* of *optical fibers* is given approximately by the relation:

$$n \approx \epsilon^{1/2}$$

where ϵ is the electric permittivity of the glass. Because it is relative to free space, a vacuum, or air, it is dimensionless for specific materials, about 2 or 3 for glass, and up to 10 or 12 for the special dielectric materials used in capacitors to increase their capacity and provide for electrical separation of electrodes. Also see *electric permittivity; refractive index*.

relative magnetic permeability. The incremental or absolute *magnetic permeability* of a material, μ_r, normally represented as μ and implied to be relative to *free space*, that is, compared with that of a vacuum, μ_0. The magnetic permeability of free space in SI (International System) units, absolute or incremental, is $4\pi \times 10^{-7}$ (webers per square *meter*) per (ampere per meter), or webers per ampere-meter, which is the same as henrys per meter. For an *optical fiber*, the magnetic permeability is very nearly equal to that of free space. Thus, μ_r for an optical fiber is very close to unity, and, because it is relative to free space, a vacuum, or air, it is dimensionless for specific materials. Also see *magnetic permeability; refractive index*.

relative refractive index. The *refractive index* of a substance relative to another substance. Thus, if two glasses have refractive indices, that is, refractive indices relative to a vacuum, of $n_1 = 2.100$ and $n_2 = 1.781$, then the refractive index of substance 1 relative to substance 2 is n_1/n_2, or 1.179.

release time. See *switch release time*.

remanence. The magnetization, that is, the magnetic polarization, that remains in a magnetized material after the applied *magnetic field* that magnetized the material is removed.

repair. The restoration of an item to serviceable condition by correction of a specific failure or unserviceable condition. See *mean time to repair (MTTR).*

repeater. A device that amplifies an input *signal* or, in the case of pulses, recovers, amplifies, reshapes, retimes, or performs a combination of any of these functions on an input signal, and *transmits* the signal. It may be either a one-way or a two-way repeater. See *optical repeater.*

repeatered optical link. An *optical link* that has *optical repeaters* along its cable to recover, amplify, and perhaps reshape and time, *optical signals.*

repeaterless optical link. A *fiber optic link,* usually in a *long-haul communication network* in which *repeaters* are not used because of the short length, the high *bandwidth-distance factor* of the fiber, or the low *attenuation rate* of the fiber. Repeaterless optical links thousands of kilometers long are expected to be in place before the year 2000. Optical links for short distances, such as for intercity or *local-area-network* use, in which repeaters were never needed or anticipated, are simply called *optical links* rather than repeaterless optical links, even though they have no repeaters.

repetition rate. See *pulse-repetition rate.*

resolution. See *borescope target resolution; distance resolution.*

resolving power. A measure of the ability of a lens or *optical system* to form separate and distinct images of two objects with small angular separation. Because of *diffraction* at the *aperture,* no optical system can form a perfect image of a point, but produces instead a small disk of light, called an "airy disk," surrounded by alternately dark and bright concentric rings. See *chromatic resolving power; theoretical resolving power.*

resonant cavity. A geometric space that may be empty or contain a fluid or solid material, that is bounded by two or more *reflectors,* and in which *electromagnetic waves* can *reflect* back and forth, thus producing *standing waves* of high intensity at certain *wavelengths.* For example, standing waves might be produced in a ruby crystal *laser* with two plane or spherical mirrors, forming a resonant cavity, with the crystal itself between the mirrors, the molecules of which can be excited by an inert-gas lamp, thus generating and *emitting* a narrow *beam* of *monochromatic light* of high *power* and high *irradiance,* that is, high *optical power density* or intensity, in the direction of the crystal axis.

response. See *impulse response; spectral response.*

response function. See *impulse-response function.*

response quantum efficiency. The ratio of the number of countable output events to the number of incident *photons* that occur when energy in the form of *electromagnetic waves,* such as *lightwaves, gamma radiation, x-rays,* and cosmic rays, is incident upon a material, often measured as electrons emitted per incident *quantum,* that is, for *lightwaves,* per incident photon. *Optical* response quantum efficiency indicates the efficiency of conversion or utilization of optical energy. It is an indication of the number of events produced for each quantum incident on the sensitive surface of a photodetector. It is a function of the *wavelength, incidence angle, polarization,* and other factors.

responsivity. 1. In a *photodetector,* such as an *avalanche photodiode (APD),* the ratio of its electrical current or voltage to the *optical power input.* It is generally expressed in amperes of electrical current output per watt of optical power input, or in volts per watt. The value will change with the *wavelength* of the incident *radiation.* 2. In a *light source,* such as a *light-emitting diode (LED)* or *laser,* the ratio of its *optical power* output to the driving current input, for example, 3mW/μA at 0.810 μ *(microns).* Sometimes the term is improperly used as a synonym for *sensitivity.* See *photodetector responsivity; spectral responsivity.* Also see *sensitivity.*

restoral. See *mean time to service restoral (MTTSR).*

restricted edge-emitting diode (REED). An *edge-emitting LED (ELED),* that is, a *light-emitting diode (LED)* in which *light* is *emitted* over a small portion of an edge, permitting improved *coupling* with *dielectric waveguides,* such as *optical fibers* and *integrated optical circuits (IOCs).*

retention. See *fiber retention.*

return loss. In a *fiber optic system, optical power* that is *reflected* back toward the *source* of optical power by a component, such as a *fiber optic splice, connector, coupler, attenuator,* or *rotary joint.* Return loss may be expressed in absolute power units, such as microwatts, or in *dB* with reference to the incident optical power input to the component. Thus, return loss is the optical power *reflected,* rather than power that is *transmitted, absorbed, scattered,* or *radiated.* However, *backscatter* is included in the return loss measurement and should be considered when the actual return loss is calculated.

ribbon. See *fiber optic ribbon.*

ribbon cable. A *fiber optic cable* in which the *optical fibers* are held in grooves and laminated within a flat semirigid strip of plastic or other material that holds them and protects them. Fiber optic cables with large numbers of fibers can be produced by stacking two or more ribbons together, adding *buffers, strength members, fillers,* and perhaps gels to inhibit water penetration, and *jacketing* or sheathing the whole assembly. A ribbon cable may be either a *tight-jacketed* or a *loose-tube* type of fiber optic cable, depending on how tightly or loosely the fibers are held by the grooves in the ribbons. Thus, the fibers are laminated within the ribbon.

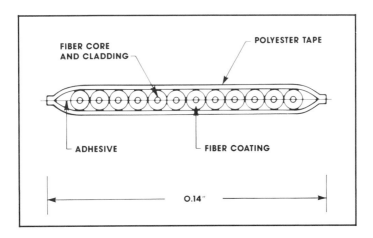

R-6. A cross section of a **ribbon cable** core with 12 *optical fibers* in each *fiber optic cable core*. The ribbon cores are identified by a unique ribbon number. The optical fibers in each ribbon are uniquely identified with a fiber *color* code. (Courtesy AT&T.)

right-hand circular polarization. *Circular polarization* of an *electromagnetic wave* in which the *electric field vector* rotates in a clockwise direction as the wave advances in the direction of *propagation* and as seen by an observer looking in the direction of propagation. Thus, in a *waveguide,* the electric field vector of a circularly polarized wave always remains perpendicular to the direction of propagation, as well as perpendicular to the *magnetic field vector.* Also see *right-hand helical polarization.*

right-hand helical polarization. *Polarization* of an *electromagnetic wave* in which the *electric field vector* rotates in a clockwise direction as it advances in the direction of *propagation* and as seen by an observer looking in the direction of propagation. The tip of the electric field vector advances like a point on the thread of a right-hand screw when entering a fixed nut or tapped hole, thus describing a helix in the shape of the thread itself. Thus, in a *waveguide,* the electric field vector of a helically polarized wave always remains skewed to the direction of propagation. Also see *right-hand circular polarization.*

right-hand polarized electromagnetic wave. An *elliptically* or *circularly polarized electromagnetic wave* in which the direction of rotation of the *electric field vector* is clockwise as seen by an observer looking in the direction of *propagation* of the wave. Also see *left-hand polarized electromagnetic wave.*

rigid borescope. A portable *fiber optic borescope* that has an *aligned bundle* (coherent bundle) in a stiff *fiber optic cable*, or a train of lenses mounted in a rod, with sufficient rigidity to cross unsupported gaps in a path to an objective. There is usually an eyepiece for viewing the image, and it may have a controllable articulating tip on the *objective* end to enable inspection at various angles, such as up to 90 degrees from the cable *optical axis*. The borescope is hand-held and energized via a fiber optic cable from a *light source*. The rigid borescope usually consists of a basic borescope unit and ancillary equipment required for operation. Also see *flexible borescope*.

ringer. See *optical fiber ringer*.

ring network. 1. A type of *network* configuration consisting of a closed loop medium. **2.** In *network* topology, a network configuration in which each *node* is directly connected to two and only two adjacent nodes, that is, to two nodes by means of a single *branch* to each. Thus, one and only one path connects all nodes, and there are no *end-point nodes*. Nor are they any connections or *ports* to other networks. (See Figs. R-7—R-8)

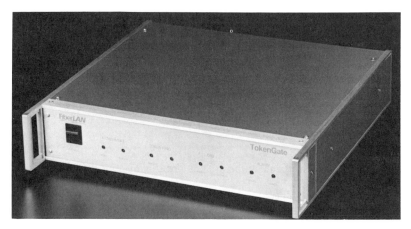

R-7. The TokenGate™ allows token ring *data* to be transported over DS2 (6.3 Mbps) telecommunications facilities, such as *fiber optic networks*. It acts as a token ring repeater. In the *transmit* direction the token ring *signal* is encoded into a DS2 signal. In the receive direction, the reverse occurs. Thus, local token **ring networks** can be interconnected via *optical fiber*, microwave, or other links. (Courtesy FiberLAN, a Siecor Company).

rise time. 1. The time it takes for the amplitude of a *pulse* to increase from a specified value, usually near the lowest or zero value, to a specified value near the peak value. Values of rise time are often specified as the time interval between the 10 and 90% values of the peak value of a pulse. When other than 10 to 90% values are used, they should be specified. **2.** The time it takes for the magnitude of a *pulse*,

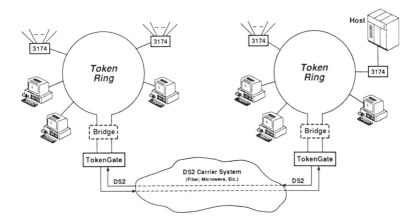

R–8. Token **ring networks** can be connected via the TokenGate and optical fiber, microwave, or other carrier systems. (Courtesy FiberLAN, a Siecor Company.)

such as the amplitude, the *phase* shift, the *frequency* shift, or other parameter used to represent a pulse, to increase from one value to another to represent binary digits. Synonymous with *pulse rise time*. Also see *fall time*.

rms pulse broadening. See *root-mean-square (rms) pulse broadening*.

rms pulse duration. See *root-mean-square (rms) pulse duration*.

rod. See *optical mixing rod*.

rod-in-tube technique. A process in which a rod placed in a tube is used as an *optical fiber preform* in the manufacture and *drawing* of *optical fibers*.

root-mean-square (rms) deviation. A quantity that characterizes a function, $f(x)$, by the relation:

$$\sigma_{\text{rms}} = [1/M_0 \left| {}^{+\infty}_{-\infty} (x - M_1)^2 f(x)\ dx]^{1/2}$$

$$M_0 = \left| {}^{+\infty}_{-\infty} f(x)\ dx, \quad M_1 = 1/M_0 \left| {}^{+\infty}_{-\infty} xf(x)\ dx$$

M_o is the normalization, which, in probability and statistics, is 1. Also see *impulse-response function; root-mean-square (rms) pulse broadening; root-mean-square (rms) pulse duration; spectral width*.

root-mean-square (rms) pulse broadening. The temporal *root-mean-square (rms) deviation* of the impulse-response function of a component or system. Also see *impulse-response function; root-mean-square (rms) pulse duration; root-mean-square (rms) deviation; spectral width*.

root-mean-square (rms) pulse duration. A special case of *root-mean-square devia-tion* in which the independent variable is time, and $f(t)$ is the *pulse* waveform. Also see *impulse-response function; root-mean-square (rms) pulse broadening; root-mean-square (rms) deviation; spectral width.*

rotary joint. See *fiber optic rotary joint.*

rotating polarization. *Polarization* of an *electromagnetic wave,* such that the *polari-zation plane* rotates with an angular displacement that is a function of the distance along the *ray* or *propagation* path.

rotation. See *optical rotation.*

rotor. See *optical rotor.*

RTM. *Reference test method.*

runner-cut. On the surfaces of glass and other ground or polished materials, a curved scratch, such as might be caused by a grinding or polishing wheel.

S

safety margin. When designing *optical links,* a *power-loss (dB)* value used to allow for unexpected *losses* and ensure performance criteria are met, such as the bit rate *(data signaling rate)* and *bit-error ratio (BER).* The safety margin should not include expected losses and degradations, e.g., *laser* aging, cable aging, *reflections,* and repairs. These effects should be included in the appropriate parameters of specific devices. Synonymous with *power margin; power safety margin.*

Sagnac fiber optic sensor. An *interferometric sensor* in which a *lightwave* is *split* and passed in opposite directions around the same rigid rotating *optical fiber loop* by means of mirrors to a single *photodetector. Phase* cancellation and enhancement, and hence *irradiance,* that is, *light intensity,* changes when the angular velocity is varied, thereby obtaining an output *frequency* that is proportional to the angular acceleration. The output *signal* from the photodetector can be integrated once for obtaining angular velocity and twice for angular displacement.

sampling theorem. For a waveform to be sampled and then effectively or satisfactorily reconstituted from the sampled data, it is necessary that the sampling rate be equal to or greater than twice the highest significant *frequency* component of the *wave* being sampled, that is, the sampling rate must satisfy the *Nyquist interval* or *Nyquist criterion.* See also *Nyquist bandwidth; Nyquist rate.*

satellite. See *direct-broadcast satellite (DBS).*

satellite optical link. A *beamed optical transmission channel,* usually consisting of a *laser source* shining a *lightwave* in the form of a narrow beam on an *avalanche photodetector,* operating between a ground station and a satellite station. The source is *modulated* in accordance with an intelligence-bearing *signal.*

scanning technique. See *near-field scanning technique.*

scatter. The process whereby the direction, *frequency,* or *polarization* of *electromagnetic waves,* such as *lightwaves,* is changed when the waves encounter particles or discontinuities, such as *microcracks,* abrupt changes in *refractive index,* and impurities in the *propagation medium* in which the waves are *propagating,* particularly

278

when the particles and discontinuities have sizes on the order of 1 *wavelength* of the propagating waves. If the particles and discontinuities are larger, the diffusion phenomenon is one of *spectral reflection* and *refraction* rather than scattering. The term is frequently used to imply a disordered change in the properties of the incident waves, such as the distribution of energy or the direction of *propagation* of *electromagnetic power.* See *backscatter; forward-scatter.* Also see *scattering.*

scattering. The change in direction of *electromagnetic waves,* such as *lightwaves* or *photons,* after striking a small particle, particles, or discontinuity in the material in which the *rays,* or photons, are *propagating.* Scattering may also be caused by the non*homogeneity* of the *propagation medium.* See *backscattering; Brillouin scattering; configuration scattering; material scattering; nonlinear scattering; radiation scattering; Raman scattering; Rayleigh scattering.* See *waveguide scattering.* Also see *scatter.*

scattering center. A site in the structure, particularly the microstructure, of a *propagation medium* at which *electromagnetic waves,* such as *lightwaves,* are *scattered.* Examples of scattering centers include *vacancy defects,* such as missing atoms or molecules in the somewhat orderly though amorphous structure of the propagation medium; interstitial defects, such as misalignment of atoms or molecules; inclusion defects, such as trapped impurity atoms, molecules, and ions, including gas molecules, hydroxide ions, iron ions, and water molecules; *microcracks;* abrupt changes in *refractive index,* perhaps caused by pressure points; and *microbends.*

scattering coefficient. In the *propagation* of *electromagnetic waves* in material media, the part of the *attenuation term,* that is, *attenuation constant,* contributed by *scattering* in the exponent of the expression that describes *Bouger's law. Absorption* also contributes to the *attenuation term.* Both parts of the attenuation term, that is, the scattering part and the absorption part, depend on impurities and the irregularity of the molecular structure of the intrinsic material.

scattering loss. In the *propagation* of *electromagnetic waves* in material media, such as *lightwaves* in an *optical fiber,* that part of the *power loss* that is due to *scattering* within the *propagation medium* and the power loss caused by irregularity or roughness of *reflecting* surfaces in the propagation medium.

scattering method. See *transverse-scattering method.*

Schottky effect. See *shot noise.*

Schottky noise. See *shot noise.*

scrambler. See *fiber optic scrambler; mode scrambler.*

secondary coating. A *coating* applied to the *primary coating* of an *optical fiber* for added protection during handling, *cabling,* installation, and use.

secondary radiation. Particles or *photons* produced by the action of primary *radiation* on matter, such as Compton recoil electrons, delta *rays,* secondary cosmic rays, and secondary electrons.

section. See *optical station/regenerator section.*

selective absorption. The process by which a substance *absorbs* only certain *frequencies,* that is, certain *wavelengths* or *colors,* in a *beam* of incident *electromagnetic radiation* and *reflects* or *transmits* all others. Some substances are *transparent* to certain wavelengths, allowing them to be transmitted, while absorbing others. Some substances reflect certain wavelengths and absorb or transmit all others. The color of a transparent object is usually the color it transmits while the color of an opaque object is usually the color it reflects. Some materials reflect and transmit the same color, absorbing all others.

self-adjusting. In *fiber optic communication networks,* pertaining to communication *networks* in which an adequate number of redundant *fiber optic links* are installed to accommodate *lightwave data* streams in both directions and to provide for automatic alternate routing if a link fails, thereby eliminating the necessity of repairing *fiber optic cables* before service can be restored. A digital access and *fiber optic cross-connect* system is used to reroute lightwave *signals.* Synonymous with *self-healing.*

self-healing. See *self-adjusting.*

self-heterodyne. In *fiber optics,* pertaining to a *fiber optic communication network* in which the reference oscillator for *coherent* communication is derived from the same *light source* as used for generating the *signal* that is *detected.*

self-lasing fiber. See *fiber laser.*

Sellmeier equation. An equation that expresses the *group delay* per unit length for an *optical fiber,* given by the relation:

$$\tau(\lambda) = A + B\lambda^2 + C\lambda^{-2}$$

where *A, B,* and *C* are empirical fit parameters, or by the relation:

$$\tau(\lambda) = \tau_0 + S_0(\lambda - \lambda^2_0/\lambda)^2/8$$

where τ is the group delay per unit length of fiber, λ is the *wavelength,* τ_0 is the relative group delay minimum at λ_0, λ_0 is the *zero-dispersion wavelength,* and S_0 is the *zero-dispersion slope.* These functional forms may be assumed when measured group delay *data* near the zero-dispersion wavelength of a *dispersion-unshifted fiber* is numerically fitted for the purposes of calculating the *chromatic dispersion coefficient,* given by the relation:

$$D(\lambda) = d\tau(\lambda)/d\lambda = S_0\lambda(1 - \lambda_0^4/\lambda^4)/4$$

where all the variables are as defined above. Also see *group delay.*

semiconductor. See *n-type semiconductor; p-type semiconductor.*

semiconductor diode laser. See *laser diode.*

semiconductor laser. See *laser diode.*

sensitivity. In a device that develops an output *signal* as a result of an input signal, such as a *transducer* or *optical fiber communication system* receiver, the minimum input signal strength required to achieve a specified quality of performance, such as to produce a specified output signal having a specified *signal-to-noise ratio* or maintain a specified *bit-error ratio (BER).* For example, the input signal may be expressed as power in dBm or as *electric field strength* in microvolts per *meter,* with input *impedance* stipulated. The output signal may be expressed in a power or other unit, such as current, voltage, or lumens. Sometimes the term is improperly used as a synonym for *responsivity.* See *polarization sensitivity; receiver optical sensitivity.* Also see *responsivity.*

sensor. A device or means for extending the natural senses, such as a device that measures and indicates temperature, pressure, speed, humidity, force, chemical composition, or other physical variable or phenomenon. Examples of sensors include thermometers, gyroscopes, barometers, tachometers, and speedometers. *Optical fibers* are being used as the sensing element in various types of sensors. See *birefringence sensor; bobbin-wound sensor; coated-fiber sensor; critical-angle sensor; critical-radius sensor; distributed-fiber sensor; Fabry-Perot fiber optic sensor; fiber end-to-end separation sensor; fiber longitudinal-compression sensor; fiber optic sensor; fiber strain-induced sensor; fiber tension sensor; fiber transverse-ompression sensor; fiber axial-alignment sensor; fiber axial-displacement sensor; frustrated total (internal) reflection sensor; interferometric sensor; Kerr-effect sensor; Mach-Zehnder fiber optic sensor; Michelson fiber optic sensor; microbend sensor; moving-grating sensor; optical cavity mirror sensor; optical fiber acoustic sensor; photoelastic sensor; Pockels-effect sensor; polarization sensor; Sagnac fiber optic sensor.*

separation sensor. See *fiber end-to-end separation sensor.*

serrodyne. See *optical serrodyne.*

service. See *plain old telephone service (POTS).*

service restoral. See *mean time to service restoral (MTTSR).*

services digital (or data) network. See *integrated-services digital (or data) network (ISDN).*

session. A length of time, a temporary grouping of equipment, or a set of interactions among humans, among machines, or among humans and machines. Examples of sessions include the time interval between logging-on and logging-off a computer remote terminal, a connection between two terminals that allows them to communicate for a span of time, a temporary connection between a control panel and a ship or aircraft propulsion system controller to enable real-time online control of the system or one of its subsystems, and a conference telephone call. Also see *session layer.*

session layer. In *open-systems architecture,* the layer that provides the functions and services that may be used to establish and maintain connections among elements of a *session,* to maintain a dialog of requests and responses between the elements of a session, and to terminate the session. Also see *session; Open-Systems-Interconnection (OSI) Reference Model (RM).*

set. See *optical fiber connector set; optical loss test set (OLTS).*

sheath. See *optical fiber jacket.*

shell. In a *fiber optic connector,* the primary connector housing. See *backshell.*

shift. To change the *frequency* or *phase* of a *signal* or the change itself. See *Goos-Haenchen shift.*

shifted fiber. See *dispersion-shifted fiber; dispersion-unshifted fiber.*

short-haul communication network. A communication network designed for handling communication traffic usually over distances less than 20 km (kilometers) and within the area covered by a telephone *switching center* or central office, such as metropolitan areas, large cities, or an entire county. Short-haul systems are characterized by not having long-distance trunks between towns and cities nor more than one switching center or central office. Like *long-haul communication networks, short-haul communication networks* may have high-quality equipment for high-fidelity and high-definition *analog* (voice) and *digital transmission, integrated-services digital (data) networks (ISDN),* high transmission capacity, and automatic switching for handling calls and messages without operator assistance. It is anticipated that by the year 2000 over 600 billion voice circuit-kilometers will be on *fiber optic* long- and short-haul networks and only 16 billion on microwave, 160 billion on satellite, and 3 billion on coaxial cable. Also see *long-haul communication network.*

short wavelength. In *fiber optics,* pertaining to *optical radiation* that has a *wavelength* less than a nominal 1 μm *(micron).*

shot effect. See *shot noise.*

shot noise. Noise, that is, random surges of voltage, current, electrical power, or *optical power* caused in electronic and *optoelectronic* devices by random variations in the number and velocity of electrons emitted by heated cathodes, photosensitive surfaces subjected to incident *radiation,* and *p-n junctions* in *semiconductors.* The random noise occurs because electric currents consist of discrete electric charges and optical power consists of a stream of discrete photons, both of which are capable of irregular and erratic movements, such as clustering as well as reinforcing and cancelling one another. Also, shot noise from the electronic circuits in a *light source* can produce equivalent surges in *optical power output,* which will produce corresponding surges in *photodetector* electrical power output. Synonymous with *shot effect; Schottky effect; Schottky noise.*

signal. The code or *pulse* that represents intelligence, message, or control function conveyed over a communication system. See *analog signal; digital signal; two-level optical signal.*

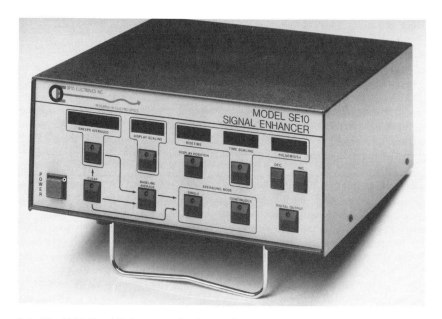

S–1. The SE10 Signal Enhancer, a **signal** averaging system that improves the *signal-to-noise ratio* of signals in the subnanosecond time and *gigahertz frequency* domains. The maximum improvement is 220X. All connections are made with standard coaxial *cables.* (Courtesy Opto-Electronics, Inc.).

signal format. A structure for the *transmission* of logical states that constitutes a method of transferring information from one point to another.

signaling rate. See *data signaling rate; range designation of data signaling rates.*

signal parameter. Any of the characteristics of a *signal,* such as the *frequency, pulse length, pulse amplitude, pulse duration, polarization* type, *modulation* type, strength, power, *irradiance,* or polarity.

signal-to-noise ratio. See *photodetector signal-to-noise ratio.*

signature. See *spectrum signature.*

significant instant. In a time-plot of a *signal,* such as a time-plot of a *pulsed electromagnetic wave,* an instant at which a particular type of a usually repetitive event occurs, such as a transition to a different *electric field strength, power level, polarization, frequency,* or *phase* in a *modulated wave.*

silica. See *fused silica.*

silica fiber. See *hard-clad silica (HCS) fiber; plastic-clad silica (PCS) fiber.*

silica process. See *doped-deposited-silica (DDS) process.*

simulator. See *equilibrium mode simulator.*

single heterojunction. In a *laser diode,* a *junction* involving two energy-level shifts and two *refractive-index* shifts. The junction provides increased confinement of *radiation* direction, improved control of radiative recombination, and reduced nonradiative (thermal) recombination.

single-mode fiber. An *optical fiber* in which only one *bound mode* can *propagate* at a given *wavelength* and *numerical aperture.* The lowest order bound mode may be a pair of orthogonally *polarized electric* and *magnetic fields.* For a fiber to support only one mode, the *core diameter* should be about three or four wavelengths of the operating *radiation,* and the *numerical aperture* must be adjusted accordingly. Thus, the ability to *guide, support,* or *propagate* only one mode will depend on the core radius, the *numerical aperture,* and the wavelength, that is, the *normalized frequency* (V-parameter or V-value) must be less than 4. Synonymous with *monomode fiber.* Also see *multimode fiber.*

single-mode launching. The insertion of an *electromagnetic wave* into a *waveguide* in such a manner that only one *propagation mode* is *coupled* into, and hence *transmitted,* by the guide. Single-mode launching can be accomplished by controlling the *incidence angle, beam diameter, skew ray* angle, *source*-to-*waveguide* gap, that is, *longitudinal offset,* and other parameters. Propagation of the mode will depend on waveguide dimensions, wavelength, and *refractive indices* of the material constituting the guide.

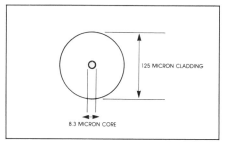

Cross Section Of Single Mode Fiber

SPECIFICATIONS

Cladding diameter   *125.0 ±2.0 micron*
Core diameter   *8.3 micron (nominal)*
Core eccentricity   *Less than or equal to 1.0 micron*
              (typically 0.3 micron)
Attenuation range   *0.35-0.5 dB/km at 1310 nm*
 (dB/km-maximum)   *0.21-0.3 dB/km at 1550 nm*
Zero dispersion wavelength   *1310 ±10 nm*
Maximum dispersion range   *1285 to 1330 nm,   3.2 ps/nm-km*
Refractive index delta   *0.37%*
Coating diameter   *245 +19 −13 micron*
Mode field diameter   *8.8 ±0.7 micron*
Cut-off wavelength   *Max. 1330 nm (5 meter reference length)*
Attenuation vs. wavelength   *Less than or equal to 0.1 dB/km*
              in the wavelength region from
              1285-1330 nm
Attenuation uniformity   *No loss discontinuities greater than*
              0.1 dB/km at either 1310 nm or 1550 nm

S-2. This **single-mode fiber** consists of a germanium-doped *core* and a silica *cladding*. A dual protective *coating* is applied over the cladding to cushion the fiber against *microbending losses*, provide abrasion resistance, and preserve the mechanical strength of the fiber. Fibers are proof-tested to survive installation and long-term loads. (Courtesy AT&T.)

single-node network. A *network* in which all stations, *user-end instruments,* or other devices are interconnected via one *node,* such as a single-star network.

skew ray. In an *optical fiber,* a *ray* that does not intersect the *optical axis.* Thus, it is a ray that is not parallel to the fiber *optical axis,* traverses a helical path around the axis, does not intersect it, and does not lie in a *meridional* plane.

slab-dielectric waveguide. An *electromagnetic waveguide* consisting of a *dielectric propagation medium* of rectangular cross section. The width and thickness of the guide, *refractive indices,* and *wavelength* determine the *modes* the guide will support and hence *transmit* over a long distance, that is, more than a hundred wavelengths or beyond the *equilibrium length.* The guide may be *cladded,* protected, distributed, and electronically controllable. Slab-dielectric waveguides are used in *integrated op-*

tical circuits (IOCs) for geometrical convenience, in contrast to the round *optical fibers* used in *fiber optic cables* for long-distance transmission. Their principle of operation is the same.

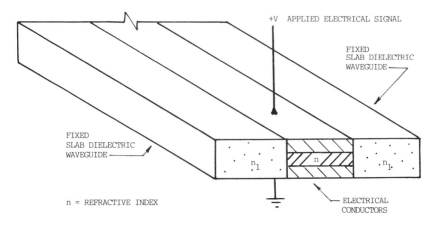

S–3. An electronically controllable *coupler* (ECC) between two **slab-dielectric (planar) wave-guides.**

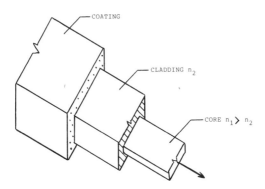

S–4. A **slab-dielectric waveguide** a few *microns* wide and thick, that might be formed, deposited, etched, or otherwise mounted on an *integrated optical circuit* chip.

slab interferometry. A method used to measure the *refractive-index profile* of a *dielectric waveguide,* such as an *optical fiber,* by using an *interferometer* that scans across the end face, perpendicular to the *optical axis,* of a thin slab, that is, a short length, of the waveguide. Also see *transverse interferometry.* Synonymous with *axial slab interferometry; axial interference microscopy.*

SLD. Superluminescent diode. See *superluminescent LED.*

sleeve splicer. See *precision-sleeve splicer.*

slope. See *chromatic dispersion slope; zero-dispersion slope.*

Snell's law. When *electromagnetic waves,* such as *lightwaves,* pass from a given *propagation medium* to a medium of greater density, its path is deviated toward the normal; when passing into a medium of lesser density, its path is deviated away from the normal to the surface at the point of incidence, density being proportional to the *refractive index.* Snell's law is given by the relation:

$$\sin \theta_1 / \sin \theta_2 = n_2 / n_1$$

where θ_1 is the *incidence angle,* θ_2 is the *refraction angle,* n_2 is the refractive index of the medium containing the *refracted ray,* and n_1 is the refractive index of the medium containing the incident ray. Also, the incidence angle and the *reflection angle* are equal. All angles are measured with respect to the normal.

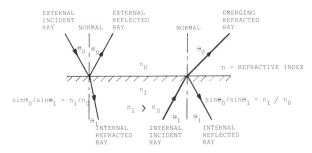

S-5. *Reflection* and *refraction* at an *optical interface* in accordance with **Snell's laws.** The *incidence* and *reflection angles* are equal. The ratio of the sines of the incidence and refraction angles is equal to the reciprocal of the ratio of the *refractive indices* of the incident and refraction media.

Sonet. See *synchronous optical network (Sonet) standard.*

source. See *fiber optic light source; isotropic source; line source; optical fiber source; optical source; standard optical source; video optical source.* (See Figs. S-6—S-7, p. 288)

source coupling efficiency. The maximum efficiency with which the *optical power,* that is, the integrated *irradiance,* luminous intensity, or *optical power density* of a *Lambertian radiator,* such as a *light-emitting diode (LED),* is coupled to an *optical fiber.* The efficiency is given by the relation:

$$\Gamma_{sc} = A_f (NA)^2 / 2 A_s n_0^2$$

where A_f is the cross-sectional area of the *fiber core; NA* is the *numerical aperture,* which is the sine of the *acceptance angle* or half of the apex angle of the *acceptance*

S-6. The M160 Fiber Optic Power Meter and S170 Dual Wavelength Fiber Optic Source. The *power meter* measures *optical power* from 10 mW to 10 nW (+10 to −50 *dB*) at 0.85, 1.30, 1.55 nm. The **source** has microlens *LEDs* at 0.85 and 1.30 μ (*microns*). (Courtesy fotec incorporated).

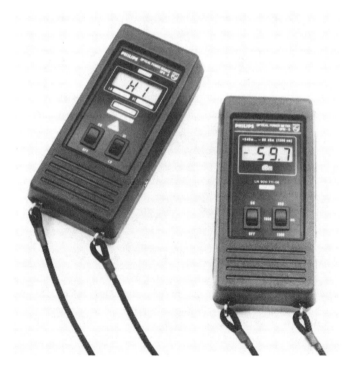

S-7. The OPS-5 hand-held *optical power* **source** (left) and the OPM-6 hand-held optical power meter. (Courtesy Philips Telecom Equipment Company, a division of North American Philips Corporation).

cone, and dependent upon the *refractive indices* of the core and the *cladding; A$_s$* is the cross-sectional area of the *source-emitted light beam;* and n_0 is the refractive index of the *propagation medium* between the light source and the optical fiber face.

source power efficiency. In an electrically powered *light source,* the ratio of the *optical power output* to the electrical input power.

source-to-fiber coupled power. The *optical power* that is actually *coupled* from a *light source* into the interior of an *optical fiber,* given by the relation:

$$P_c = P_0(\theta_a/\theta_e)T$$

where P_0 is the source *optical power output; θ_a* is the *acceptance cone* apex solid angle, or collection angle, of the fiber; θ_e is the *emission* solid angle, *beam divergence,* or *exit* (solid) *angle* of the light source; and *T* is the *transmission coefficient* of the *optical system,* mostly consisting of the source-to-fiber *interface. Angular misalignment loss* and *lateral offset loss* must also be taken into account.

source type. See *optical source type.*

space coherence. See *spatial coherence.*

space-division multiplexing. A method of deriving two or more simultaneous *channels* by dividing available *transmission media,* such as a *fiber optic cable,* into separate circuits. In contrast to a single coaxial cable, a fiber optic cable can have hundreds of *optical fibers* or *fiber bundles,* each fiber or fiber bundle constituting a separate *optical link,* with little or no *crosstalk* among fibers or bundles. The *signals* in each fiber or bundle can be *frequency-division multiplexed,* that is, separate *bandwidth* channels, and *time-division multiplexed,* that is, separate time slots for each channel, even before they are *wavelength-division multiplexed,* that is, separated by *color.*

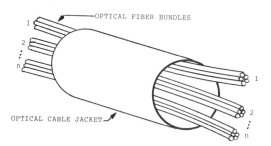

S–8. **Space-division multiplexing** by placing many *optical fibers* in one *fiber optic cable.* Each fiber can serve as a *waveguide* that can be *frequency-, wavelength-,* or *time-division multiplexed.*

spatial coherence. 1. In *electromagnetic waves,* pertaining to a fixed *phase* relationship between various points in the wave *propagating* in a given region of space or in a given *propagation medium,* such that recurring relationships can be predicted everywhere in the wave with that space. 2. In *fiber optic cables* containing an *aligned bundle* or bundles, pertaining to a fixed relationship between the *optical fibers* at one end of a bundle relative to the fibers at the other end, such that each fiber has the same relative position with respect to all other fibers at both ends. Thus, an image focused on one end of a bundle will appear, with only limited distortion, at the other end of the bundle. Synonymous with *space coherence.* Also see *aligned bundle.*

spatial resolution. See *distance resolution.*

specific detectivity. A figure of merit used to characterize *photodetector* performance, defined as the reciprocal of the *noise-equivalent power (NEP),* normalized to a unit area and a unit *bandwidth.* The specific detectivity, D^*, is given by the relation:

$$D^* = (A\Delta f)^{1/2}/\text{NEP}$$

where A is the area of the photosensitive surface of the detector that is exposed to *radiation,* Δf is the effective noise bandwidth, and NEP is the noise equivalent power; or by the relation:

$$D^* = D[A(\Delta f)]^{1/2}$$

where D is the detectivity, that is, the reciprocal of the noise-equivalent power; A is the area of the photosensitive surface of the detector exposed to radiation; and Δf is the effective noise bandwidth. Synonymous with *normalized detectivity; D-star.*

speckle effect. A mottling effect produced by *color distortion* due to *interference* in *light transmission* caused by the *propagation medium* itself, such as an *optical fiber.*

speckle noise. See *modal noise.*

speckle pattern. An *irradiance* pattern, that is, a *power-density* pattern, produced by the mutual *interference* of partially *coherent light beams* that are subject to minute temporal and spatial fluctuations. In a *multimode fiber,* a speckle pattern results from a superposition of *mode* field patterns. If the relative modal *group velocities* change with time, the speckle pattern will also change with time. If, in addition, *differential mode attenuation (DMA)* occurs, *modal noise* will also occur.

spectral absorptance. The absorptance of *electromagnetic radiation* by a material evaluated at one or more *wavelengths.*

spectral absorption coefficient. In the *transmission* of *electromagnetic waves* in a *propagation medium,* the *attenuation term, a,* in the relation:

$$I = I_0 e^{-ax}$$

where *I* is the *radiance* or the *irradiance,* that is, the *optical power density* or intensity of *radiation* at point *x,* and I_0 is the initial radiance or irradiance at the point from which *x* is measured, when the *attenuation* is caused only by *absorption,* and not by *refraction, diffraction, scattering, dispersion,* diffusion, *divergence, geometric spreading,* or other causes.

spectral bandwidth. See *spectral width.*

spectral density. In a finite *bandwidth* of *electromagnetic radiation* consisting of a continuous *spectrum* or of a distributed spectrum of *frequencies,* the *optical power* distribution expressed in watts per *hertz* of the given bandwidth.

spectral dispersion. See *material dispersion.*

spectral emittance. The *radiant emittance* plotted as a function of *wavelength.*

spectral irradiance. The *irradiance* per unit *wavelength* interval at a given nominal median wavelength, usually expressed as power per unit area, such as watts per square meter, per unit wavelength interval, for example, $W/(m^2\text{-nm})$. It may be calculated as (1) the irradiance contained in an elementary range of wavelengths at a given wavelength, divided by (2) the range of wavelengths.

spectral line. In the *optical spectrum,* an extremely narrow range of wavelengths of *light,* that is, *monochromatic electromagnetic radiation,* generated when an atomic or molecular particle undergoes a transition from a higher to a lower energy level, such as when an electron in an atom drops from a higher, or excited, energy state to a lower, or ground, state. Excited states are created when an atom or molecule absorbs energy, such as the energy in *photons* or other forms of electromagnetic energy. In such *quantum* mechanical systems, only certain energy levels are possible. Only certain transitions can occur, giving rise to *absorption* and *emission* of only certain quanta of energy, as expressed by *hf,* where *h* is *Planck's constant* and *f* is the *frequency* of the absorbed or emitted photon. Therefore, only certain discrete *wavelengths* of *optical radiation* exist in a given spectral line.

spectral linewidth. A measure of the distribution of *wavelengths,* that is, the wavelength composition of the *optical radiation,* in a *spectral line.* A quantitative measure of spectral linewidth is the difference between the shortest wavelength and the longest wavelength in the spectral line. Also see *spectral width.*

spectral-loss curve. In *fiber optics,* a plot that shows *attenuation* as a function of *wavelength* of *light propagating* in an *optical fiber.* Spectral *loss* curves should be normalized before meaningful comparisons can be made.

spectral power. See *peak spectral power.*

spectral radiance. The *radiance* per unit *wavelength* interval at a given nominal wavelength, expressed in watts per steradian per unit area per wavelength interval. For example, spectral radiance can be expressed as $(W/ster)/(m^2/\mu)$. It may be calculated as (1) the radiance contained in an elementary range of wavelengths at a given wavelength, divided by (2) the range of wavelengths.

spectral reflectivity. The *reflectivity* of a surface evaluated as a function of *wavelength.*

spectral response. 1. In *fiber optics,* a graph representing the specific *wavelength* versus *detected* power curve of a *photodetector.* The spectral response curve is typically drawn with *wavelength* on the ordinate axis and spectral responsivity on the abscissa. It shows the wavelength dependence of the detector. **2.** The *optical power output* of a *light source* as a function of driving current or driving power, that is, the *responsivity.* It may be indicated as watts/ampere, lumens/(ampere-second), or lumens/coulomb.

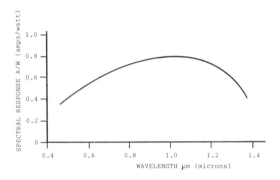

S–9. The **spectral response** curve of a typical *photodiode.*

spectral responsivity. The *responsivity* per unit *wavelength* interval at a given nominal or median wavelength, such as watts/ampere-*nanometer* at 1.31 μ *(microns).*

spectral transmittance. *Transmittance* evaluated at one or more *wavelengths.*

spectral wavelength. See *peak spectral wavelength.*

spectral width. The *wavelength* interval in which a *radiated* spectral quantity is a specified fraction of its maximum value. The fraction is usually taken at 0.50 of the maximum power level, or 0.707 of maximum (3 dB) current or voltage level. In a *lightwave,* the wavelength interval is the difference between the wavelengths at which the power level at those wavelengths is 3 dB down from the maximum power level. For example, the spectral width of a *laser* is an order of magnitude less than that of a *light-emitting diode (LED)*. *Optical fiber* wavelengths are of the order of 1 μ *(micron)*. Spectral widths of LEDs are of the order of 20 to 50 nm *(nanometers)*, whereas lasers are of the order of 1 or 2 nm. Dispersion of an *optical pulse* in a fiber might be expressed in nanoseconds per microsecond of launched pulse per nanometer of spectral width of the source. The difference of the *electromagnetic frequencies* corresponding to the two limiting wavelengths would express the same spectral width in *hertz*. If the spectral width of a laser is 1 nm, this would correspond to a spectral bandwidth of 3×10^{11} Hz. If the nominal wavelength is 1 μ, the nominal frequency is 3×10^{14} Hz. The ratio of width-to-nominal frequencies is 10^{-3}, and the ratio of width-to-nominal wavelengths is also 10^{-3}. Synonymous with *spectral bandwidth*. Also see *impulse-response function; monochromatic; root-mean-square (rms) pulse broadening; root-mean-square (rms) pulse duration; root-mean-square (rms) deviation; spectral linewidth; spectral width.*

S–10. The A500 Fiber Optic Spectrum Analyzer for characterizing the *spectral* output of a *source* for both *peak wavelength* and **spectral width**. It features *digital* readout of *wavelength* and operates in the 0.60 to 1.70 μ *(micron)* range. (Courtesy fotec incorporated).

spectral window. In a given *propagation medium* or material, a *wavelength* region of relatively high *transmittance*, surrounded by regions of low transmittance, that is, a trough in a plot of *optical attenuation*, caused by the material, versus wave-

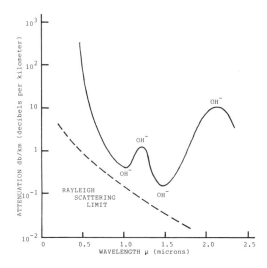

S–11. *Attenuation* versus *wavelength* for a *low-loss optical fiber,* showing the peaks and **spectral windows** (valleys) of hydroxyl ion *absorption.*

length. The best region of the *optical spectrum* for *transmission* is between 1.1 and 1.6 μ *(microns),* with spectral windows at 1.1, 1.3, and 1.55 μ. The 0.8- to 0.9-μ region is also favorable for transmission and is currently in wide use because components operating at this wavelength are more readily available. To achieve lower *losses,* that is, reduced *attenuation rates,* particularly for *long-haul communications,* longer wavelengths, deeper into the *infrared* will be required. For *local-area networks,* such as *networks* in urban areas, *local loops,* and networks on ships and aircraft, use of the *far-infrared region* of the optical spectrum may not be necessary. Synonymous with *transmission window.*

spectrometer. A spectroscope equipped with an angle scale capable of measuring the angular deviation of *radiation* of different *wavelengths.* The spectrometer may also be used to measure angles between surfaces of *optical* elements.

spectroscope. An instrument capable of dispersing *radiation* into its component *wavelengths* and observing, or measuring, the resultant spectrum.

spectrum. A continuous range or group of *frequencies,* or groups of ranges, of *waves* that have something in common, for example, all *frequencies* that comprise a string of similar, equally spaced, retangular *pulses;* all the frequencies that comprise *visible light;* a *band* of radio frequencies; or all the frequencies in a sound wave, such as a musical chord. See *electromagnetic spectrum; line spectrum; optical spectrum; visible spectrum; visual spectrum.*

spectrum analyzer. See *lightwave spectrum analyzer.*

spectrum designation of frequency. A method of referring to a range or *band* of communication *frequencies.* In American practice the designator is a two- or three-letter abbreviation for the name. In ITU practice, the designator is numeric. These frequency ranges, or bands, and their designators are given in the TABLE OF FREQUENCY RANGES.

Table of Frequency Ranges

FREQUENCY RANGE (Lower Limit Exclusive, Upper Limit Inclusive)	AMERICAN DESIGNATOR	ITU FREQUENCY BAND DESIGNATOR
Below 300 Hz	ELF (Extremely Low Frequency)	2
330–3000 Hz	ULF (Ultra Low Frequency)	3
3–30 kHz	VLF (Very Low Frequency)	4
30–300 kHz	LF (Low Frequency)	5
300–3000 kHz	MF (Medium Frequency)	6
3–30 MHz	HF (High Frequency)	7
30–300 MHz	VHF (Very High Frequency	8
300–3000 MHz	UHF (Ultra High Frequency)	9
3–30 GHz	SHF (Super High Frequency)	10
30–300 GHz	EHF (Extremely High Frequency)	11
300–3000 GHz	THF (Tremendously High Frequency) Frequency)	12

spectrum signature. The *frequencies* that make up the *waves* emanating from a particular *source,* such as the frequencies (or *wavelengths*) in *lightwaves* emanating from a *light-emitting diode (LED),* or the frequencies contained in a sound wave emanating from a given ship.

specular reflection. *Reflection* from a smooth surface, curved or straight, so that there is negligible diffusion as *Snell's laws* of reflection and *refraction* are microscopically obeyed over a uniformly directed surface. Thus, only clear images that are sharp are obtained in specular reflection.

specular transmission. The *propagation* of *lightwaves* in a *propagation medium* such that *diffusion attenuation* is negligible, and smooth changes in *refractive indices* result in *refraction* of *rays* at the microscopic level and the preservation of clear images during propagation. Specular transmission is usually desired in *optical waveguides,* such as *optical fibers, bundles,* and *cables.*

speed. See *optical speed.*

splice. See *cable splice; fiber optic splice; fusion splice; mechanical splice; optical fiber splice; optical waveguide splice; waveguide splice.*

splice closure. See *cable splice closure.*

splice housing. See *fiber splice housing.*

splice loss. An *insertion loss* caused by a *splice* in an *optical fiber.* The loss results from *angular misalignment, longitudinal offset, lateral offset,* lateral eccentricity, differences in *core diameter* (larger to smaller), *Fresnel reflection* at the *interface* or interfaces due to *refractive index* mismatch, and other causes. Splicing loss can vary from a small fraction of a *decibel,* about 0.1 dB, to several dB. Synonymous with *splicing loss.*

S-12. The Orionics FW-305 *Optical Fiber Splicer* featuring *multimode* and *single-mode fiber splicing,* 60X view screen, automatic piezoelectric *alignment,* and an average **splice loss** less than 0.10 dB, including many *splices* measuring 0 *dB.* (Courtesy Orionics Products, Controls Division, AMETEK, Inc.).

splice organizer. See *fiber-and-splice organizer.*

splicer. See *loose-tube splicer; precision-sleeve splicer.*

splice tray. See *fiber optic splice tray.*

splicing. The joining of two *optical fibers* or two *fiber optic cables.* See *field-splicing; fusion-splicing.* See Figs. S-13 and S-14.

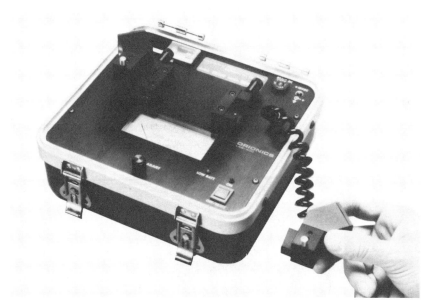

S-13. The Orionics LAS-400 Local Alignment System. Local injection and detection fixtures coupled with a *splice* indication meter optimizes *alignment* of *opical fibers,* allows immediate evaluation of *splice* quality, and can be used with most *fusion* **splicing** and bonding equipment. (Courtesy Orionics Products, Controls Division, AMETEK, Inc.).

splicing loss. See *splice loss.*

splitter. See *beam splitter; fiber optic splitter.*

spontaneous emission. 1. *Electromagnetic radiation,* such as *radiation* from a *light-emitting diode (LED)* or an *injection laser* operating below the *lasing threshold, emitted* when the energy level of an internal *quantum* system, such as electron energy levels in a population of molecules, drops to a lower energy level without regard to the simultaneous presence of similar radiation. **2.** The *emission* of *electromagnetic radiation,* such as *light,* that does not bear an amplitude, *phase, frequency,* time, or other relationship with an applied *signal* and is therefore a spurious, random, or noiselike form of *radiation.*

spot. See *launch spot.*

spot illumination. See *fiber optic spot illumination.*

spreading. See *geometric spreading.*

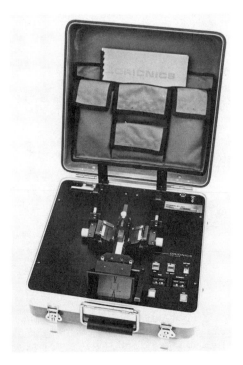

S-14. The Orionics FW-303 *Optical Fiber Splicer* 100–400 μ *(micron)* for **splicing** *multimode optical fibers,* featuring 60X view screen, V-groove fast *alignment,* and battery or external power source, including 12 V DC. (Courtesy Orionics Products, Controls Division, AMETEK, Inc.).

spurious emission. *Electromagnetic radiation* having *frequencies* that are outside the necessary *emission bandwidth,* the level of which may be reduced without affecting the corresponding *transmission* of information. Such emissions include harmonic emission, parasitic emission, and intermodulation products.

SRD. Superradiant diode. See *superluminescent LED (SRD).*

standard. See *fiber distributed-data-interface (FDDI) standard; synchronous optical network (Sonet) standard.*

standard operating condition. In the design of an *optical station/regenerator section,* a nominal environmental condition over which specified parameters must maintain their stated performance ratings, e.g., terminal input voltage range, room ambient temperature range, relative humidity range, electrical interfaces, *optical line rate,* and other parameters that may be specified as standard. Also see *extended operating condition.*

standard optical source. A reference *optical source* to which *emitting* and *detecting* devices are compared for calibration purposes. In the United States, standard optical sources must be traceable to the National Institute of Standards and Technology (NIST).

standing wave. In contrast to a *propagating wave,* a wave that exists only between two points or surfaces in a spatial dimension, such as between the two ends of a *transmission* line, opposite surfaces of an *optical cavity,* or the fixed end of an elastic string, but does not propagate beyond the fixed limits. For example, a crest of a standing wave simply increases, decreases, and reverses in a transverse direction while remaining at the same spatial point on the line, in the cavity, or on the string. However, the amplitude has maxima and minima along a longitudinal direction, with a *wavelength* depending on the *frequency,* the distance between fixed points, the characteristics of the vibrating or oscillating entity, and the characteristics of the environment. In a resonant cavity, a standing *electromagnetic wave* can be generated by two waves propagating in opposite directions, each *reflecting* from opposite walls of the cavity.

star coupler. In *fiber optics,* a *coupler* that distributes *optical power* from one input *port* to two or more output ports; that combines optical power from two or more input ports and distributes it to a larger number of output ports; or that combines optical power from two or more input ports and distributes it to a smaller number of output ports. See *reflective star coupler.*

star-mesh network. 1. In *network* topology, a *mesh network* in which one or more *nodes* serve also as the central node of a *star network.* Inversely, a star-mesh network is a communication network configuration in which the central nodes of two or more star networks are connected by a *bus,* such as by a partially or fully connected mesh network bus. **2.** A *mesh network* in which a large number of *local loops* or *data links* are connected to mesh *nodes* that serve as network control, *processing,* and *switching centers.* Synonymous with *distributed star-coupled bus; multiply connected star network.*

star network. In *network* topology, the interconnection of *nodes* such that there is a direct path between each node and a central node that serves as a control node. The central node is connected directly to all *end-point nodes,* which usually are the locations of equipment, such as a data station, ship system, computer, display device, switchboard, or other *user end-instrument.*

station cable. See *optical station cable.*

station facility. See *optical station facility.*

station optical cable. See *optical station cable.*

station/regenerator section. See *optical station/regenerator section.*

statistical fiber optic cable facility loss. See *statistical optical cable facility loss.*

statistical loss-budget constraint. In the design of an *optical station/regenerator section* using statistical methods, the constraint imposed by the relation:

$$\mu_{G-L} > 2\sigma_{G-L}$$

where

$$\mu_{G-L} = \mu_G - \mu_L \text{ and } \sigma_{G-L}{}^2 = \sigma_G{}^2 + \sigma_L{}^2$$

where μ_G is the *statistical (mean) terminal/regenerator system gain (dB)*, μ_L is the *statistical (mean) optical cable facility loss (dB)*, σ_G is the standard deviation of the statistical terminal/regenerator system gain (dB), and σ_L is the standard deviation of the *fiber optic cable* within the *optical cable facility* of the optical station/regenerator system section, assuming distributions are Gaussian, *splice losses* are uncorrelated with *optical fiber* losses, cable losses are uncorrelated from reel to reel, loss allowance for *fiber optic transmitter wavelength* drift is correlated in different reels, averages and sigmas are representative over time, sample sizes are sufficient to warrant use of Gaussian statistical theory, and product distributions are reasonably Gaussian in shape. Also see *loss-budget constraint.*

statistical optical cable facility loss. In a *single-mode optical station/regenerator section,* the mean *loss (dB)* in the *optical cable facility,* i.e., in the outside plant, given by the relation:

$$\mu_L = \ell_t(\mu_c + \mu_{cT} + \mu_{S\lambda}) + N_S(\mu_S + \mu_{ST})$$

and the standard deviation is given by the relation:

$$\sigma_L = [\ell_t\ell_R(\sigma^2{}_{cmx} + \sigma^2{}_{cT}) + N_S(\sigma_S{}^2 + \sigma_{ST}{}^2)]^{1/2}$$

where ℓ_t is the total *jacket* length of *spliced*-fiber cable (km); ℓ_R is the reel length of fiber cable (km); μ_c is the mean end-of-life cable *attenuation rate* (dB/km) at the transmitter *nominal central wavelength;* $\mu_{S\lambda}$ is the increase in mean cable attenuation rate above μ_c measured at the *wavelength* within the transmitter central wavelength range at which the largest mean-plus-two-sigma loss occurs; μ_{cT} and σ_{cT} are the mean and standard deviation of the effect of temperature on cable loss (dB/km) at the worst-case temperature conditions over the expected cable operating temperature; σ_{cmx} is the standard deviation of cable loss (dB/km) determined at the wavelength at which the largest mean-plus-two-sigma cable loss occurs, including an allowance for uncertainty in cable-loss measurements; N_S is the number of splices in the length of the cable in the optical cable facility, including the splice at the *optical station facility* on each end and allowances for cable repair splices; μ_{ST} and σ_{ST} are the mean and standard deviation of the effect of temperature on splices (dB/splice) at the worst-case temperature conditions over the cable operating temperature range, that must be specified for underground, buried, and aerial applications; and μ_S and σ_S are the mean and standard deviation of splice loss (dB/splice). Synonymous with *statistical fiber optic cable facility loss.*

statistical terminal/regenerator system gain. In a *single-mode optical station/regenerator section,* the *transmitter optical power* less the receiver *sensitivity* (power) and the optical power requirements of the *terminal* or *regenerator station facilities,* such as the *dispersion power penalty,* the *reflection power penalty,* the overall *safety margin,* and only the power losses in cables and connectors within the station facility at both ends. The mean statistical terminal/regenerator system gain is given by the relation:

$$\mu_G = P_T - P_R - P_D - R_P - M - \mu_{WDM} - \ell_{SM}\mu_{SM} - N_{con}\mu_{con}$$

and the standard deviation is given by the relation:

$$\mu_G = [\sigma^2_{PT} + \sigma^2_{PR} + \sigma^2_{WDM} + \sigma^2_{PD} + N_{con}\sigma^2_{con}]^{1/2}$$

where P_T and σ_{PT} are the mean and standard deviation transmitter power, the *optical power* (dBm) into the single-mode station cable on the line side of the transmitter unit *fiber optic connector* or *splice,* specified as P_{T1} and σ_{PT1} for *standard operating conditions* or P_{T2} and σ_{PT2} for *extended operating conditions;* P_R and σ_{PR} are the mean and standard deviation of the receiver *sensitivity,* the input *optical power (dBm)* to the receiver on the line side of the *fiber optic receiver* connector or splice, specified as P_{R1} and σ_{PR1} for standard operating conditions or P_{R2} and σ_{PR2} for extended operating conditions, that are necessary to achieve the manufacturer-*specified bit-error ratio (BER)* and includes the performance degradations caused by manufacturing variations introduced by temperature and aging drifts, maximum transmitter power penalty resulting from the use of a transmitter with a worst-case *extinction ratio,* and maximum transmitter power penalty resulting from use of a transmitter with a worst-case *rise* and *fall time,* but does not include the power penalty associated with *dispersion;* P_D and σ_{PD} are the mean and standard deviation of the *dispersion power penalty (dB);* R_P is the *reflection power penalty (dB);* M is the overall *safety margin (dB);* σ_{WDM} and μ_{WDM} are the mean and standard deviation of the all-inclusive *loss* (dB) associated with *wavelength-division-multiplexing* equipment at both ends, including effects of temperature, humidity and aging; ℓ_{SM} is the total length (km), and μ_{SM} is the mean *attenuation rate* (dB/km) for the single-mode station/regenerator cable; N_{con} is the number of single-mode connectors, not including the transmitter and receiver unit connectors; and μ_{con} and σ_{con} are the mean and standard deviation of the single-mode connector *insertion loss.* Also see *statistical optical cable facility loss.*

steady-state condition. See *equilibrium modal power distribution.*

step-index fiber. An *optical fiber* with a uniform *homogeneous isotropic core* and *cladding refractive-index profile,* with the core *refractive index* being greater than the cladding refractive index. There may be more than one layer of cladding each with a different refractive index. For the step-index fiber, the *profile parameter, g,* is infinite. Also see *graded-index (GI) fiber.*

step-index optical waveguide. A *dielectric waveguide* that has a *core* of uniform *refractive index* and one or more *claddings,* each with a different uniform refractive index.

step-index profile. A *refractive-index profile* characterized by a uniform *refractive index* within the *core,* a sharp decrease in refractive index at the core-*cladding interface,* and a uniform refractive index in one or more claddings. This corresponds to a *power-law index profile* with the *profile parameter* approaching infinity. See *equivalent step-index (ESI) profile.*

stepwise-variable optical attenuator. A device that *attenuates* the *irradiance* or *radiance* of *lightwaves* in discrete steps, each of which is selectable by some means, such as by changing sets of *filter* cells, rather than over a continuously variable range. For example, if fixed attenuation cells of 0, 1, 2, 6, 10, dB *(decibels)* are used three at a time, all attenuations from 0 to 14 dB are achievable in steps of 1 dB. The attenuator may be inserted in an *optical link* to control the irradiance, that is, the intensity of *light* at the receiver *photodetector.* Also see *continuously variable optical attenuator.*

stimulated emission. 1. *Electromagnetic radiation,* such as *radiation* from an *injection-locked laser* operating above the *lasing threshold, emitted* when the energy level of an internal *quantum* system, such as electron energy levels in a population of molecules, drops to a lower energy level when induced to do so by the simultaneous presence of *radiation* at the same *wavelength.* **2.** In a *laser,* the *emission* of *light* caused by a *signal* applied to the laser such that the response is directly proportional to and in *phase coherence* with the *electromagnetic field* of the stimulating signal. This coherence or correlation, between the applied signal and the response is the key to the usefulness of the laser as a *fiber optic light source* for *optical systems.*

stimulated emission of radiation. See *light amplification by stimulated emission of radiation (laser).*

stop-band. The *frequency band* of *electromagnetic waves* that a *filter* does not allow to pass. All frequencies higher than the highest in the stop-band and lower than the lowest in the stop-band are allowed to pass. If the stop-band is a single frequency or an extremely narrow band of frequencies, the filter is called a *notch filter.*

stria. A defect in *optical* materials consisting of a sharply defined streak of material having a slightly different *refractive index* than the main body of the material. Striae usually cause *wave*like distortions in the images of objects seen through the material, exclusive of similar *distortion* caused by variations in thickness or curvature. Striae are usually caused by temperature variation, or poor mixing of ingredients, causing *optical density,* that is, *refractive-index,* variations from place to place in the material.

strain-induced sensor. See *fiber strain-induced sensor.*

strength. See *electric field strength; field strength; magnetic field strength; optical fiber tensile strength; twist strength.*

stretching. See *optical pulse stretching.*

stripe laser diode. A *laser diode* usually fabricated as a multiheterostructured monolithic element with deposited metallic stripes for electrical conductors to create necessary excitation, *electric fields,* and contacts.

stripper. See *cladding-mode stripper; fiber optic mode stripper.*

structures. See *fiber optic supporting structures.*

stuffing process. See *molecular stuffing process.*

subscriber loop. See *local loop.*

superfiber. An *optical fiber* that has the best attributes of all the specialty fibers, such as high strength, long shelf-life, precision *polarization* control, precision *dispersion shifting,* high hermeticity, high-level *radiation hardness,* low *attenuation rate,* benign *cabling,* and low cost.

superluminescence. The phenomenon of amplification of *spontaneous emission* in a gain medium, such as the medium in an *avalanche photodiode,* characterized by *spectral line* narrowing and directionality. This process is generally distinguished from *lasing* action by the absence of feedback and hence the absence of well-defined *modal distribution* of *optical power.* Synonymous with *superluminescent effect; superradiance.*

superluminescent effect. See *superluminescence.*

superluminescent LED (SRD). A *junction*-type semiconducting device that *emits optical radiation* caused by *superluminescence.*

superradiance. See *superluminescence.*

supporting structures. See *fiber optic supporting structures.*

suppressor. See *optical pulse suppressor.*

surface. See *optical surface; reference surface.*

surface center. See *reference-surface center.*

surface-emitting LED. A *light-emitting diode (LED)* that *emits radiation* normal to the plane of the *junction,* that is, emission is perpendicular to the surface plane, rather than from the edges of the surface plane. The surface-emitting LED has a lower output *irradiance* and a lower *coupling efficiency* to an *optical fiber* or *integrated optical circuit (IOC)* than the *edge-emitting LEDs* and the *injection lasers.* Surface- and edge-emitting LEDs provide several milliwatts of *optical power* in the 0.8-1.3-μ *(micron) wavelength* range at drive currents of 100 to 200 mA. *Diode lasers* at these currents provide tens of milliwatts of optical power. Synonymous with *front-emitting LED; Burrus diode; Burrus LED.*

surface noncircularity. See *reference-surface noncircularity.*

surface tolerance field. See *reference-surface tolerance field.*

surface wave. An *electromagnetic wave* that *propagates* along the *interface* surface between different *propagation media* in accordance with the geometrical shape of the surface and the properties of the propagation media near that surface, such as the *electric permittivity, magnetic permeability,* and *electrical conductivity,* whose values define the *refractive indices* of the media. The wave is guided by the interface surface between the two different media, such as by a refractive-index gradient in the media. The *electromagnetic field components* of the surface wave may exist throughout space, but become negligible within a short distance from the interface. The surface wave propagates along the interface between two different propagation media without *radiating* away from the interface surface. *Optical* energy is not converted from the surface wave field to another form of energy, and the wave does not have a component directed normal to the interface surface. In *optical fiber transmission, evanescent waves* are surface waves. In radio transmission, ground waves are surface waves that propagate close to the surface of the earth, the earth having one *refractive index* and the atmosphere another, thus constituting an interface surface.

susceptibility. See *ambient light susceptibility.*

swim. Slow, graceful, undesired movement of display elements, groups, or images about their mean or normal position on the display surface of a display device, for example, the slow movement of the display image on the *faceplate* of a *fiberscope* that might be caused when the *objective* or object is moved slowly relative to the *aligned bundle* at the transmitting end, or the slow movement of the display on the screen of a CRT when the deflection plate bias voltage drifts slowly. Swim is the same as *jitter,* except that swim is usually slower and of larger amplitude than jitter, and usually refers to images rather than *pulses.* Also see *jitter.*

switch. See *fiber optic switch; latching switch; nonlatching switch; optical switch; thin-film optical switch.*

switch bounce time. Either the *switch operating bounce time* or the *switch release bounce time.*

switching center. In *optical fiber communication systems,* an installation in which switching equipment is used to interconnect *optical links* and circuits in a *network* for message-switching or circuit-switching operation. A *switching center* is usually located at a *node* in the network.

switch operating bounce time. In a *fiber optic switch,* the time between the instant an applied actuating force or *signal* reaches a specified value and the instant the *transmittance* permanently changes 90% of its maximum value minus the *switch operating time.* Also see *switch release bounce time.*

switch operating time. In a *fiber optic switch,* the time between the instant an applied actuating force or *signal* reaches a specified value and the instant the *transmittance* changes 90% of its maximum value.

switch release bounce time. In an *optical switch,* the time interval between the cessation of an applied actuating force or *signal* and the time that the optical transmittance has permanently changed by 90% of its maximum value, minus the *switch release time.*

switch release time. In a *fiber optic switch,* the time between the instant an applied actuating force or *signal* ceases and the instant the *transmittance* has changed by 90% of its maximum value.

synchronous optical network (Sonet) standard. A *fiber optic communication network* standard, developed by the American National Standards Institute (ANSI), in which a set of *optical transmission rates* and formats are established so that operating telephone companies can synchronize public telephone *networks* and efficiently interconnect high-speed *fiber optic links* using an 8-bit byte format, rather than the single bit-stream format. For example, the byte format will enable the telephone network to be efficiently synchronized to an 8-bit frame so that *optical signals* at 45 Mbps (megabits/second) need not be *demultiplexed* down to 1.5 Mbps for routing signals around the network. Thus, *data* can be transmitted at any *data signaling rate* on any line. Also see *fiber distributed-data-interface (FDDI) standard.*

system. See *advanced television system; enhanced-definition television (EDTV) system; fiber optic telemetry system; high-definition television (HDTV) system; imaging system; improved-definition television (IDTV) system; integrated fiber optic system; metric system; optical system; telemetry system.*

system gain. See *statistical terminal/regenerator system gain; terminal/regenerator system gain.*

system information rate. In *optical transmission systems,* the bit-repetition rate in bits/second internal to equipment that when acted upon by the line-coding algorithm results in the *optical line rate.*

systems architecture. See *open-systems architecture.*

systems interconnection. See *Open-Systems-Interconnection (OSI) Reference Model (RM).*

T

tactical fiber optic cable. A *fiber optic cable* specially designed and constructed to meet the stringent environmental and mechanical needs of the military. The cables are tested for cyclic flexing, radial compression, cold bend, freezing water, flammability, impact, and other tests in accordance with military specifications. (See Figs. T–1 and T–2.)

tap. 1. To remove a part of the *signal* energy from a line, such as an *optical fiber* carrying *optical pulses.* Tapping may be authorized for purposes of establishing a signal line, or it may be clandestine, though clandestine tapping of optical fiber is

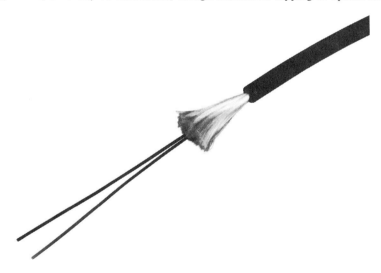

T–1. A *single-mode* two-fiber **tactical fiber optic cable** specifically designed to meet military needs. In this cable two radiation-qualified *single-mode buffered optical fibers* are twisted together forming the center. Aramid yarns are stranded around the fibers in two opposing layers to provide a torque-balanced design. Directly over the aramid is a flame-retardant polyurethane *jacket* containing four reinforcing elements evenly spaced in a helical lay. (Courtesy AT&T.)

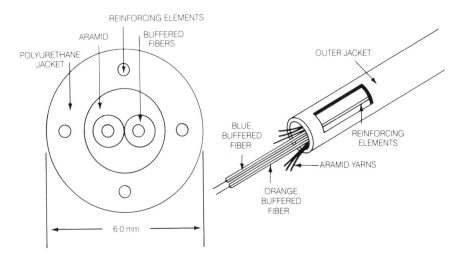

T–2. Cross section and cutaway of the *single-mode* two-fiber **tactical fiber optic cable**. (Courtesy AT&T.)

difficult to accomplish without *detection*. **2.** The physical connection to a *signal* line that removes part of the signal energy in a line, such as a *branch* or *coupler*.

taper. See *optical taper*.

tapered fiber. An *optical fiber* whose cross section increases or decreases with distance along the *optical axis* of the fiber.

tapoff. See *optical tapoff*.

target resolution. See *borescope target resolution*.

T-coupler. See *tee coupler*.

TE. *Transverse electric.*

technique. See *cutback technique; near-field scanning technique; rod-in-tube technique*.

tee coupler. In *fiber optics,* a *coupler* that connects three *ports* and operates in such a manner that *optical power* into any one port is distributed to the other, or power into any two ports can be combined and sent out on the third. In effect, it is a three-port *star coupler*. Synonymous with *T-coupler*. (See Fig. T-3).

telemetry. The branch of science and technology devoted to measuring a quantity or quantities, such as motion, pressure, temperature, humidity, *radiation,* sound,

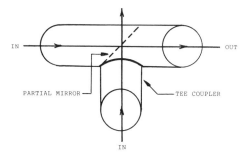

T-3. A *fiber optic* **tee coupler**. The coupler consists of two *optical fibers* joined to form a T-shaped junction at which is placed a *partial mirror*. The partial mirror allows two input *signals,* each on a different fiber, to be transferred to a single output fiber. For example, an *optical signal* input from the left is *transmitted* by the partial mirror to the output *port* at the right. An optical signal from below is *reflected* by the partial mirror also to the output port at the right. The partial mirror is placed at 45° to the *optical axes* of the fibers. The part of the signal from below that is transmitted by the partial mirror is lost, or it may be transmitted away via another fiber, where it will *mix* with the reflected part of the signal from the left.

smoke, flame, or toxic fumes, *transmitting* the results of the measurements to a distant station, and there interpreting, indicating, or recording the measured values.

telemetry system. A system, usually consisting of (1) a *sensor,* that is, a *data source* that measures the absolute values and variations of a physical variable, such as acceleration, pressure, temperature, fluid flow, humidity, *radiation,* sound, smoke, flame, or toxic fumes; (2) a data sink that receives, converts, displays, or otherwise processes the data, such as a *photodetector* and indicator; and (3) a *data link* that connects the source to the sink. Usually the sensor is at a remote location and the sink is at another location where the data can be interpreted, displayed, recorded, or used. In some systems, *signals* can be sent from the sink or other location to the sensor to reset, calibrate, test, or otherwise control the sensor. See *fiber optic telemetry system.*

telephone service. See *plain old telephone service (POTS).*

television system. See *advanced television system; enhanced-definition television (EDTV) system; high-definition television (HDTV) system; improved-definition television (IDTV) system.*

TEM mode. *Transverse electromagnetic mode.*

TE mode. *Transverse electric mode.*

Tempest-proofed fiber optic cable. *Fiber optic cable* that has been tested and certified to U.S. National Security Agency and U.S. Department of Defense standards in regard to *emission power levels* of *optical radiation.*

template. See *four-concentric-circle near-field template; four-concentric-circle refractive-index template.*

temporal coherence. See *time coherence.*

tensile strength. See *optical fiber tensile strength.*

tension sensor. See *fiber tension sensor.*

terahertz. A unit of *frequency* equal to 10^{12} Hz *(hertz).*

term. See *attenuation term; phase term.*

terminal/regenerator system gain. In a *single-mode fiber optic transmission system,* the *transmitter optical power* less the *receiver sensitivity* (power) and the optical power requirements of the *terminal* or *regenerator station facilities,* such as the *dispersion power penalty,* the *reflection power penalty,* the overall *safety margin,* and only the power *losses* in cables and connectors within the station facility at both ends. The terminal/regenerator system gain is given by the relation:

$$ G = P_T - P_R - P_D - R_P - M - U_{WDM} - \ell_{SM}U_{SM} - N_{con}U_{con} $$

where P_T is the transmitter power, P_R is the receiver sensitivity (power), P_D is the dispersion power penalty, R_P is teh reflection power penalty, M is the overall safety margin, U_{WDM} is the worst-case value of all losses associated with *wavelength-division multiplexing* equipment at both ends. ℓ_{SM} is the fiber length in the stations, U_{SM} is the worst-case end-of-life loss (dB/km) of the *single-mode* cable at both stations, N_{con} is the number of *fiber optic connectors* within the stations (inside plants), and U_{con} is the loss (dB) per connector. G is used to overcome all the losses, $L,$ in the *optical cable facility* (outside plant), which lies between the two stations of the *optical station/regenerator section.* When designing a system, G must be equal to or greater than L for the section to function satisfactorily, that is, the relation:

$$ G \geq L $$

must hold. $L,$ the sum of all the losses in the optical cable facility (outside plant) of the section expressed in dB, is given by the relation:

$$ L = \ell_t(U_c + U_{cT} + U_\lambda) + N_S(U_S + U_{ST}) $$

where ℓ_t is the total *jacket* length of *spliced-fiber cable* (km), U_c is the worst-case end-of-life cable *attenuation rate* (dB/km) at the transmitter *nominal central wavelength,* U_λ is the largest increase in cable *attenuation rate* that occurs over the transmitter central wavelength range, U_{cT} is the effect of temperature on the end-of-life cable attenuation rate at the worst-case temperature conditions over the cable operating temperature range, N_S is the number of *splices* in the length of the cable in the optical cable facility, including the splice at the *optical station facility* on each end and allowances for cable repair splices, U_S is the loss (dB/splice) for each splice, and U_{ST} is the maximum additional loss (dB/splice) caused by temperature variation. Of necessity, P_T and P_R are expressed in dB referenced to a common power unit, such as milliwatts, because they are single-ended in regard to optical power, but all the losses that contribute to L are double-ended and thus are simply expressed in dB. See *statistical terminal/regenerator system gain.* Also see *optical cable facility loss.*

termination. See *optical waveguide termination.*

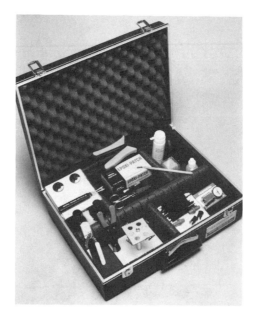

T–4. Field **Termination** Kit for use with all major *fiber optic connector* designs. (Courtesy 3M Fiber Optic Products, EOTec Corporation).

terminus. A device used to terminate an electrical conductor or *dielectric waveguide,* such as an *optical fiber,* that provides a means of positioning and holding the conductor or *waveguide* within a *connector.* In *fiber optics,* the term "contact" is deprecated. (See Fig. T–5, p. 312) See *fiber optic terminus.*

testing. See *qualification testing.*

test method. See *alternative test method (ATM); fiber optic test method (FOTM); reference test method (RTM).*

test procedure. See *fiber optic test procedure (FOTP).*

test set. See *optical-loss test set (OLTS).* (See Figs. T–6 – T–10, pages 312–314)

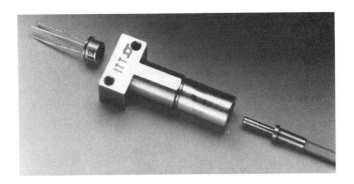

T-5. This fiber optic device receptacle saves weight and simplifies the logistics of fiber-to-device applications in an avionics *fiber optic link*. The device receptacle accepts the MIL-C-38999 *optical* pin contact (MIL-T-29504/4). It contains a "Little Ceasar" clip and cable sealing grommet that retain the **terminus** and allow easy installation and removal. (Courtesy ITT Cannon, Military/Aerospace Division, Fiber Optic Products Group).

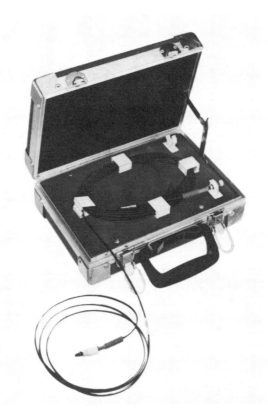

T-6. This *Fiber Optic Pulse* Suppressor allows for the **test** and measurement of *connections* and *splices* close to the *ODTR*. (Courtesy 3M Fiber Optic Products, EOTec Corporation).

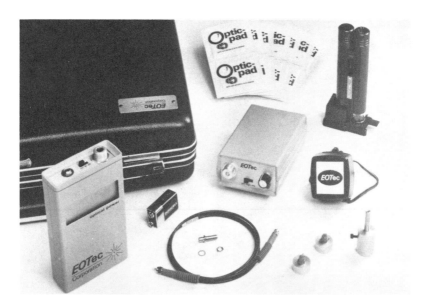

T-7. This Fiber Optic Test Kit contains everything needed to **test** *fiber optic cable* assembly *loss* and *transmitter optical power output*. (Courtesy 3M Fiber Optic Products, EOTec Corporation).

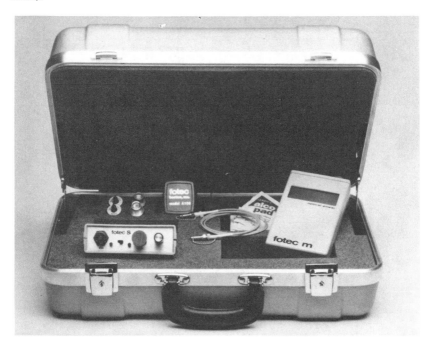

T-8. The T320 Fiber Optic Test Kit for telco *local loop* and central office use. The **test set** operates at 1.30 and 1.55 μ (*microns*). (Courtesy fotec corporation).

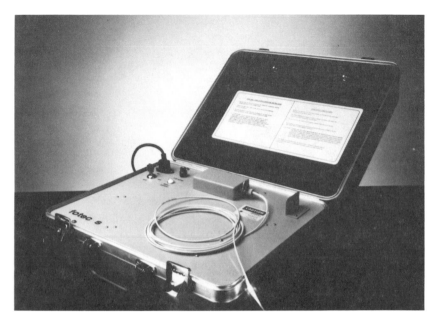

T-9. The S650 Fiber Optic Cable Fault Locator, a portable *fiber optic cable* **tester** and fault locator that uses a HeNe *laser* to inject *visible light* into the cable for a visual indication of *optical fiber* trace and location of faults, such as cable breaks, areas of *microbending loss,* and bad *connectors.* It has a five-hour rechargeable battery for a full work shift. The S650 can also operate from a 12-V DC source for vehicular operation. (Courtesy fotec incorporated).

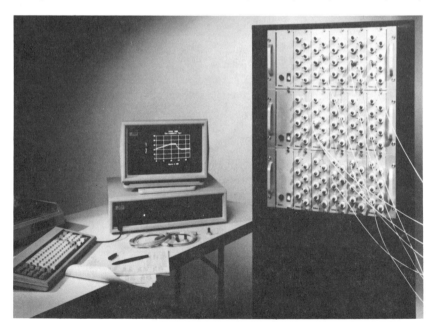

T-10. The T800 Multichannel Fiber Optic Test Systems can be configured to **test** large numbers of *fiber optic* components, including *cables, harnesses, couplers,* and many other single and multifiber assemblies, such as *splices* and *connectors.* (Courtesy fotec incorporated).

TE wave. *Transverse electric wave.*

theorem. See *sampling theorem.*

theoretical resolving power. The maximum possible *resolving power* determined by *diffraction,* frequently measured as an angular resolution determined from the relation:

$$A = 1.22\lambda/D$$

where A is the limiting resolution angle in radians, λ is the *wavelength* of the *light* at which the resolution is determined, and D is the effective diameter of the *aperture.*

thermal noise. The noise generated by thermal agitation of electrons in a conductor. The noise power, P, in watts, is given by the relation:

$$P = kT(\Delta f)$$

where k is *Boltzmann's constant,* T is the conductor temperature in kelvins, and Δf is the *bandwidth* in *hertz.* Thermal noise is distributed equally throughout the *electromagnetic frequency spectrum.* Synonymous with *Johnson noise.*

thickness. See *optical thickness.*

thin film. See *optical thin film.*

thin-film optical modulator. A device made of multilayered films of *dielectric* materials and other materials in which *electrooptic, electroacoustic, magnetooptic, magnetostrictive,* or other effects are used to *modulate* a *lightwave.* Thin-film optical modulators are used as component parts of *integrated optical circuits (IOCs).*

thin-film optical multiplexer. A *multiplexer* consisting of layered *dielectric* materials and other materials in which *electrooptic, electroacoustic, magnetooptic, magnetostrictive,* or other effects are used to multiplex *lightwave signals.* Thin-film optical multiplexers are used as component parts of *integrated optical circuits (IOCs).*

thin-film optical switch. A switch in which *electrooptic, electroacoustic, magnetooptic, magnetostrictive,* or other effects are used to switch *lightwaves* in order to perform logic functions, such as AND, OR, and NEGATION. The thin dielectric films usually support only one *mode* and are used as component parts of *integrated optical circuits (IOCs).*

thin-film optical waveguide. An *optical waveguide* made of *dielectric* or *semiconductor* material consisting of thin layers of materials with differing *refractive indices.* The lower refractive-index material on the outside serves as the substrate as well as the *cladding.* Thin-film optical waveguides usually support only one *mode* and operate from *laser sources* for effective *launching* of *lightwaves* into the thin films. Various combinations of thin-film waveguides, *lasers, modulators, switches, directional couplers, filters,* and other components may be *light-coupled* to form *integrated optical circuits (IOCs),* which may be *optically* or electrically powered.

thin-film waveguide. See *periodically distributed thin-film waveguide.*

third-harmonic nonlinear material. See *third-order nonlinear material.*

third-order nonlinear material. In *fiber optics,* pertaining to certain *transparent* materials, such as special glasses, polymers, and organic materials, that generate a third harmonic *frequency* when energized with *optical signals.* The generation of third harmonic frequencies means switching speeds of fewer than 100 fs *(femtoseconds).* The process is also *loss*less and thus can be repeated many times. An exit *wave* can be frequency-mixed, and output *beams* can be considerably different from input beams. For example, at the exit point of an *integrated optical circuit waveguide,* different signals can be switched to different *ports.* Synonymous with *third-harmonic nonlinear material.*

threshold. See *absolute luminance threshold; lasing threshold.*

threshold current. In a *laser,* the driving current corresponding to the *lasing threshold.*

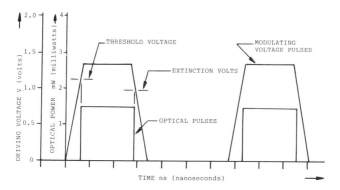

T-11. *Modulation* of a *light source,* such as a *laser* or *LED,* with an applied modulating electrical signal. The source does not *emit* until the **threshold current** or voltage is reached, and the source ceases to emit when the *extinction* voltage is reached. The threshold voltage is higher than the extinction voltage.

threshold frequency. In *optoelectronic* devices, the frequency of incident *radiant* energy below which there is no *photoemissive effect* created in the illuminated material. Thus, if the *photon frequency* is less than the *threshold frequency,* it will not have sufficient energy to overcome the *work function* of the material it strikes. If no photon in a beam has sufficient energy to overcome the work function of the material it strikes, there will be no photoemissive effect no matter how many nor at what rate photons strike the material. Also see *work function.*

THz. *Terahertz.*

tight-jacketed cable. In *fiber optics,* a *fiber optic cable* in which the *coated optical fibers* are not free to assume their own position within the *jacket,* but are completely constrained by the jacket.

time. See *acquisition time; coherence time; fall time; group delay time; real time; rise time; switch operating bounce time; switch operating time; switch release time; switch release bounce time.*

time between failures. See *mean time between failures (MTBF).*

time between outages. See *mean time between outages (MTBO).*

time coherence. In *waves,* such as *acoustic* and *electromagnetic waves,* pertaining to the existence of a correlation between the *phases* of two waves or between the phases of one wave at one point in space at two or more instants of time. Thus, the phase relationships are such that they can be calculated and predicted everywhere over a time span. Synonymous with *temporal coherence.*

time-coherent light. *Lightwaves* in which the amplitude, *phase, polarization,* or other characteristic at any point in space and time can be predicted given the values of the same characteristics at any previous time at that same point.

time-division multiplexing. A method of deriving two or more apparently simultaneous *channels* from a given *frequency spectrum* in a *propagation medium* connecting two or more points by assigning discrete time intervals in sequence to each of the individual channels, that is, a form of parallel-to-serial conversion. During a given time interval, the entire available frequency spectrum can be used by the channel to which it is assigned. In general, time-division multiplexing systems use *pulse transmission.* The multiple-pulse train may be considered to be the interleaved pulse trains of the individual channels. The individual channel pulses may be *modulated* either in an *analog* or a *digital* manner. Thus, individual signals from separate *sources* share the time of a circuit by being assigned time slots on the circuit. In *fiber optic data links,* a single *optical fiber* can share its time among a large number of sources. For example, if each source is assigned 1 μs (microsecond) out of each ms (millisecond), 1000 sources can be accommodated by the single fiber. Of course, the individual sources can be *frequency-division multiplexed,* giving rise to many more channels. The *lightwaves* in the optical fiber can also be *wavelength-division multiplexed,* that is, a different *color* lightwave for each channel, giving rise to even more channels. (See Figs. T–12 and T–13, p. 318.)

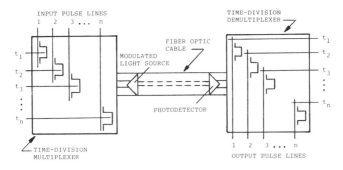

INPUT PULSE LINES
1 2 3 ... n

t_1
t_2
t_3
$\vdots$
t_n

TIME-DIVISION
MULTIPLEXER

MODULATED
LIGHT SOURCE

FIBER OPTIC
CABLE

PHOTODETECTOR

TIME-DIVISION
DEMULTIPLEXER

t_1
t_2
t_3
$\vdots$
t_n

1 2 3 ... n
OUTPUT PULSE LINES

T-12. **Time-division multiplexing** on an *optical link*.

T-13. The TDMR/560™, a 560 *Mbps* system that provides drop/insert transport for DS3 (44.736 Mbps) *channels*. The system has a capacity of 12 DS3 channels. In the TDMR/560, multiplexing *nodes* are interconnected via counter-rotating *optical fiber* rings to accomplish the **time-division multiplexing**. (Courtesy FiberLAN, a Siecor Company).

318

time-domain reflectometer. See *optical time-domain reflectometer (OTDR).*

time-domain reflectometry. See *optical time-domain reflectometry.*

time jitter. The short-term variation or instability in the duration of a specified interval or in the occurrence time of a specific event, such as a *signal* transition from one significant condition to another. When viewed with reference to a fixed time frame, the signal transition moves back and forth on the time axis, that is, the *significant instant* shifts.

time to failure. See *mean time to failure (MTTF).*

time to repair. See *mean time to repair (MTTR).*

time to service restoral. See *mean time to service restoral (MTTSR).*

tip length. See *borescope tip length.*

TM. *Transverse magnetic.*

TM mode. *Transverse magnetic mode.*

TM wave. *Transverse magnetic wave.*

tolerance. See *diameter tolerance.*

tolerance area. See *cladding tolerance area; core tolerance area.*

tolerance field. The region between two curves used to specify the tolerance in the size of a component. When specifying the size of *optical fiber cladding,* the tolerance field is the annular region between the two concentric circles of the diameter $D + \Delta D$ and $D - \Delta D$. The first is the smallest circle that circumscribes the outer surface of the *homogeneous cladding;* the second is the largest circle that fits within the outer surface of the homogeneous cladding. When specifying the *core* size, the tolerance field is the annular region between the two concentric circles of diameter $d + \Delta d$ and $d - \Delta d$. The first is the smallest circle that circumscribes the *core area;* the second is the largest circle that fits within the core area. The cladding circles may not necessarily be concentric with the core circles. See *reference-surface tolerance field.*

tool. See *fiber-cutting tool.*

total harmonic distortion. When a single *frequency signal* of specified power is applied to the input of a system or component, such as a length of *optical fiber,* a *transmitter,* or a receiver, the ratio of the sum of the powers of all harmonic *frequencies* to the power of the fundamental frequency *signal* at the output. This ratio is measured under specified conditions and is expressed in *decibels.* The harmonic frequencies do not include the fundamental frequency, that is, the input frequency. The *distortion* is caused by the *nonlinearity* of the component or system.

total internal reflection. See *total reflection.*

total reflection. *Reflection* that occurs when an *electromagnetic wave* strikes an *interface* surface between *propagation media* with different properties at *incidence angles* greater than the *critical angle.* At these incidence angles, there is no *refracted wave,* except that at the critical angle, there is a wave that *propagates* along the interface surface. In an *optical fiber,* the incident *ray* is in the *core,* and the *refractive index* of the core is higher than that of the *cladding.* When the incidence angle is greater than the critical angle, the incident ray will be totally *reflected* back into the medium of incidence, that is, back into the core. Synonymous with *total internal reflection.*

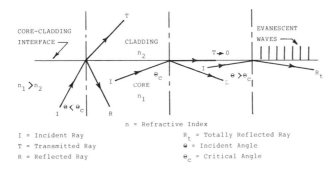

T–14. **Total (internal) reflection** occurs when the angle θ is greater than the *critical angle,* $\theta_c = \sin^{-1} n_2/n_1$. When total internal reflection occurs, *evanescent waves* are produced in the *cladding* close to the *core.* The diagram shows incident, *transmitted,* and *reflected waves* for *incidence angles* less than, equal to, and greater than the critical angle at the core-cladding *interface* of an *optical fiber.*

total reflection sensor. See *frustrated total (internal) reflection sensor.*

trajectory. See *ray trajectory.*

transceiver. A device consisting of a *transmitter* and *receiver* in a common housing, often for use in a fixed *data transmission link,* or for portable or mobile use. It employs common circuit components for transmitting and receiving *signals.* Each transceiver is usually designed for simplex operation. A *fiber optic* transceiver has electrical terminals and optical *pigtails* for signals and a lead for electrical power. (See Figs. T–15 – T–17.)

transceiver dispersion. See *maximum transceiver dispersion.*

T-15. The RS232 Optical Transceiver for converting standard RS232 *ports* to *fiber optic transmission*. The fiber optic **transceiver** is normally powered from *signals* found in the RS232 *interface*. OPTIMATE™ *connectors* from a *fiber optic cable* assembly are attached. The transceiver provides 17 dB of *optical power budget*. (Courtesy AMP Incorporated).

T-16. A single-*fiber* bidirectional **transceiver**. The optical subassembly incorporates a fused *single-mode* or *multimode coupler* or *wavelength-division multiplexer (WDM) emitters* and/or *detectors* in an athermally stable package with a single-*fiber pigtail*. (Courtesy Aster Corporation).

INTEGRATED OPTICAL SUBASSEMBLY

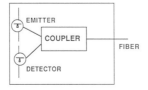

T-17. Schematic diagram for the **transceiver**. (Courtesy Aster Corporation.)

transducer. A device capable of transforming energy from one form to another, usually with such fidelity that if the original energy represents information, the transformed energy can represent the same information or its time derivative. For example, a microphone that converts electrical currents to *modulated lightwaves,* a *photodetector* that converts modulated lightwaves to electrical currents, or a piezoelectric crystal that produces a voltage proportional to the time derivative of the pressure wave that impinges upon it. See *optoacoustic transducer.*

transfer function. A mathematical statement expressing the transfer characteristics of a system, subsystem, equipment, or component. Thus, the transfer function is such that when it operates on the input the output will be fully determined or described. Thus, the transfer function is the ratio of the output function to the input function of the device. For an *optical fiber,* the transfer function may be the ratio of the *optical power output* to the optical power input as a function of *modulation frequency.* In general, for a *linear device* the transfer function and the *impulse response* are related through the Fourier transform pair, commonly given by the relations:

$$H(f) = \int_{-\infty}^{+\infty} h(t)e^{i2\pi ft}\, dt$$

where $H(f)$ is the ratio of optical power output to optical power input as a function of modulation frequency, f is the frequency, t is time, and $h(t)$ is given by the relation:

$$h(t) = \int_{-\infty}^{+\infty} H(f)e^{-2\pi ft}\, df$$

$H(f)$ is often normalized to $H(0)$ and $h(t)$ to the relation:

$$h_n(t) = \int_{-\infty}^{+\infty} h(t)\, dt$$

which by definition is $H(0)$. Synonymous with *baseband response function; frequency-response function.* See *optical fiber transfer function (OFTF).*

transfer network. See *data transfer network (DTN); fiber optic data transfer network.*

transient attenuation rate. 1. In a *multimode dielectric waveguide,* such as an *optical fiber,* a change in the *attenuation rate* with distance along the guide between the entrance face and the *equilibrium length,* that is, the equilibrium modal-power-distribution length. The change can be an increase or a decrease in the *attenuation rate* with distance. Within this length, the attenuation rate changes with distance because of the over- or underexcitation of the *lossy high-order propagating modes.* Beyond this length, the attenuation rate does not change with distance along the guide. Thus, this change, that is, "transient" condition, occurs as a function of distance. **2.** At a point between the entrance face of a *multimode dielectric waveguide,* such as an *optical fiber,* and the *equilibrium length,* that is, the equilibrium modal-power-distribution length, a temporary change in the attenuation rate at the point, usually occurring during the *rise*-and-*fall time* of *optical pulses* or by temporal changes in environmental conditions. This transient occurs as a function of time.

transimpedance. See *optical transimpedance.*

transition fiber. An *optical* element of such geometric shape at input and output ends that it enables *coupling* between elements of different cross-sectional shape, such as coupling between an element with a rectangular cross section and an element with a round cross section. For example a transition fiber may be used to couple a *thin-film optical waveguide* to a round *optical fiber,* the transition fiber having a smooth transition from a rectangular to a round cross section.

transmission. 1. The transfer or transport of electrical, *optical,* or any other form of power from one location to another over electrical conductors, such as wires, or *waveguides,* such as *optical fibers.* Purposes of transmission include transfer of energy from one point to another or transfer of *signals* that represent information. **2.** The dispatching of a *signal,* message, or other form of information by means of wire, *optical fiber,* radio, heliograph, semaphore, horns, or any other means, using one or more modes, such as telegraphy, telephony, or facsimile. See *bidirectional transmission; specular transmission.*

transmission coefficient. 1. The ratio of the *transmitted field strength* to the *incident field strength* when an *electromagnetic wave* is incident upon an *interface* surface between *dielectric propagation media* with different *refractive indices.* For example, if, at oblique incidence, the *electric field component* of an incident *plane-polarized electromagnetic wave* is parallel to the interface surface, the transmission coefficient is given by the relation:

$$T = 2m_2 \cos A/(m_2 \cos A + m_1 \cos B)$$

where m_1 and m_2 are the reciprocals of the refractive indices of the incident and transmitted media, respectively, and A and B are the *incidence* and *refraction angles,* respectively. This is one of the *Fresnel equations.* The sum of the *reflection coefficient* and the transmission coefficient is not necessarily unity. **2.** The ratio between the amplitude of the *transmitted wave* and the amplitude of the incident wave. **3.** A number indicating the probable performance of a portion of a *transmission* circuit. The value of the transmission coefficient is inversely related to the quality of the link or circuit. Synonymous with *refraction coefficient.* Also see *reflection coefficient.*

transmission loss. The decrease in *power level* of a *signal* during *transmission* from one point to another within a component, from the output of one component to the input of another component, or from one point to another in a *propagation medium.* The *loss* is usually expressed in dB *(decibels),* which requires that the loss must first be expressed as a ratio of the power levels at the two points. The number of decibels of transmission loss is given by the relation:

$$10 \log_{10} (P_0/P_i)$$

where P_0/P_i is the ratio of the output power at one point to the input power at the other point, such as the output power divided by the input power of a transmission line, connector, or other component. Also see *insertion loss.*

transmission medium. See *propagation medium.*

transmission window. See *spectral window.*

transmissivity. The intrinsic *transmittance* per unit length of a material at a given *wavelength,* excluding the *reflectance* from inner or outer surfaces. The term is no longer in common use. Also see *transmittance.*

transmittance. In *optics,* when an *electromagnetic wave,* such as a *lightwave,* is incident upon an *interface* surface between *propagation media* with different *refractive indices,* the ratio of *transmitted power* to incident power. The conditions under which the transmittance and *reflectance* occur should be stated, such as the *wavelength* composition and *polarization* of the incident wave, the geometrical distribution of the *reflecting* surface or surfaces, and the composition of the media on both sides of the reflecting surface. In *optics,* the transmittance is often expressed as a percentage or as transmittance density, the base-10 logarithm of the reciprocal of the transmittance. In communications, it is usually expressed in *decibels* (dB). See *coupler transmittance; spectral transmittance.* Also see *transmissivity.*

transmittance change. See *induced attenuation.*

transmittance density. See *optical density.*

transmitted power. The energy per unit time *propagated* through or across a specified area in a direction perpendicular to the area, such as might be *transmitted* through a wire or *fiber optic cable.* Because the instantaneous power can vary with time at a point, various values of power can be measured, such as the peak envelope power; the average power over an area or over time; the *radiance,* that is, the power *radiated* in a given solid angle; the *irradiance,* that is, the *power density;* the total power radiated by a *source;* or the power in a specified portion or *band* of the *frequency spectrum,* such as, for an *electromagnetic wave,* only the power in the *infrared region* of the spectrum.

transmitter. A device that generates, usually *modulates,* and dispatches a *signal* for communication, control, or other purpose, such as a unit consisting of a *modulator,* a *laser,* and an *optical fiber pigtail* for connection to a *fiber optic* cable; or a unit consisting of a microphone, amplifier, modulator, and antenna for transmitting radio *signals.* See *electrooptic transmitter; fiber optic transmitter; optoelectronic transmitter; optical transmitter.*

T-18. The OPTIMATE™ Continuity Checker for aiding the set-up and *testing* of *optical fibers.* The compact **transmitter** and receiver attaches to either end of a *cable* run and gives instant readout of *attenuation* within a 20 *dB* range. A graphic *LED* display indicates *signal strength* while a variable-pitch audible indicator can be activated when visibility is limited. (Courtesy AMP Incorporated).

transmitter central wavelength range. In *fiber optics,* the range of *wavelengths* between the minimum and maximum wavelength limits of the total worst-case variation of *fiber optic transmitter nominal central wavelengths* caused by manufacturing, temperature, aging, and any other factors when operated under standard or extended operating conditions.

transmitter information descriptor. A unique descriptor from which information about a transmitter can be determined, such as the manufacturer, terminal equipment association, system design application *(single-mode, multimode),* operating *wavelength,* output *power level, source* type, temperature controller, FCC classification (Class I, Class II, etc), and manufacturer product change designation (issue, revision).

transmitter optical connector. The *optical connector* provided at the output of an *optical transmitter* that is attached to the transmitter *pigtail.* The transmitter optical connector description should include the manufacturer, type (biconic, FC, etc.) model number, classification *(multimode, single-mode),* and mating connector model number.

transmitter optical power. The worst-case minimum value of *optical power (dBm) coupled* into a *fiber optic cable* on the line side of the *fiber optic transmitter connector* under specified standard or extended operating conditions using standard measurement procedures. The worst-case minimum value combines manufacturing, temperature, and aging variations in a worst-case fashion.

transparency. A property of an entity that allows another entity to pass through it without altering either one or both of the entities, such as (1) the quality of a material that allows *lightwaves* to pass through without *absorption* or *scattering,* (2) the property that allows the *transmission* of *signals* without changing their electrical characteristics or coding beyond the limits specified by the design, (3) a component in a communication system whose operation is independent of the codes used in representing the information being *transmitted,* or (4) the film on which an image is placed so that the image can be projected on a screen without alteration of the image.

transport layer. In *open-systems architecture,* the layer that directly provides functions and facilities for the actual movement of *data* among *network* elements, such as among user *end-instruments;* among land vehicle, ship, aircraft, or spacecraft systems, subsystems, and control panels; and among network *nodes. Optical fiber systems* can be used to implement the transport layer and the *physical layer.* Also see *Open-Systems-Interconnection (OSI) Reference Model (RM).*

transverse-compression sensor. See *fiber transverse-compression sensor.*

transverse electric (TE) mode. In an *electromagnetic wave propagating* in a *waveguide,* a *mode* whose *electric field vector* is normal to the direction of propagation and whose *magnetic field vector* is not normal to the direction of propagation. In an *optical fiber,* transverse electric and *transverse magnetic modes* correspond to *meridional rays.* Also see *transverse magnetic (TM) mode.*

transverse electric (TE) wave. An *electromagnetic wave propagating* in a *propagation medium,* or in *free space,* in such a manner that the *electric field vector* is directed entirely transverse, that is, perpendicular, to the direction of propagation. Hence there is no electric field vector *component* in the direction of propagation, but there is a component of the *magnetic field vector* in the direction of propagation. Thus, in a *waveguide,* the electric field vector is entriely transverse and does not have a longitudinal component, but the magnetic field does have a longitudinal component. In *optical systems* the *propagation medium* is usually a waveguide with a uniform cross section in the longitudinal direction. Also see *transverse magnetic (TM) wave.*

transverse electromagnetic (TEM) mode. In an *electromagnetic wave,* a *mode* whose *electric field vector* and *magnetic field vector* are both normal, that is, perpendicular, to the direction of *propagation.*

transverse electromagnetic (TEM) wave. An *electromagnetic wave* whose *electric field vector* and *magnetic field vector,* although varying in time at every point in the space occupied by the wave, whether in *free space* or in a material *propagation medium,* are contained in a local plane at that point. The orientation of the plane is independent of time. Thus, in general, the orientation of the local plane is different for different points in space or material media, the exception to this being the special case of a *uniform plane-polarized electromagnetic wave,* in which case all the *polarization planes* are parallel.

transverse interferometry. A method used to measure the *refractive-index profile* of a *dielectric waveguide,* such as an *optical fiber,* by placing a short length of the guide in an *interferometer* and illuminating it in a direction transverse to the *optical axis.* Normally a computer is used to interpret the *interference pattern* obtained when making the measurement. Also see *slab interferometry.*

transverse magnetic (TM) mode. In an *electromagnetic wave,* a *mode* whose *magnetic field vector* is perpendicular to the direction of *propagation* and whose *electric field vector* is not perpendicular to the direction of propagation. In an *optical fiber,* transverse magnetic modes and *transverse electric modes* correspond to *meridional rays.* Also see *transverse electric (TE) mode.*

transverse magnetic (TM) wave. An *electromagnetic wave propagating* in a *propagation medium* or in *free space* in such a manner that the *magnetic field vector* is directed entirely normal, that is, perpendicular, to the direction of *propagation.* Hence there is no magnetic field vector *component* in the direction of propagation, but there is an *electric field vector* component in the direction of propagation. Thus, in a *waveguide,* the magnetic field vector is entirely transverse and does not have a longitudinal component, but the electric field does have a longitudinal component. For *optical systems,* the *propagation medium* is usually a *waveguide* with a uniform cross section in the longitudinal direction. Also see *transverse electric (TE) wave.*

transverse mode. See *lowest order transverse mode.*

transverse offset loss. See *lateral offset loss.*

transverse propagation constant. The *propagation constant* evaluated along a direction perpendicular to the *waveguide* longitudinal axis. The transverse propagation constant for a given mode can vary with the transverse coordinates.

transverse scattering method. A method for measuring the refractive index profile of an *optical waveguide,* such as an *optical fiber* or *preform,* by illuminating the guide *coherently* and transversely to its longitudinal axis, and examining the *far-field irradiance pattern.* A computer is used to interpret the pattern of *scattered light* obtained.

trap. See *optical fiber trap.*

trapped electromagnetic wave. An *electromagnetic wave* that enters a layer of material on both sides of which is a layer of material with a lower *refractive index,* such that, if the wave is *propagating* parallel or nearly parallel to the surfaces of the layers, the wave will be forced to remain in the layer it has entered as long as the *incidence angles* that the *rays* make with the local surfaces they encounter are greater than the *critical angle,* thus obtaining *total* (internal) *reflection.*

trapped mode. See *bound mode.*

trapped ray. See *guided ray.*

tray. See *fiber optic splice tray.*

tree network. In *network* topology, a network configuration in which there is one and only one path between any two *nodes.*

triangular-cored optical fiber. An *optical fiber* consisting of a *core* whose cross section is shaped like a triangle with bowed convex sides placed in a hollow *jacket* or *cladding* tube, such that only the vertices of the core touch the inner walls of the jacket or cladding. Air surrounds the core and performs the function of the cladding.

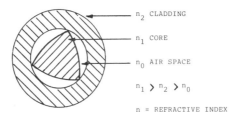

n_2 CLADDING

n_1 CORE

n_0 AIR SPACE

$n_1 > n_2 > n_0$

n = REFRACTIVE INDEX

T–19. A **triangular-cored optical fiber** with airspace between *core* and *cladding.*

tube. See *bait tube.*

tunable laser. A *laser* whose spectral *emission wavelength,* that is, the wavelengths that make up its *spectral width,* can be varied. The spectral range of some *lasers* can be varied continuously across a broad spectral range, such as an organic dye or optical parametric-oscillator laser. Others are stepwise-tunable, such as carbon dioxide lasers and other molecular lasers. Emission of these lasers can be tuned to one of several wavelengths, that is, one of several *spectral lines.*

tunneling mode. See *leaky mode.*

tunneling ray. See *leaky ray.*

TV fiber optic link. A *video signal link* that accepts an electrical video signal from a television camera or video tape, converts it into an *optical signal,* transmits it over a *fiber optic cable,* and reconverts or reconstitutes it back to an electrical signal for use by a television monitor or set.

twist strength. For an *optical* element, such as an *optical fiber, cable, splice, bundle,* or *pigtail,* a measure of the ability of the element to withstand alternate torsional flexing without degrading performance, such as increasing the *attenuation rate* of an optical fiber.

two-level optical signal. A *modulated lightwave carrier* that is at either one of two different *optical power levels,* that is, *irradiance, optical power density,* or *electric field strength* is a specific value at any given instant. For example, a *lightwave* carrier that is turned on and essentially off, depending on whether a 1 or a 0 is being transmitted at any given instant.

Twyman-Green interferometer. An *interferometer* in which an observer sees a contour map of an emergent *electromagnetic wavefront* based on the *wavelength* of the *light* that is entering the system.

type. See *optical detector type; optical source type.*

U

u. Often used in lieu of μ to indicate *microns,* that is, 10^{-6} m *(meters).* The symbol um is also used to indicate microns, that is, micrometers.

ultralow-loss fiber. **1.** *Optical fiber* that has an *attenuation rate* approaching that caused only by *Rayleigh scattering.* **2.** *Optical fiber* that has a potential *optical attenuation rate* less than 0.01 dB/km *(decibels*/kilometer), perhaps as low as 0.002 dB/km, the attenuation being limited only by Rayleigh scattering, perhaps operating at a window in the 2- to 5-μ-(micron)-*wavelength* region of the *optical spectrum,* that is, wavelengths in the *near-* and *middle-infrared* regions of the *electromagnetic spectrum,* such as might be obtained from certain halogen glasses like zirconium fluoride glasses.

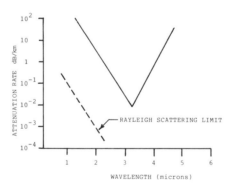

U-1. *Attenuation rate* versus *light source nominal central wavelength* for a typical **ultralow-loss fiber.**

ultraviolet (UV). Pertaining to the region of the *electromagnetic spectrum* in which *wavelengths* are shorter than those of the *visible spectrum,* but longer than *x-rays.* The ultraviolet region extends from about 1 nm *(nanometer),* which is the beginning of the optical spectrum, to about 400 nm, that is, the beginning of the visible region, or 0.001 to 0.4 μ *(microns).* Because the *frequency* of ultraviolet *radiation* is higher

330

than that of the visible region, the *rays,* that is, the *photons,* can kill some living organisms, such as some germs and bacteria, and can have a damaging effect on human cells if exposure time and *irradiance* are sufficiently high. Because of the higher frequencies, *attenuation rates* in glass are higher for the shorter-than-visible ultraviolet wavelengths and, because of the lower frequencies, attenuation rates are lower for the *infrared;* hence the interest in the use of infrared in *long-haul optical fiber communication systems.* Also see *ultraviolet fiber optics.*

ultraviolet fiber optics. *Fiber optics* involving the use of *ultraviolet (UV) light* in *optical* components designed to handle UV light. Uses include medical engineering, medicine, measurements, materials testing, sterilization, photochemistry, genetics, security, and secondary-*emission* (fluorescence) applications. Also see *ultraviolet.*

ultraviolet light. *Radiation* in the *ultraviolet* region of the *electromagnetic spectrum,* namely *electromagnetic waves* that have *wavelengths* between 1 and 400 nm *(nanometers)* or between 0.001 and 0.4 μ *(microns).* The *lightwave* region of the *electromagnetic spectrum* lies between 0.3 to 3 μ, which is the region in which the techniques and components designed for the *visible spectrum,* 0.4 to 0.8 μ, also apply. Thus, only the narrow region between 0.3 and 0.4 μ might be considered ultraviolet light, and the remaining portion of the ultraviolet region, from 0.001 to 0.3 μ, might be considered as ultraviolet *radiation.*

unaligned bundle. A bundle of *optical fibers* in which the spatial relationship among the fibers may vary along the bundle, and the spatial relationship among the ends of the fibers at one end of the bundle is not the same as at the other end. Thus, an unaligned bundle, that is, an incoherent bundle, cannot be used to *transmit* an image, but it can be used for signaling or illumination. Synonymous with *incoherent bundle.*

unbound mode. In an *electromagnetic wave propagating* in a *waveguide,* a *mode* that is not a *bound mode.* In *optical fibers,* unbound modes include *leaky modes,* that is, *radiation modes.* Also see *bound mode; cladding mode; leaky mode; leaky ray.*

undercompensated optical fiber. An *optical fiber* whose *refractive-index profile* is adjusted so that the *high-order propagating modes* arrive at the end of the fiber after the *low-order modes,* as in the case of the uncompensated fiber, that is, a fiber in which no deliberate effort is made to compensate for *dispersion.* Higher order modes propagate faster than lower order modes in a *propagation medium* with a given *refractive index.* In an undercompensated fiber they are made to travel longer paths in higher index media and hence are delayed so they arrive at the end after the lower order modes, which are made to travel shorter paths in lower refractive-index media. Higher order modes have higher eigenvalues in the solution of the *wave equations.* Higher order modes also have higher *frequencies* and shorter *wavelengths* than lower order modes. Also see *compensated optical fiber; overcompensated optical fiber.*

uniformly distributive coupler. A *fiber optic coupler* in which *optical power* input to each input *port* is distributed equally among the output ports, and the *optical power* input to each output port is not distributed equally among the input ports. Also see *nonuniformly distributive coupler.*

uniform plane-polarized electromagnetic wave. A *plane-polarized electromagnetic wave* whose *electric* and *magnetic field vector* amplitudes are independent of the coordinate direction perpendicular to the direction of *propagation,* that is, they are independent of the transverse coordinates, but are dependent on the longitudinal coordinates in the direction of propagation.

uniform refractive-index profile. See *linear refractive-index profile.*

unshifted fiber. See *dispersion-unshifted fiber.*

user end-instrument. A device connected to the terminals of a circuit and used to convert electrical, *optical,* acoustic, or other types of *signals,* into usable information, or vice versa, such as a telephone, control panel, computer, or remote *data* terminal. *Optical data transfer networks* in land-based communication systems, aircraft communication and control systems, shipboard communication and control systems, *local-area networks,* and other *optical fiber communication systems* are designed to allow communication among users by means of their end-instruments, such as intercoms, telephones, and data terminals. From the point of view of the communication system, equipment, such as computers, aircraft and ship propulsion systems, and sensor systems may also be considered as user end-instruments connected by a communication or control network.

UV. *Ultraviolet.*

V

vacancy defect. In the somewhat ordered array of atoms and molecules of *optical propagation media,* such as the glass used in *optical fibers,* a site at which an atom or molecule is missing in the array. The defect will be a *scattering* center and will cause *diffuse reflection,* heating, and *absorption,* all of which will contribute to *optical attenuation.*

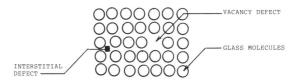

V-1. Interstitial and **vacancy defects** in the glass used for making *optical fibers* can result in high *attenuation rates* because of the *reflection, refraction,* and *scattering* caused by the defects.

VAD. *Vapor-phase axial deposition.*

VAD process. *Vapor-phase axial-deposition process.*

vapor deposition. See *outside vapor deposition (OVD).*

vapor-deposition process. See *chemical-vapor-deposition (CVD) process; modified chemical-vapor-deposition (MCVD) process; plasma-activated chemical-vapor-deposition (PACVD) process.*

vapor-phase axial-deposition (VAD) process. A *chemical-vapor-deposition (CVD) process* in which *dopants* in vapor form are axially deposited on substrates, such as glass tubes, to make a *preform* from which *optical fibers* may be *pulled.*

vapor-phase-oxidation process. See *axial vapor-phase-oxidation (AVPO) process; chemical-vapor-phase-oxidation (CVPO) process; inside vapor-phase-oxidation (IVPO) process; outside vapor-phase-oxidation (OVPO) process.*

variable optical attenuator. See *continuously variable optical attenuator; stepwise-variable optical attenuator.*

velocity. See *group velocity; phase-front velocity; phase velocity.*

vector. See *electric field vector; magnetic field vector; Poynting vector.*

Verdet constant. The constant of proportionality, that is, the V, in the equation that defines the *Faraday effect,* which is given by the relation

$$\Theta = V \int B \cdot dl = V \int \mu H \cdot dl$$

where Θ is the rotation angle of the *polarization plane* of *plane-polarized light propagating* in a *medium,* V is the Verdet constant, B is the magnetic flux density in the medium, μ is the *magnetic permeability* of the medium, and H is the *magnetic field strength* at the point in the medium at which the path-length differential, dl, is taken. The integrand is the dot (scalar) product. Thus, if the magnetic field is parallel to the propagation path, such as occurs when an optical fiber is parallel to a magnetic field, the magnitude of the rotation angle is directly proportional to the product of the path length and the magnetic flux density, or for nonferrous substances such as glass, the product of the path length and the magnetic field strength, because the magnetic permeability of glass is unity. Typical units for the Verdet constant are degrees/centimeter-tesla, degrees/meter-oersted, or radians/meter-tesla, depending on the units used for the variables in the solution integral of the Faraday equation. Also see *magnetooptic effect.*

video. In communication systems, such as *optical fiber communication systems,* pertaining to a capability or characteristic to *transmit* or process *signals* having *frequencies* that lie within that part of the frequency *band* that is employed for the representation of pictures that change in accordance with events occurring in real time. Video frequency bands range nominally from 100 kHz to several MHz and are generally used for television *transmission.* (See Figs. V–2 and V–3.)

video disk. See *optical video disk (OVD).*

video optical detector. In a *video-signal optical fiber transmission* system, a device that converts an optical video signal in an *optical fiber* or other optical *propagation medium* to an electrical video signal in a wire or other electrical propagation medium.

video optical source. In a *video-signal optical fiber transmission* system, a device that converts an electrical video signal in a wire or other electrical *propagation medium* to an optical signal in an *optical fiber* or other optical propagation medium.

view field. See *borescope view field.*

V-2. This FM Video Link allows *fiber optic transmission* of black-and-white or *color* **video signals** up to 10,000 ft. (Courtesy 3M Fiber Optic Products, EOTec Corporation).

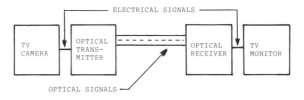

V-3. A **video** (TV) closed-circuit *fiber optic link.*

viewing angle. See *borescope axial viewing angle.*

visible radiation. *Electromagnetic radiation* with *wavelengths* in the *visible spectrum.*

visible spectrum. The portion of the *electromagnetic frequency spectrum* to which the human retina is sensitive and by which humans see. The visible spectrum is considered to extend from a *wavelength* of 400 to 800 nm *(nanometers),* that is, 0.4 to 0.8 μ *(microns),* in air, which corresponds to a *frequency band* of 3.75–7.50 × 10^{14} Hz. The visible spectrum lies between the *ultraviolet* and the *infrared* regions, the visible wavelengths being longer than the ultraviolet and shorter than the infrared.

visual spectrum. The band of *color* produced when *white light* is decomposed into its *wavelength* components by means of *dispersion.* For example, the rainbow produced by a cloud of water droplets suspended in air or the spread of colors produced by a prism or cut diamond when illuminated by white light are examples of visual spectra.

vitreous silica. See *fused silica.*

V-number. See *normalized frequency.*

voice. In communication systems, pertaining to a capability or characteristic to *detect, transmit,* or process signals having *frequencies* that lie within that part of the frequency *band* that is employed for the representation of speech. Voice frequency bands range from about 300 to about 3400 Hz.

voltage. See *noise voltage.*

voltaic effect. See *photovoltaic effect.*

voltaic photodetector. See *photovoltaic photodetector.*

volume. See *effective mode volume; mode volume.*

V-parameter. See *normalized frequency.*

V-value. See *normalized frequency.*

W

wave. See *clockwise-polarized wave; counterclockwise-polarized wave; electromagnetic wave; evanescent wave; guided wave; Hertzian wave; indirect wave; leaky wave; left-hand-polarized electromagnetic wave; lightwave; plane (electromagnetic) wave; plane-polarized electromagnetic wave; right-hand-polarized electromagnetic wave; standing wave; surface wave; transverse electric (TE) wave; transverse magnetic (TM) wave; trapped electromagnetic wave; uniform plane-polarized electromagnetic wave.*

wave equation. The equation, derived from *Maxwell's equations,* the *constitutive relations,* and the vector algebra, that relates the *electromagnetic field* of an *electromagnetic wave* time and space derivatives with the *propagation medium electric permittivity* and *magnetic permeability* in a region without electrical charges or currents, that is, in a charge-free *dielectric* propagation medium, such as that of an *optical fiber.* The solution of the wave equation yields the *electric* and *magnetic field strengths* everywhere in the medium in which the wave is propagating. The wave equation is given by the relation:

$$\nabla^2 H - \mu\epsilon\partial^2 H/\partial t^2 = 0$$

or by the relation:

$$\nabla^2 E - \mu\epsilon\partial^2 E/\partial t^2 = 0$$

in a current and charge-free electrically nonconducting medium, that is, dielectric medium, where E is the electric field strength, H is the magnetic field strength, ϵ is the electric permittivity, μ is the magnetic permeability, usually equal to 1, and $\partial^2/\partial t^2$ is the second partial derivative with respect to time. The ∇ is the vector spatial derivative operator. When the wave equation for *uniform plane waves* is transformed into cartesian coordinates, the scalar equations for the electric and magnetic field strengths in the transverse directions (x and y for a wave propagating in the z direction) are obtained in terms of propagation distance and *propagation medium* parameters, namely the electric permittivity and magnetic permeability. These are the *Helmholtz equations.* They are ordinary differential equations. Their solutions are simple exponential functions whose exponents are the *propagation constants.* For example, the relation:

$$(d^2/dz^2)E_x - \Gamma^2 E_x = 0$$

is one of the Helmholtz equations derivable from the wave equation. A solution of this Helmholtz equation is the relation:

$$E_x = E_{xo}e^{-\Gamma z}$$

where $\Gamma = -\omega^2\mu\epsilon$, μ being the magnetic permeability, ϵ the electric permittivity, and $\omega = 2\pi f$, where f is *frequency.* In *transparent dielectric media,* $\mu = 1$. Thus, the *relative refractive index* is given by the relation:

$$n_r = \epsilon_r^{1/2}$$

where ϵ_r is the electric permittivity relative to a vacuum or approximately to air. The wave equation, the derived Helmholtz equations, and their solutions apply to *electromagnetic waves* propagating in *optical waveguides.* Their eigenvalues give rise to many possible *modes.*

wavefront. A theoretical surface normal to an *electromagnetic ray* as it *propagates* from a *source,* the surface passing though those parts of all the *radiating waves* that have the same *phase,* that is, all points on a wavefront surface have the same *optical path length* from the source. Hence, it is the locus of all points of the wave where the *electric* and *magnetic field components,* that is, *electric* and *magnetic field vectors,* have the same phase at the same time. For parallel rays, the wavefront is a plane. For waves diverging from or converging on a point, the wavefront is spherical. The wavefront at a given point in space is perpendicular to the direction of propagation. The electric and magnetic field vectors at a point in space define a plane, that plane being tangent to the wavefront surface at that point.

waveguide. A device capable of guiding an *electromagnetic* or sound *wave* along a prescribed path, such as an *optical fiber* or a *transmission* line consisting of a hollow or *dielectric*-filled metallic conductor, generally rectangular, elliptical, or circular in shape, within which electromagnetic waves can propagate. See *circular dielectric waveguide; closed waveguide; dielectric waveguide; diffused optical waveguide; multimode waveguide; open waveguide; optical waveguide; periodically distributed thin-film waveguide; planar diffused waveguide; slab-dielectric waveguide; step-index optical waveguide; thin-film optical waveguide.*

waveguide connector. See *optical waveguide connector.*

waveguide coupler. See *optical waveguide coupler.*

waveguide delay distortion. In an *electromagnetic wave,* such as a *lightwave propagating* in a *waveguide,* the *distortion* in received *signals* caused by the differences in group-delay time for each *wavelength* or each *propagation mode,* causing an increase in *pulse duration,* that is, pulse broadening or spreading, or causing *distortion* of *analog signals,* as the wave propagates along the guide.

waveguide dispersion. For each *mode* of an *electromagnetic wave* or for an *optical pulse propagating* in a *waveguide,* the *dispersion* caused by the dependence of the *phase* and *group velocities* on *wavelength* as a consequence of the geometric properties of the waveguide. For circular dielectric waveguides, such as *optical fibers,* the dependence is on the ratio a/λ, where a is the core radius, that is, one-half the *core diameter,* and λ is the wavelength.

waveguide isolator. A *passive attenuator* in which *transmission losses* are much greater in one direction than in the other, thus causing *absorption* of end *reflections, backscatter,* and *noise.* Certain such nonreciprocal properties can be obtained from some ferrite materials.

waveguide scattering. *Scattering* other than material scattering in a *waveguide,* such as an *optical fiber.* Waveguide scattering is caused by variations of geometry and variations in the *refractive-index profile* of the waveguide.

waveguide splice. A permanent joint between the *transmission* elements of two *waveguides* so that *signals* may pass from one waveguide to the other with minimal *loss,* for example, a joint between the *cores* and *claddings* of two *optical fibers.* See *optical waveguide splice.*

waveguide termination. See *optical waveguide termination.*

wave impedance. In an *electromagnetic wave propagating* in a *propagation medium,* the ratio of the *electric field strength* to the *magnetic field strength* at the point and instant of observation. The characteristic *impedance* of a *plane-polarized electromagnetic wave* in *free space* is 377 ohms. From *Maxwell's equations,* the characteristic impedance in a *linear, homogeneous, isotropic, dielectric,* and electric-charge-free *propagation medium* is given by the relation:

$$Z = (\mu/\epsilon)^{1/2}$$

where μ is the *magnetic permeability* and ϵ is the *electric permittivity.* For free space, $\mu = 4\pi \times 10^{-7}$ H/m (henries per meter) and $\epsilon = (1/36\pi) \times 10^{-9}$ F/m (farads per meter), from which 120π ohms, that is 377 ohms, is obtained. For many dielectric media, such as glass, the electric permittivity, that is, the dielectric constant, ranges from about 2 to 7. However, the *refractive index* is the square root of the electric permittivity. The characteristic impedance of dielectric materials is $377/n$, where n is the refractive index. Also see *refractive index.*

wavelength. 1. For a sinusoidal *wave,* the distance, usually expressed in *meters,* between points of corresponding *phase* of two consecutive cycles. The *wavelength* is also given by the relation:

$$\lambda = v/f$$

where v is the *propagation* velocity and f is the *frequency*. **2.** For any periodic type of *wave,* the distance between corresponding points, such as positive-peak values, in two consecutive cycles of the wave. If the propagation velocity and repetition rate of the wave are known, the wavelength can be calculated from the relation:

$$\delta = s/r$$

where s is the speed and r is the repetition rate. See *cable cutoff wavelength; cutoff wavelength; long wavelength; nominal central wavelength; peak spectral wavelength; short wavelength; zero-dispersion wavelength; zero-material-dispersion wavelength.*

wavelength-dependent attenuation rate characteristic. A plot of the variation of the *attenuation rate* (dB/km) of *optical fiber* as a function of the *wavelength* λ of the energizing *light source.* The plot usually also shows the minimum, nominal, and maximum wavelength in the *spectral width* of the source. The characteristic is used to determine the increase in *attenuation rate* that has to be allowed for in the design of an *optical station/regenerator section.* The increase in attenuation rate to be used in the design is the maximum value, which may occur at the minimum or maximum wavelength, or anywhere in between, relative to the *nominal central wavelength.*

wavelength-division multiplexing (WDM). In *fiberoptics* when referring to *lightwaves,* multiplexing that is similar to *frequency-division multiplexing,* a multiplexing system in which the available *transmission wavelength* range is divided into narrow *bands* and each is used as a separate channel. Because an *optical fiber* can transmit more than one wavelength of *light* at the same time, each wavelength can be separately *modulated* and used as a separate transmission *channel* as long as a combination of dispersive components, such as prisms, and *photodetectors* on the receiving end are wavelength-sensitive or spatially distributed for demultiplexing. In *fiber optics,* such frequency-division multiplexing (FDM) is called wavelength-division multiplexing (WDM), because *lightwaves* are more often described and measured in terms of wavelength rather than *frequency,* and perhaps more importantly, because the use of the term WDM also avoids confusion with the possible use of FDM in assembling *baseband signals,* such as FDM-*pulse* patterns, that are to be transmitted over an *optical link* at each of the operating optical wavelengths. Because optical frequencies are so high, these multiplexing schemes can be accomplished during assigned time slots. Thus, *time-division multiplexing (TDM),* along with *space-division multiplexing (SDM),* can take place concurrently on an optical link. Thus,WDM, FDM, TDM, and SDM can all be used concurrently. (See Figs. W–1 and W–2.)

wavelength measurement. See *central wavelength measurement.*

wavelength range. See *transmitter central wavelength range.*

W-1. A **wavelength-division multiplexer**. (Courtesy Aster Corporation).

WAVELENGTH DIVISION MULTIPLEXER

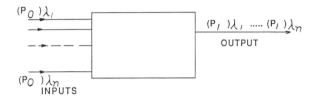

WAVELENGTH DIVISION DEMULTIPLEXER

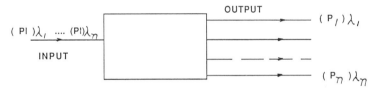

W-2. *Single-mode* or *multimode* **wavelength-division multiplexer (WDM)** and wavelength-division demultiplexer (WDD). (Courtesy Aster Corporation.)

wave number. The reciprocal of the *wavelength* of a single-*frequency* sinusoidal *wave,* such as a single-frequency *uniform plane-polarized electromagnetic wave* or *monochromatic lightwave.* The wave number is used for waves in or near the *visible spectrum,* because wavelength is more readily measured than *frequency,* but it is frequency, proportional to wave number, that is directly related to energy. For example, *photon* energy is given by the relations:

$$PE = hf = hkc$$

where h is *Planck's constant,* k is the wave number, and c is the velocity of *light.* The wave number turns out to be the number of wavelengths per unit distance in the direction of *propagation.* Also see *wave parameter.*

wave parameter. In periodic *waves,* such as *electromagnetic waves,* the wave parameter is given by the relationship:

$$p = 2\pi/\lambda$$

where π is approximately 3.1416 and λ is the wavelength. Also see *wave number.*

WDM. *Wavelength-division multiplexing.*

weakly guiding fiber. An *optical fiber* in which the difference between the maximum and the minimum *refractive index* is small, usually less than 1%.

wet-mateable connector. A connector that can be mated and demated underwater, usually many times and under high pressure, without deleterious effects on the quality of the connection. For example, a *fiber optic connector* that can be connected and disconnected many times on the ocean bottom is a wet-mateable connector.

white light. *Electromagnetic waves* having a spectral distribution, that is, energy distribution, among the *wavelengths* such that the *color* sensation to the average human eye is identical to that of average noon sunlight.

wideband. Pertaining to a *frequency band* that is wide relative to the particular frequency or frequency band under consideration. For example, a band 60 *kHz* wide at a midband frequency of 50 *MHz* would be wide, because the *bandwidth* of 60 kHz is greater than 0.1% of the midband frequency, whereas at a midband frequency of 10 GHz the same 60-kHz band would be considered narrow.

width. See *spectral width.*

width distortion. See *pulse-width distortion.*

window. See *spectral window.*

work function. The amount of energy required to free an electron from an atom in a given material, usually measured in electron-volts. If a *photon* has an energy greater that the work function of the material it strikes, it will be capable of freeing an electron from that material. The freed electron will possess kinetic energy equal to the difference between the energy of the photon and the material work function. The work function of a material is given by the relation:

$$W = hf_0$$

where h is *Planck's constant* and f_0 is the *threshold frequency* required to release an electron from the material. Also see *photoelectric equation.*

working diameter. See *borescope working diameter.*

working length. See *borescope working length.*

worst-case chromatic dispersion over wavelength range. See *worst-case dispersion coefficient.*

worst-case dispersion coefficient. In *single-mode optical station* and *optical cable facilities,* the *worst-case* (end-of-life) *dispersion coefficient* (in ps/(nm-km)) that corresponds to the greater absolute value of (1) the difference between the lower limit of the *transmitter (light source) central wavelength range* and the lower limit of the fiber *zero-dispersion-wavelength* range, and (2) the difference between the upper limit of the transmitter light source central wavelength range and the upper limit of the fiber zero-dispersion-wavelength range. The ranges are caused by manufacturing tolerances, temperature and humidity variations, and aging. Synonymous with *worst-case chromatic dispersion over wavelength range.*

woven fiber optics. The application of textile-weaving and *fiber optic* techniques for producing special-effect devices, such as reproducible-image guides, image dissectors, *color* image panels, high-speed alphanumeric display surfaces, and *distributed-fiber sensors.*

W-type fiber. An *optical fiber* that has two or more concentric layers of *cladding,* in which the *fiber core* usually has the highest *refractive index* and the innermost cladding has the lowest refractive index. This fiber has several advantages, such as reduced *bending losses,* over conventionally cladded fibers that have a *stepped-* or *graded refractive index* that decreases all the way from the core *optical axis* to the outer surface. For example, *cladding modes* in the innermost cladding can be stripped off as in a cladding-mode stripper.

wye coupler. In *fiber optics,* a *coupler* with one input *port* and two output ports. Synonymous with *Y-coupler.*

X Y Z

x-ray. A ray of *electromagnetic waves* that has a *wavelength* between 0.001 and 1 $\times$ 10^{-9} m (*meters*), between 0.001 and 1 nm (*nanometer*), or between 0.000001 and 0.001 μ (*microns*) and therefore a *frequency* in *free space* between 0.3 to 300 $\times$ 10^{18} Hz (*hertz*). The *wavelengths* of x-rays are shorter than those of *ultraviolet light*. Because the energy of a *photon* is given as *Planck's constant* times the frequency, the energy of an x-ray photon is sufficient to penetrate deeply into solids, ionize gases, and destroy molecular structures, such as those of living tissues. They can have sufficient energy to pass through a substance with negligible *attentuation,* except for highly dense substances, such as lead and gold. X-rays must be used with great care. They are used to produce images of materials they pass through in transit to sensitive film. The x-ray photos thus produced indicate varitions in *transmittance* of the material. Also see *optical spectrum.*

Y-coupler. See *wye coupler.*

Zehnder fiber optic sensor. See *Mach-Zehnder fiber optic sensor.*

zero dispersion. Pertaining to a *waveguide* in which an *electromagnetic wave* is *propagating* such that all *modes* in the wave have the same end-to-end *propagation delay.* The *refractive index* is different for each of the various *wavelengths* contained in the *spectral width* of the *source.* The *refractive-index profile* of the guide can be adjusted so that the geometric path taken by each mode is shorter geometric paths, resulting in all modes is the same, which reduces the *signal distortion* by *waveguide dispersion* to nearly zero. The geometric distance is the actual measured distance over which a given mode propagated. If the proper operating wavelength is chosen, *material dispersion* can also be reduced to nearly zero, resulting in a zero-dispersion waveguide at a particular wavelength. In a *single-mode optical fiber,* there is a singular case in which the total *dispersion* (ps/(nm-km)) is identically equal, or nearly equal, to zero. For a given appropriate refractive index profile, zero dispersion will occur at a specific wavelength. The wavelength at which zero dispersion occurs will vary with manufacturing tolerances, temperature and humidity variations, and aging. Also see *zero-material-dispersion wavelength.*

zero-dispersion slope. In a *dielectric waveguide,* such as an *optical fiber,* the slope (ps/(nm^2-km)) of the *global dispersion characteristic* at the *wavelength* at which *zero dispersion* occurs. The zero-dispersion slope is used to calculate the upper and lower limiting values of the *global-dispersion coefficient.* The zero-dispersion slope will vary with manufacturing tolerances, temperature and humidity variations, and aging. It is the value of the *chromatic dispersion slope* at the *zero-dispersion wavelength* of the waveguide, usually expressed by the relation:

$$S_0 = S(\lambda_0)$$

where S_0 is the zero-dispersion slope, $S(\lambda)$ is the chromatic dispersion slope as a function of *wavelength,* and λ_0 is the zero-dispersion wavelength of the waveguide.

zero-dispersion wavelength. For a *lightwave propagating* in a *single-mode optical fiber* with an appropriate *refractive-index profile,* the *wavelength* at which *zero* or nearly zero *dispersion* occurs. The zero-dispersion wavelength will have a maximum and minimum value among fibers and over time because of manufacturing tolerances, temperature and humidity variations, and aging. It is the wavelength at which the fiber *chromatic dispersion parameter,* $D(\lambda)$, or the *material dispersion coefficient,* $M(\lambda)$, is zero. In *single-mode* fibers, the zero-dispersion wavelength is the wavelength at which the total *dispersion* is zero.

zero-material-dispersion wavelength. In an *optical waveguide,* the *electromagnetic weavelength* at which *theree* will be no *material dispersion.* The wavelength usually occurs at that point in the *electromagnetic frequency spectrum* at which the electronic *band-edge absorption,* extending from where the *ultraviolet* ceases and *infrared* absorption begins. The *absorptions* occur in the *visible spectrum* and are primarily caused by oxides of silicon, sodium, boron, calcium, germanium, and other elements, and by hydroxyl ions. For *zero dispersion* in an optical waveguide, the *refractive-index profile* must be such that the guide is compensated for zero dispersion at this *wavelength.*

zone. See *dead zone.*

Bibliography

Abebe, M., C. A. Villarruel, and W. K. Burns. Reproducible fabrication method for polarization preserving single-mode fiber couplers. *Journal of Lightwave Technology* **6** (7): 1191–1198 (July 1988).

Aiki, M. Low-noise optical receiver for high-speed optical transmission. *Journal of Lightwave Technology* **LT-3** (6): 1301–1306 (December 1985).

Alexander, S. B. Design of wide-band optical heterodyne balanced mixer receivers. *Journal of Lightwave Technology* **LT-5** (4): 523–537 (April 1987).

Allan, W. B. *Fiber Optics: Theory and Practice.* New York: Plenum, 1973.

American National Standards Institute. *American National Standard Dictionary for Information Systems.* Washington, DC: Computer and Business Equipment Manufacturers Association, 1988.

AMP, Incorporated. *Designers Guide to Fiber Optics.* Harrisburg, Pennsylvania, 1982.

Andrews, R. A., A. F. Milton, and T. G. Giallorenzi. Military applications of fiber optics and integrated optics. *IEEE Transactions on Microwave Theory and Techniques* **MTT-21**: 763–769 (1973).

Arie, A. and M. Tur. The effects of polarization control on the transfer function and the phase induced intensity noise of a fiber optic recirculating delay line. *Journal of Lightwave Technology* (Special Issue of the 1988 Optical Fiber Communications/Optical Fiber Sensors Conference) **6** (10): 1566–1574 (October 1988).

Atternas, L., L. Thylen. Single-layer antireflection coating of semiconductor lasers: Polarization properties and the influence of the laser structure. *Journal of Lightwave Technology* **7** (2): 426–430 (February 1989).

Auracher, F. A. Photoresist coupler for optical waveguides. *Optical Communication* **11**: 191–195 (1974).

Bachus, E. J., R. P. Braun, C. Caspar, H. M. Foisel, E. Grossman, B. Strebel, and F. J. Westphal. Coherent optical multicarrier systems. *Journal of Lightwave Technology* **7** (2): 375–384 (February 1989).

Baker, D. G. *Monomode Fiber-Optic Design with Local-Area and Long-Haul Network Applications.* New York: Van Nostrand Reinhold Company, 1987.

Barnoski, M. K., ed. *Fundamentals of Fiber Optic Communications.* New York: Academic Press, 1976.

——, ed. *An Introduction to Integrated Optics.* New York: Plenum, 1974.

Bartree, T. C. *Digital Communications.* Indianapolis, Indiana: Howard W. Sams & Company, 1986.

——. *Data Communications, Networks, and Systems.* Indianapolis, Indiana: Howard W. Sams & Company, 1985.

Berian, A. G. Life-reliability stressed at cable & connector workshop. *Sea Technology:* 47–48 (June 1987).

Bordon, E. E., W. L. Anderson. Dispersion-adapted monomode fiber for propagation of nonlinear pulses. *Journal of Lightwave Technology* **7** (2): 353–357 (February 1989).

Botez, D. Laser diodes are power-packed. *IEEE Spectrum:* 43–53 (June 1985).

Brain, M. and T.-P. Lee. Optical receivers for lightwave communication systems. *Journal of Lightwave Technology* **LT-3** (6): 1281–1300 (December 1985).

Brody, H. The rewiring of America. *High Technology Business:* 34–37 (February 1988).

Bucholtz, F., K. P. Koo, and A. Dandridge. Effect of external perturbations on fiber-optic magnetic sensors. *Journal of Lightwave Technology* **6** (4): 507–512 (April 1988).

Burns, W. K., R. P. Moeller, C. A. Villarruel, and M. Abebe. All-fiber gyroscope with polarization-holding fiber. *Optics Letters* **9** (12): 570–572 (December 1984).

Burrus, C. A. Radiance of small-area high-current-density electroluminescent diodes. *Proceedings of the IEEE* **60**: 231–232 (1972).

—— and R. D. Standley. Viewing refractive-index profiles and small-scale inhomogeneities in glass optical fibers: Some techniques. *Applied Optics* **13**: 2365 (1974).

——, E. L. Chinnock, D. Gloge, W. S. Holden, T. Li, and D. B. Keck. Pulse dispersion and refractive index profiles of some low-noise multimode fibers. *Proceedings of the IEEE* **61**: 1498 (1973).

Cancellieri, G. and U. Ravaioli. *Measurements of Optical Fibers and Devices: Theory and Experiments.* Dedham, Massachusetts: Artech House, Inc., 1984.

Cardama, A. and E. T. Kornhauser. Modal analysis of coupling problems in optical fibers. *IEEE Transactions on Microwave Theory and Techniques* (Special Issue on Integrated Optics and Optical Waveguides) **MTT-23** 162–169 New York: IEEE Press (1975).

Cartlege, J. C. Impulse response of a step index optical fiber excited by a Lambertian source: Addendum. *Applied Optics* **16**: 3082 (December 1977).

Chang, W. S. C. Integrated optics at 10.6 μm wavelength. *IEEE Transactions on Microwave Theory and Techniques* (Special Issue on Integrated Optics and Optical Waveguides) **MTT-23**: 31–44 (1975).

Chase, K. M. On wave propagation in inhomogeneous media. *Journal of Mathematics and Physics* **13**: 360 (1972).

Chen, K.-L. and D. Kerps. Coupling efficiency of surface-emitting LED's to single-mode fibers. *Journal of Lightwave Technology* **LT-5** (11): 1600–1604 (November 1987).

Cheng, Y. H., T. Okoshi, O. Ishida. Performance analysis and experiment of a homodyne receiver insensitive to both polarization and phase fluctuations. *Journal of Lightwave Technology* **7** (2): 368–374 (February 1989).

Cherin, A. H. *An Introduction to Optical Fibers.* New York: McGraw-Hill Book Company, 1983.

Chiang, K. S. Stress-induced birefringence fibers designed for single-polarization single-mode operation. *Journal of Lightwave Technology* **7** (2): 436–441 (February 1989).

Chu, P. L., T. Whitbread. An on-line fiber drawing tension and diameter measurement device. *Journal of Lightwave Technology* **7** (2): 255–262 (February 1989).

Cochrane, P., M. Brain. Future optical fiber transmission technology and networks. IEEE Communications Magazine 26 (11): 45–60 (November 1988).

Cotter, D. Stimulated Brillouin scattering in monomode optical fiber. Journal of Optical Communications 4 (1): 10–19 (January 1983).

CSELT (Centro Studi e Laboratori Telecomunicazioni). *Optical Fiber Communication.* Torino (Turin), Italy, 1980.

Culshaw, B. *Optical Fiber Sensing and Signal Processing.* London: Peter Peregrinus Ltd., 1984.

Dakin, J. P., W. A. Gambling, H. Matsumura, D. N. Payne, and H. R. D. Sunak. Theory of dispersion in lossless multimode optical fibers. *Optics and Communications* **7**: 1 (1973).

Darcie, T. E. Differential frequency-deviation multiplexing for lightwave networks. *Journal of Lightwave Technology* **7** (2): 314–322 (February 1989).

Davis, C. M., E. F. Carome, M. H. Weik, S. Ezekiel, and R. E. Einzig. *Fiberoptic Sensor Technology Handbook*. Herndon, Virginia: Optical Technologies, Inc., 1986.

Dimarcello, F. V., D. L. Brownlow, R. G. Huff, and A. C. Hart, Jr. Multikilometer lengths of 3.5-GPa (500 kpsi) prooftested fiber. *Journal of Lightwave Technology* **Lt-3** (5): 946–948 (October 1985).

Drexhage, M. G. and C. T. Moynihan. Infrared optical fibers. *Scientific American:* 110–116 (November 1988).

Drögenmüller, K. A compact optical isolator with a plano-convex YIG lens for laser-to-fiber coupling. *Journal of Lightwave Technology* **7** (2): 340–346 (February 1989).

Dynamic Systems, Incorporated. *Fiber Optic Sensor Abstracts*. Reston, Virginia, 1982.

EIA. *EIA-440A: Fiber Optic Terminology*. Electronic Industries Association, 2001 Eye Street, NW, Washington, D.C. 20006, 1988.

Eickhoff, W. and E. Weidel. Measuring method of the refractive index profile of optical glass fibers. *Optical and Quantum Electronics* **7**: 109 (1975).

Elrefaie, A. F., R. E. Wagner, D. A. Atlas, and D. G. Daut. Chromatic dispersion limitations in coherent lightwave transmission systems. *Journal of Lightwave Technology* **6** (5): 704–709 (May 1988).

Fermann, M. E., S. B. Poole, D. N. Payne, and F. Martinez. Comparative measurement of Rayleigh scattering in single-mode optical fibers based on an OTDR technique. *Journal of Lightwave Technology* **6** (4): 545–551 (April 1988).

Foschini, G. R. and G. Vannucci. Using spread-spectrum in a high-capacity fiber-optic local network. *Journal of Lightwave Technology* **6** (3): 370–379 (March 1988).

Freeman, R. L. *Telecommunication Transmission Handbook*. New York: John Wiley & Sons, 1981.

Freude, W. and A. Sharma. Refractive-index profile and modal dispersion prediction for a single-mode optical waveguide from its far-field radiation pattern. *Journal of Lightwave Technology* **LT-3** (3): 628–634 (June 1985).

———, H.-H. Yao, and Z.-J. He. Propagation constant and waveguide dispersion in single-mode fibers measured from the far-field. *Journal of Lightwave Technology* **6** (2): 318–321 (February 1988).

———, E. K. Sharma, and A. Sharma. Propagation constant of single-mode fibers measured from mode-field radius and from bending loss. *Journal of Lightwave Technology* **7** (4): 225–228 (February 1989).

Friebele, E. J., E. W. Taylor, G. Turguet de Beauregard, J. A. Wall, and C. E. Barnes. Interlaboratory comparison of radiation-induced attenuation in optical fibers. Part I: Steady-state exposures. *Journal of Lightwave Technology* **6** (2): 165–171 (February 1988).

Friedrich, N. T. Fiber optic trends: Linking LANs with fiber: Cost and other factors. *Photonics Spectra:* 89–92 (September 1988).

Fukuda, M. Lasers and LED reliability update (invited paper). *Journal of Lightwave Technology* (Special Issue on the 1988 Optical Fiber Communications/Optical Fiber Sensors Conference) **6** (10): 1488–1495.

Gagliardi, R. M. and S. Karp. *Optical Communications*. New York: John Wiley & Sons, 1976.

———. *A User's Manual for Optical Waveguide Communications* PB 252 901. U.S. Department of Commerce. Washington, D.C.: U.S. Government Printing Office (March 1976).

Gallawa, R. L. Optical waveguide technology for modern urban communications. *IEEE Transactions on Communications* **COM-23**: 131–142 (1975).

Gardner, W. B. and D. Gloge. Microbending loss in coated and uncoated optical fibers. *Digest of Topical Meeting on Optical Fiber Transmission,* Williamsburg, Virginia **WA-1-WA-4** (January 1975).

——, E. L. Chinnock, and T. P. Lee. GaAs twin-laser setup to measure mode and material dispersion in optical fibers. *Applied Optics* **13**: 261–263 (1974).

——, ed. *Optical Fiber Technology.* New York: IEEE Press, 1976.

——. Optical power flow in multimode fibers. *Bell System Technical Journal* **51**: 1767 (1972).

——. Propagation effects in optical fibers. *IEEE Transactions on Microwave Theory and Technology* (Special Issue on Integrated Optics and Optical Waveguides) **MTT-23**: 106–120 (1975).

Garmire, E. Simple solutions for modeling symmetric step-index dielectric waveguides. *Journal of Lightwave Technology* (Special Issue on Integrated Optics) **6** (6): 1105–1108 (June 1988).

Garrett, I., G. Jacobsen, E. Bodtker, R. J. S. Pedersen, and J. S. Kan. Weakly coherent optical systems using lasers with significant phase noise (invited paper). *Journal of Lightwave Technology* (Special Issue on the 1988 Optical Fiber Communications/Optical Fiber Sensors Conference) **6** (10): 1520–1526 (October 1988).

Garth, S. J. Birefringence in bent single-mode fibers. *Journal of Lightwave Technology* **6** (3): 445–449 (March 1988).

Gaspar, J. and H. Soffer. Electro-optics on the modern battlefield. *Defense Electronics:* 139–153 (October 1986).

Geittner, P., H.-J. Hagemann, H. Lydtin, and J. Warnier. Hybrid technology for large SM fiber preforms. *Journal of Lightwave Technology* (Special Issue on the 1988 Optical Fiber Communications/Optical Fiber Sensors Conference) **6** (10): 1451–1454 (October 1988).

Giallorenzi, T. G., R. A. Steinberg, and R. G. Priest. New electrode design for polarization insensitive optic switches. *Applied Optics* **16**: 2166 (1977).

——, J. A. Bucaro, A. Dandridge, G. H. Sigel, Jr., J. H. Cole, S. C. Rashleigh, and R. G. Priest. Optical fiber sensor technology. *IEEE Journal of Quantum Electronics* **QE-18** (4): 626–664 (April 1982).

Gilpin, J. B., J. Roschen, S. R. Hampton, and M. J. Kniffin. A high performance 8 to 12 μm (Hg,Cd)Te 72-element detector array. *Society of Photo-Optical Instrumentation Engineers* (SPIE) **132**: 141–144 (1978).

Gloge, D. Weakly guiding fibers. *Applied Optics* **10**: 2252–2258 (October 1971).

Goodfellow, R. C., B. T. Debney, G. J. Rees, and J. Buus. Optoelectronic components for multigigabit system. *Journal of Lightwave Technology* **LT-3** (6): 1170–1179 (December 1985).

Griffioen, W. The installation of conventional fiber-optic cables in conduits using the viscous flow of air. *Journal of Lightwave Technology* **7** (2): 297–302 (February 1989).

Gruber, J., P. Marten, R. Petschacher, and P. Russer. Electronic circuits for high bit rate digital fiber optic communications systems. *IEEE Transactions on Communications* (Special Issue on Fiber Optics) **COM-26** (7): 1088 (July 1978).

Habbab, I. M. I. and L. J. Cimini, Jr. Polarization switching techniques for coherent optical communications. *Journal of Lightwave Technology* (Special Issue on the 1988 Optical Fiber Communications/Optical Fiber Sensors Conference) **6** (10): 1537–1548 (October 1988).

Harger, R. O. *Optical Communication Theory.* Benchmark papers in Electrical Engineering and Computer Science **18**. New York: Academic Press, 1977.

Harris, A. J., P. A. Shrubshall, and P. F. Castle. Wavelength demultiplexing using bends in a single-mode optical fiber. *Journal of Lightwave Technology* **6** (1): 80–86 (January 1988).

Hecht, J. *Understanding Fiber Optics.* Indianapolis, Indiana: Howard W. Sams & Company, 1987.

——. Optical computers. *High Technology* 44–49 (February 1987).

——. Pushing the limits of fiberoptic technology. *Lasers and Optronics:* 68–72 (January 1988).

Henry, C. H. and B. H. Verbeek. Solution of the scalar wave equation for arbitrarily shaped dielectric waveguides by two-dimensional Fourier analysis. *Journal of Lightwave Technology* 7 (2): 308–313 (February 1989).

Hentschel, C. *Fiber Optics Handbook: An Introduction and Reference Guide to Fiber Optic Technology and Measurement Techniques.* Böblingen, Federal Republic of Germany: Hewlett-Packard GmbH, 1983.

Hill, A. M. and D. B. Payne. Linear crosstalk in wavelength-division-multiplexed optical-fiber transmission systems. *Journal of Lightwave Technology* LT-3 (3): 643–651 (June 1985).

Hudson, M. C. and P. J. Dobson. Fiberoptic cable technology. *Microwave Journal:* 46–53 (July 1979).

Hudson, M. L. and F. L. Thiel. The star coupler: A unique interconnection component for multimode optical waveguide communications systems. *Applied Optics* 14: 2540–2545 (1974).

IEC. IEC-731: *Optical Communication Terminology.* International Electrotechnical Commission, draft under preparation, 1988.

IEEE. *IEEE Proceedings* (Special Issue on Optical Communications) 58: 1410–1786 (1970).

——. *IEEE Std 812-1984: Definitions of Terms Relating to Fiber Optics.* The Institute of Electrical and Electronics Engineers, Inc., 445 Hoes Lane, Piscataway, New Jersey 08854-4150, 1984.

——. *IEEE Transactions on Microwave Theory and Techniques.* (Special Issue on Integrated Optics and Optical Waveguides) MTT-23 New York: IEEE Press (1975).

Iino, A. and J. Tamura. Radiation resistivity in silica optical fibers. *Journal of Lightwave Technology* 6 (2): 145–149 (February 1988).

Inoue, H., K. Hiruma, K. Ishida, T. Asai, and H. Matsumura. Low loss GaAs optical waveguides. *Journal of Lightwave Technology* LT-3 (6): 1270–1276 (December 1985).

Institution of Electrical Engineers. *First European Conference on Optical Fibre Communication.* Conference Publication Number 132. London: Institution of Electrical Engineers, 1975.

International Organization for Standardization. *Information Processing Vocabulary. International Standard 2382.* New York: American National Standards Institute, 1988.

Iwamoto, Y. and H. Fukinuki. Recent advances in submarine optical fiber cable transmission systems in Japan. *Journal of Lightwave Technology* LT-3 (5): 1005–1015 (October 1985).

Jackel, J. L. and J. J. Johnson. Nonsymmetric Mach-Zehnder interferometers used as low-drive voltage modulators. *Journal of Lightwave Technology* 6 (8): 1348–1351 (August 1988).

Jaeger, N. A. F. and L. Young. High-voltage sensor employing an integrated optics Mach-Zehnder interferometer in conjunction with a capacitive divider. *Journal of Lightwave Technology* 7 (2): 229–235 (February 1989).

Jung, H. S., O. Eknoyan, and H. F. Taylor. Enhancement of refractive index in $Ti:LiTaO_3$ optical waveguides by Zn diffusion. *Journal of Lightwave Technology* 7 (2): 390–392 (February 1989).

Kaiser, P., J. Midwater, and S. Shimada. Status and future trends in terrestrial fiber systems in North America, Europe, and Japan. *IEEE Communications* 25 (10): 28–36 (October 1987).

Kapany, N. S. *Fiber Optics.* New York: Academic Press, 1967.

Kapron, F. P. The evolution of optical fibers. *Microwave Journal:* 111–120 (April 1985).

Karstensen, H. and R. Frankenberger. High-efficiency two lens laser diode to single-mode fiber coupler with a silicon plano convex lens. *Journal of Lightwave Technology* **7** (2): 244–249 (February 1989).

Kasper, B. L., J. C. Campbell. Multi-gigabit/s avalanche photodiode lightwave receivers. *IEEE Journal of Lightwave Technology* **LT-5** (10): 1138–1142 (October 1987).

Katoh, S., I. Yamamoto, and T. Nakamura. New cable layer Koyo Maru advanced cable ship navigation aid system. *Naval Engineering Center Res. & Develop.* 77: 80–87 (April 1985).

Kazovsky, L. G. *Transmission of Information in the Optical Waveband.* New York: Halsted Press, 1978.

———. Balanced phase-locked loops for optical homodyne receivers: Performance analysis, design considerations, and laser linewidth requirements. *Journal of Lightwave Technology* **LT-4** (2): 182–195 (February 1986).

———, R. Welter, A. F. Elrefaie, and W. Jessa. Wide-linewidth phase diversity homodyne receivers. *Journal of Lightwave Technology* (Special Issue on the 1988 Optical Fiber Communications/Optical Fiber Sensors Conference) **6** (10): 1527–1536. (October 1988).

———. Phase- and polarization-diversity coherent optical techniques. *Journal of Lightwave Technology* **7** (2): 279–292 (February 1989).

Keck, D. B. Observation of externally controlled mode coupling in optical waveguides. *Proceedings of the IEEE* **62**: 649–650 (1974).

Kerker, M. *Scattering of Light and Other Electromagnetic Radiation.* New York: Academic Press, 1969.

Kersey, A. D., M. J. Marrone, A. Dandridge, and A. B. Tveten. Optimization and stabilization of visibility in interferometric fiber-optic sensors using input-polarization control. *Journal of Lightwave Technology* (Special Issue on the 1988 Optical Fiber Communications/Optical Fiber Sensors Conference) **6** (10): 1599–1609 (October 1988).

Khoe, G. D., H. G. Kock, D. Küppers, J. H. F. M. Poulissen, and H. M. de Vrieze. Progress in monomode optical-fiber interconnection devices. *Journal of Lightwave Technology* **LT-2** (3): 217–226 (June 1984).

Kikuchi, K., T. Okoshi, M. Nagamatsu, and H. Henmi. Degradation of bit-error rate in coherent optical communications due to spectral spread of the transmitter and the local oscillator. *Journal of Lightwave Technology* **LT-2** (6): 1024–1033 (December 1984).

Kimura, T. Factors affecting fiber-optic transmission quality. *Journal of Lightwave Technology* **6** (5): 611–619 (May 1988).

———. Coherent optical fiber transmission (invited paper). *Journal of Lightwave Technology* (Special Issue on Coherent Communications) **LT-5** (4): 414–428 (April 1987).

——— and K. Daikoku. A proposal on optical fiber transmission systems in a low loss 1.0–1.4 μm wavelength region. *Optical and Quantum Electronics* **9**: 33–42 (1977).

Koa, C. K., C. F. Blanco, and A. Asam. Fiber cable technology. *Journal of Lightwave Technology* **LT-2** (4): 479–482 (August 1984).

Koo, K. P., A. Dandridge, F. Bucholtz, and A. B. Tveten. An analysis of a fiber optic magnetometer with magnetic feedback. *Journal of Lightwave Technology* **LT-5** (12): 1680–1685 (December 1987).

Kuester, E. F. and D. C. Chang. Propagation, attenuation, and dispersion characteristics of inhomogeneous dielectric slab waveguides. *IEEE Transactions on Microwave Theory and Techniques* (Special Issue on Integrated Optics and Optical Waveguides) **MTT-23**: 98–106 (1975).

Kull, M. and G. Manneberg. Integrated optical amplifier for fast phase modulated signals. *Journal of Lightwave Technology* **7** (2): 331–335 (February 1989).

Kumar, A., R. K. Varshney, and R. K. Sinha. Scalar modes and coupling characteristics of eight-port waveguide couplers. *Journal of Lightwave Technology* **7** (2): 293–296 (February 1989).

Kurtz, C. N. and W. Streifer. Guided waves in inhomogeneous focusing media. *IEEE Transactions on Microwave Theory and Technology* **MTT-17**: 250–263 (May 1969).

Lacy, E. A. *Fiber Optics*. Englewood Cliffs, New Jersey: Prentice-Hall, Inc., 1982.

Lau, K. Y. Short-pulse and high-frequency signal generation in semiconductor lasers. *Journal of Lightwave Technology* **7** (2): 400–419 (February 1989).

Lewin, L. Radiation from curved dielectric slabs and fibers. *IEEE Transactions on Microwave Theory and Techniques* **MTT-22**: 718–727 (1974).

Lipson, J., W. J. Minford, E. J. Murphy, T. C. Rice, R. A. Linke, and G. T. Harvey. A six-channel wavelength multiplexer and demultiplexer for single mode systems. *Journal of Lightwave Technology* **LT-3** (5): 1159–1162 (October 1985).

Loeb, M. L. and G. R. Stilwell, Jr. High-speed data transmission on an optical fiber using a byte-wide WDM system. *Journal of Lightwave Technology* **6** (8): 1306–1311 (August 1988).

Lucas, A. D. Epitaxial silicon avalanche photodiode. *Opto-electronics* **6**: 153–160 (1974).

Maeda, M. and H. Nakano. Integrated optoelectronics for optical transmission systems. *IEEE Communications* **26** (5): 45–51 (May 1988).

Mahlke, G. and P. Gössing. *Fiber Optic Cables*. Chichester, England: John Wiley & Sons Limited, 1987.

Marcuse, D. *Light Transmission Optics*. New York: Van Nostrand Reinhold Company, 1982.

———. Pulse propagation in multimode dielectric waveguides. *Bell System Technical Journal* **51**: 1199–1232 (1972).

———, ed. *Integrated Optics*. New York: IEEE Press, 1973.

———. Losses and impulse response of a parabolic-index profile. *Bell System Technical Journal* **52**: 1423–1437 (1973).

———. *Theory of Dielectric Optical Waveguides*. New York: Academic Press, 1974.

———. LED fundamentals, comparison of front and edge emitting diodes. *Journal of Qantum Electronics* **QE-13**: 819 (1977).

——— and H. M. Presby. Light scattering from optical fibers with arbitrary refractive index profiles. *Journal of the Optical Society of America* **65**: 367 (1975).

———. Launching light into fiber cores from sources located in the cladding. *Journal of Lightwave Technology* **6** (8): 1273–1279 (August 1988).

———. Reflection loss of laser mode from tilted end mirror. *Journal of Lightwave Technology* **7** (2): 336–339 (February 1989).

Matsushita, S., K. Kawai, and H. Uchida. Fiber-optic devices for local area network application. *Journal of Lightwave Technology* **LT-3** (3): 544–555 (June 1985).

Maurer, R. D. Glass fibers for optical communication. *Proceedings of the IEEE* **61**: 452–462 (April 1973).

Maurer, S. J. and L. B. Felsen. Ray methods for trapped and slightly leaky modes in multilayered or multiwave regions. *IEEE Transactions on Microwave Theory and Technology* **MTT-18**: 584–595 (September 1970).

McNulty, J. J. A 150-km repeaterless undersea lightwave system operating at 1.55 μm. *Journal of Lightwave Technology* **LT-2** (6): 787–791 (December 1984).

Meneghini, G., M. Meliga, C. De Bernardi, S. Morasca, and A. Carnera. High-accuracy characterization of titanium films for LiNbO$_3$ guided wave devices by optical densitometry. *Journal of Lightwave Technology* **7** (2): 250–254 (February 1989).

Mermelstein, M. D. Fiber-optic polarimetric DC magnetometer utilizing a composite metallic glass resonator. *Journal of Lightwave Technology* **LT-4** (9): 1376–1380 (September 1986).

Midwinter, J. E. *Optical Fibers for Transmission*. New York: John Wiley & Sons, 1979.

———. Current status of optical communications technology. *Journal of Lightwave Technology* **LT-3** (5): 927–929 (October 1985).

Miller, C. M. Loose tube splices for optical fibers. *Bell System Technical Journal* **54**: 1212 (1975).

Mochizuki, K., Y. Namihira, M. Kuwazuru, and M. Nunokawa. Influence of hydrogen on optical fiber loss in submarine cables. *Journal of Lightwave Technology* **LT-2** (6): 802–807 (December 1984).

Moeller, R. P., W. K. Burns, and N. J. Frigo. Open-loop output and scale factor stability in a fiber-optic gyroscope. *Journal of Lightwave Technology* **7** (2): 262–269 (February 1989).

Möller, K. D. *Optics.* Mill Valley, California: University Science Books, 1988.

Moore, J. A. Fiber finds a niche in process control. *Photonics Spectra:* 57–60 (December 1987).

Moslehi, B., M. R. Layton, and H. J. Shaw. Efficient fiber-optic structure with applications to sensor arrays. *Journal of Lightwave Technology* **7** (2): 236–243 (February 1989).

Murphy, E. J. Fiber attachment for guided-wave devices (invited paper). *Journal of Lightwave Technology* (Special Issue on Integrated Optics) **6** (6): 862–871 (June 1988).

Nagel, S. R. Review of the depressed cladding single-mode fiber design and performance for the SL undersea system application. *Journal of Lightwave Technology* **LT-2** (6): 792–801 (December 1984).

Nakagawa, K. and K. Nosu. An overview of very high capacity transmission technology for NTT networks. *Journal of Lightwave Technology* **LT-5** (10): 1498–1504 (October 1987).

Nakahara, T. and N. Uchida. Optical cable design and characterization in Japan. *Proceedings of the IEEE* **68** (10): 1220–1225 (October 1980).

———, H. Kumamaru, and S. Takeuchi. An optical fiber video system. *IEEE Transactions on Communications* (Special Issue on Fiber Optics) **COM-26** (7): 955 (July 1978).

Nakano, Y., H. Sudo, G. Iwane, T. Matsumoto, and T. Ikegami. Reliability of semiconductor lasers and detectors for undersea transmission systems. *Journal of Lightwave Technology* **LT-2** (6): 945–951 (December 1984).

National Technical Information Service. U.S. Department of Commerce. NTIS/PS-78/0716 *Fiber Optics* **3** (August 1978).

Neumann, E. G. and H. D. Rudolph. Radiation from bends in dielectric rod transmission lines. *IEEE Transactions on Microwave Theory and Techniques.* (Special Issue on Integrated Optics and Optical Waveguides) **MTT-23**: 142-149 (1975).

Nicholson, G. and T. D. Stephens. Performance analysis of coherent optical phase-diversity receivers with DPSK modulation. *Journal of Lightwave Technology* **7** (2): 393–399 (February 1989).

Nosu, K. and K. Iwashita. A consideration of factors affecting future coherent lightwave communication systems. *Journal of Lightwave Technology* **6** (5): 686–694 (May 1988).

———. Advanced coherent lightwave technologies. *IEEE Communications Magazine* **26** (2): 15–21 (February 1988).

Ohashi, M., N. Kuwaki, and N. Uesugi. Suitable definition of mode field diameter in view of splice loss evaluation. *Journal of Lightwave Technology* **LT-5** (12): 1676–1679 (December 1987).

Okoshi, T. Polarization-state control schemes for heterodyne or homodyne optical fiber communications. *Journal of Lightwave Technology* **LT-3** (6): 1232-1237 (December 1985).

Olshansky, R. and R. D. Mauer. Tensile strength and fatigue of optical fibers. *Journal of Applied Physics* **41**: 4497 (1976).

Panish, M. B. Heterostructure injection lasers. *Proceedings of the IEEE* **64**: 1512 (1977).

Paoli, T. C. Modulation of diode lasers. *Laser Focus* **13**: 54 (1977).

Paris, D. T. and F. K. Hurd. *Basic Electromagnetic Theory.* New York; McGraw-Hill Book Company, 1969.

Peng, S. T., T. Tamir, and H. L. Bertoni. Theory of periodic dielectric waveguides. *IEEE*

Transactions on Microwave Theory and Techniques (Special Issue on Integrated Optics and Optical Waveguides) **MTT-23:** 123–133 (1975).

Perry, M. W., G. A. Reinold, and P. A. Yeisley. Physical design of the SL repeater. *IEEE Journal on Selected Areas of Communications* **SAC-2** (6): 903–909 (November 1984).

Personick, S. D. *Fiber Optics.* New York: Plenum Press, 1985.

———. Fiber optic communications. *IEEE Communications Society Magazine* (March 1978).

———. Receiver design for digital fiber optic communication systems, II. *Bell System Technical Journal* **52:** 875–886 (1973).

Pietzsch, J. Scattering matrix analysis of 3 × 3 fiber couplers. *Journal of Lightwave Technology* **7** (2): 303–307 (February 1989).

Plastow, R. LEDs for fiber optic communication systems. *Photonics Spectra:* 109–116 (September 1986).

Pratt, W. R. *Laser Communication Systems.* New York: John Wiley & Sons, 1969.

Presby, H. M., N. Amitay, R. Scotti, and A. F. Benner. Laser-to-Fiber coupling via optical fiber up-tapers. *Journal of Lightwave Technology* **7** (2): 274–278 (February 1989).

Quan, F. J-Y and J. C. Vernon. Optical fiber will enhance Navy's ASW capabilities. *Sea Technology:* 17–21 (July 1988).

Rawson, E. G. and R. M. Metcalke. Fibernet: Multimode optical fibers for local computer networks. *IEEE Transactions on Communications* (Special Issue on Fiber Optics) **COM-26** (7): 983 (July 1978).

Rediker, R. H., T. A. Lind, and B. E. Burke. Optical wavefront measurement and/or modification using integrated optics. *Journal of Lightwave Technology* (Special Issue on Integrated Optics) **6** (6): 916–932 (June 1988).

Runge, P. K. and P. R. Trischitta. The SL undersea lightwave system. *Journal of Lightwave Technology* **LT-2** (6): 744–753 (December 1984).

Safford, E. L., Jr. *The Fiberoptics & Laser Handbook.* Blue Ridge Summit, Pennsylvania: Tab Books, Inc. 1984.

Sakaguchi, S. Drawing of high-strength long-length optical fibers for submarine cables. *Journal of Lightwave Technology* **LT-2** (6): 808–815 (December 1984).

Saleh, A. A. M., J. Stone. Two-stage Fabry-Perot filters as demultiplexers in optical FDMA LAN's. *Journal of Lightwave Technology* **7** (2): 323–330 (February 1989).

Sansonetti, P., E. C. Caquot, and A. Carenco. Design of semiconductor electrooptic directional coupler with the beam propagation method. *Journal of Lightwave Technology* **7** (2): 385–389 (February 1989).

Sasaki, Y., T. Hosaka, and J. Noda. Polarization-maintaining optical fibers used for a laser diode redundancy system in a submarine optical repeater. *Journal of Lightwave Technology* **LT-2** (6): 816–823 (December 1984).

Schickentanz, D. and J. Schubert. Coupling losses between laser diodes and multimode glass fibers. *Optical Communications* **5:** 291–292 (1972).

Schwartz, M. I., F. G. Gagen, and M. R. Santana. Fiber cable design and characterization. *Proceedings of the IEEE* **68** (10): 1214–1219 (October 1980).

Seligson, J. The orthogonality relation for TE- and TM-modes in guided-wave optics. *Journal of Lightwave Technology* **6** (8): 1260–1264 (August 1988).

Senior, J. M. *Optical Fiber Communications: Principles and Practice.* London: Prentice-Hall International, Inc., 1985.

Severin, P. J. W. A fail-safe self-routing fiber-optic ring network using multitailed receiver-transmitter units as nodes. *Journal of Lightwave Technology* **7** (2): 358–363 (February 1989).

Shama, Y., A. A. Hardy, and E. Marom. Multimode coupling of unidentical waveguides. *Journal of Lightwave Technology* **7** (2): 420–425 (February 1989).

Sharma, A., P. K. Mishra, and A. K. Ghatak. Single-mode optical waveguides and directional couplers with rectangular cross section: A simple and accurate method of analysis. *Journal of Lightwave Technology* (Special Issue on Integrated Optics) **6** (6): 119–125 (June 1988).

Smith, D. W. Techniques for multigigabit coherent optical transmission. *Journal of Lightwave Technology* **LT-5** (10): 1466–1478 (October 1987).

Smith, P. W. and A. M. Weiner. Ultrashort light pulses. *IEEE Circuits and Devices Magazine:* 3–7 (May 1988).

Snitzer, E. Cylindrical dielectric waveguide modes. *Journal of the Optical Society of America* **51**: 491–505 (May 1961).

Someda, C. G. Simple low loss joints between single-mode optical fibers. *Bell System Technical Journal* **52**: 1579–1588 (November 1973).

Sorensen, H. Designer's guide to optoisolators. *Electronic Design News* (February 5, 1976).

Stegeman, G. I., E. M. Wright, N. Finlayson, R. Zanoni, and C. T. Seaton. Third order nonlinear integrated optics (invited paper). *Journal of Lightwave Technology* (Special Issue on Integrated Optics) **6** (6): 953–970 (June 1988).

Steinberg, R. Pulse propagation in multimode fibers with frequency-dependent coupling. *IEEE Transactions on Microwave Theory and Techniques* (Special Issue on Integrated Optics and Optical Waveguides) **MTT-23**: 121–122 (1975).

Stone, J. Stress-optic effects, birefringence, and reduction of birefringence by annealing in fiber Fabry-Perot interferometers. *Journal of Lightwave Technology* **6** (7): 1245–1248 (July 1988).

Stueflotten, S. and H. H. Bruusgaard. Development and testing of a fire-resistant optical cable. *Journal of Lightwave Technology* **LT-4** (8): 1173–1177 (August 1986).

Suhir, E. Effect of initial curvature on low temperature microbending in optical fibers. *Journal of Lightwave Technology* **6** (8): 1321–1327 (August 1988).

Swain, R. A. and D. C. Poyer. Application of fiber optic technology to shipboard use: Near and far term. *Naval Engineers Journal:* 165–170 (July 1984).

Takasaki, Y. T., M. Tanaka, N. Maeda, K. Yamashita, and K. Nagano. Optical pulse formats for fiber optic digital communications. *IEEE Transactions on Communications* **COM-24**: 404–413 (April 1976).

Taylor, H. F. Polarization independent guided-wave optical modulators and switches. *Journal of Lightwave Technology* **LT-3** (6): 1277–1280 (December 1985).

Tekippe, V. J. and W. R. Wilson. Single-mode directional couplers. *Laser Focus/Electro-Optics:* 132–144 (May 1985).

Thylen, L. Integrated optics in LiNbO$_3$: Recent developments in devices for telecommunications. *Journal of Lightwave Technology* (Special Issue on Integrated Optics) **6** (6): 847–861 (June 1988).

Tingye, W. and R. A. Linke. Multigigabit-per-second lightwave systems research for long-haul applications. *IEEE Communications* **26** (4): 29–35 (April 1988).

Toda, H., K. Kasazumi, M. Haruna, and H. Nishihara. An optical integrated circuit for time-division 2-D velocity measurement. *Journal of Lightwave Technology* **7** (2): 364–367 (February 1989).

Tomlinson, W. J. and R. H. Stolen. Nonlinear phenomenon in optical fibers. *IEEE Communications* **26** (4): 36–44 (April 1988).

Tucker, R. S. High-speed modulation of semiconductor lasers. *Journal of Lightwave Technology* **LT-3** (6): 1180–1192 (December 1985).

Ungar, J., N. Bar-Chaim, and I. Ury. High-power GaAlAs window lasers. *Lasers and Applications:* 111–114 (September 1985).

Ura, S., T. Suhara, and H. Nishihara. Integrated-optic interferometer position sensor. *Journal of Lightwave Technology* **7** (2): 270–273 (February 1989).

Ury, I. Optical communications. *Microwave Journal:* 24–35 (April 1985).

U.S. Department of Defense. *MIL-STD-2196(SH) Glossary, Fiber Optics.* (12 January 1989).

U.S. Department of Defense. *Military Specification MIL-F-FFFF (Navy): Fiber, optical, shipboard, general specification for.*——. Draft under preparation. 1989.

U.S. Department of Defense. *Military Specification MIL-C-XXXXX (Navy): Cable, fiber optic, shipboard, general specification for.* Draft under preparation. 1989.

U.S. Department of Defense. *Military Specification MIL-N-NNNNN (Navy): Connectors, fiber optic, shipboard, general specification for.*——. Draft under preparation. 1989.

U.S. Department of Defense. *Military Specification MIL-S-SSSSS (Navy): Splice, fiber optic, shipboard, general specification for.*——. Draft under preparation. 1989.

U.S. Department of Defense. *Military Specification MIL-I-IIIII (Navy): Interconnection box, fiber optic, shipboard, general specification for.*——. Draft under preparation. 1989.

U.S. Department of Defense. *Military Specification MIL-C-PPPPP (Navy): Couplers, passive, fiber optic, shipboard, general specification for.*——. Draft under preparation. 1989.

U.S. Department of Defense. *Military Specification MIL-A-AAAAA (Navy): Attenuators, fiber optic, shipboard, general specification for.*——. Draft under preparation. 1989.

U.S. Department of Defense. *Military Specification MIL-W-WWWWW (Navy): Switches, fiber optic, shipboard, general specification for.*——. Draft under preparation, 1989.

U.S. Department of Defense. *Military Specification MIL-J-JJJJJ (Navy): Rotary joints, fiber optic, shipboard, general specification for.*——. Draft under preparation. 1989.

U.S. Department of Defense. *Military Specification MIL-B-BBBBB (Navy): Borescopes, fiber optic, shipboard, general specification for.*——. Draft under preparation. 1989.

U.S. Department of Defense. *Military Specification MIL-D-TTTTT (Navy): Transmitter, digital, fiber optic, shipboard, general specification for.*——. Draft under preparation. 1989.

U.S. Department of Defense. *Military Specification MIL-D-RRRRR (Navy): Receiver, digital, fiber optic, shipboard, general specification for.*——. Draft under preparation. 1989.

U.S. Department of Defense. *Military Specification MIL-A-TTTTT (Navy): Transmitter, analog, fiber optic, shipboard, general specification for.* ——. Draft under preparation. 1989.

U.S. Department of Defense. *Military Specification MIL-A-RRRRR (Navy): Receiver analog fiber optic, shipboard, general specification for* ——. Draft under preparation. 1989.

U.S. Department of Defense. *Military Specification MIL-M-FFFFF (Navy): Multiplexers, demultiplexers, multiplexers/demultiplexers (MULDEMS), frequency-division, fiber optic interfaceable, shipboard, general specification for.*——. Draft under preparation. 1989.

U.S. Department of Defense. *Military Specification MIL-M-TTTTT (Navy): Multiplexers, demultiplexers, multiplexers/demultiplexers (MULDEMS), time-division, fiber optic interfaceable, shipboard, general specification for.*——. Draft under preparation. 1989.

U.S. General Services Administration. *Federal Standard FED-STD 1037A: Glossary of Telecommunication Terms.* GSA, Specification Branch, 7th and D Streets, Washington, D.C. 20407, 1987.

U.S. National Institute of Standards and Technology. *PB82-166257, NBS Hdbk 140: Optical Waveguide Communications Glossary.* U.S. Government Printing Office, Washington, D.C. 20401, 1982.

Van der Ziel, J. P. Characteristics of 1.3-μm InGaAsP lasers used as photodetectors. *Journal of Lightwave Technology* 7 (2): 347–352 (February 1989).

Villarruel, C. A., C.-C. Wang, R. P. Moeller, and W. K. Burns. Single-mode data buses for local area network applications. *Journal of Lightwave Technology* **LT-3** (3): 472–478 (June 1985).

Weik, M. H. *Communications Standard Dictionary.* New York: Van Nostrand Reinhold Company, 1988.

——. *Standard Dictionary of Computers and Information Processing.* Rochelle Park, New Jersey: Hayden Book Company, Inc., 1977.

——. *Vocabulary for Fiber Optics and Lightwave Communications.* National Communications System Technical Information Bulletin 79-1. February 1979. National Technical Information Service, Springfield, Virginia. AD A068 510.

Worthington, P. Cable design for optical submarine systems. *Journal of Lightwave Technology* **LT-2** (6): 833–838 (December 1984).

Yamashita, K., Y. Koyamada, and Y. Hatano. Launching condition dependence of bandwidth in graded-index multimode fibers fabricated by MCVD or VAD method. *Journal of Lightwave Technology* **LT-4** (2): 601–607 (June 1985).

Yamazaki, Y., Y. Ejiri, and K. Furusawa. Design and test results of an optical fiber feedthrough for an optical submarine repeater. *Journal of Lightwave Technology* **LT-2** (6): 876–882 (December 1984).

Yariv, A. *Introduction to Optical Electronics,* 2nd ed. New York: Holt, Rinehart & Winston, Inc., 1976.

Zheng, X.-H., W. M. Henry, and A. W. Snyder. Polarization characteristics of the fundamental mode of optical fibers. *Journal of Lightwave Technology* **6** (8): 1300–1305 (August 1988).

Zucker, J. and R. B. Lauer. Optimization and characterization of AlGaAs double heterojunction LEDs for optical communication systems. *IEEE Transactions on Electron Devices* (February 1978).

Appendix

"See" refers to entries in this appendix. Italics used in definitions identifies terms defined in the main listing or this appendix.

amplifier. See *photodetector transimpedance amplifier; transimpedance amplifier.*

beat length. See *polarization beat length.*

birefringence fiber. See *high-birefringence fiber; low-birefringence fiber.*

birefringence noise. See *polarization noise.*

borescope light source. In a *fiber optic borescope,* the *light source* used to provide the *optical power* for illuminating the borescope *view field.* The source may be a separate component or an integral part of the main body of the borescope.

bundle. See *cable bundle.*

bundle jacket. See *cable bundle jacket.*

cable bundle. A number of bare *optical fibers, buffered fibers, fiber ribbons,* or *OFCCs* grouped together in the *cable core* and within a common protective layer.

cable bundle jacket. The material that forms a protective layer around a *bundle* of bare *optical fibers, buffered fibers, fiber ribbons,* or *OFCCs.* Several jacketed *cable bundles* and *strength members* may be used to form a *fiber optic cable.*

cable core component. A part of a *cable core,* such as a bare *optical fiber, buffered fiber, fiber ribbon, OFCC, buffer, strength member,* or *jacket.*

coating offset ratio. See *fiber/coating offset ratio.*

component. See *cable core component.*

core/cladding offset. In an *optical fiber,* the distance between the *core* center and the *cladding* center.

core component. See *cable core component.*

demultiplexing. See *time-division demultiplexing.*

dispersion-flattened fiber. An *optical fiber* with a low *dispersion* value over the *wavelength* region from about 1.3 to 1.6 μ *(micron).*

dispersion-unshifted single-mode fiber. 1. A *single-mode optical fiber* that has a nominal *zero-dispersion wavelength* near 1.3 μ *(micron).* 2. A *single-mode optical fiber* whose *dispersion* curve (*dispersion coefficient* versus operating wavelength) increases monotonically with a single crossing of the *zero-dispersion* axis at a *wavelength* near 1.3 μ *(micron).* Synonymous with *dispersion-unshifted fiber.*

dispersion-unshifted fiber. See *dispersion-unshifted single-mode fiber.*

electric. See *transverse electric (TE).*

eye pattern. A figure on a cathode ray tube (oscilloscope) screen obtained by applying a *digital data* stream to the vertical amplifier of the cathode ray tube and triggering the tube with the digital data system's clock pulse. The eye-shaped pattern qualitatively describes the proper functioning of the digital system, whereas the *bit error ratio (BER)* quantitatively describes the proper functioning of the system. The quality of the eye pattern, critical for achieving a low BER, is influenced by two phenomenon, namely noise, which is proportional to the receiver's *bandwidth,* and *intersymbol interference,* which arises from other bits interfering with the bit of interest. The lower level of the eye pattern and the zero level of the screen are not identical. Intersymbol interference is inversely proportional to the bandwidth. A bandwidth compromise has to be made to obtain the best possible eye pattern *aperture.*

fiber. See *dispersion-unshifted single-mode fiber; dispersion-flattened fiber; high-birefringence fiber; low birefringence fiber; polarization-preserving (PP) fiber; SEL® fiber; self-focusing optical fiber.*

fiber/coating offset ratio. At a given cross section of an *optical fiber,* the minimum *coating* thickness divided by the maximum coating thickness.

fiber-optic rotary joint. See *off-axis fiber-optic rotary joint; on-axis fiber-optic rotary joint.*

focusing optical fiber. See *self-focusing optical fiber.*

GRIN® rod. A cylindrically shaped transparent rod, such as a short length of *optical fiber,* with a *graded refractive-index profile parameter* that will cause all *light* *rays* entering within its *acceptance cone* to be focused at one point for a given operating *wavelength*. Thus, the GRIN® rod also has the capability to *collimate* and focus a *light beam* emanating from a point *source* of light. The GRIN rod was developed by Bell Laboratories. A GRIN® rod's profile parameter is approximately 2, making the refractive-index profile close to that of a parabola. Synonymous with *GRIN® lens.* Also see *SELFOC® lens; self-focusing fiber.*

high-birefringence fiber. An *optical fiber* that has a *polarization beat length* of 1 mm (millimeter) or less. The beat length, L_B, is given by the relation:

$$L_B = 2\pi/(\beta_x - \beta_y)$$

where β_x and β_y are the *propagation constants* of the two *polarized electromagnetic waves* caused by the *birefringence* of the fiber. In a high-birefringence fiber the difference between the propagation constants is large. High-birefringence fiber is designed for applications in which a stable polarization state is required and must be maintained. Also see *birefringence; low-birefringence fiber.*

holding parameter. See *polarization-holding parameter.*

h-parameter. See *polarization-holding parameter.*

h-value. See *polarization-holding parameter.*

IM. *Intensity modulation.*

integrated receiver. See *monolithic integrated receiver.*

intensity modulation (IM). The variation of *optical power* level of an optical *carrier* in accordance with an applied *modulating* signal. IM can be achieved in many ways, such as by direct modulation of a *light source* and by variation of light *intensity* after *launching* from the source. Light intensity modulation is similar to amplitude modulation of a radio frequency carrier.

jacket. See *cable-bundle jacket.*

joint. See *off-axis fiber-optic rotary joint; on-axis fiber-optic rotary joint.*

launch. See *xx/yy restricted launch.*

length. See *polarization beat length.*

lens. See *SELFOC® lens.*

light source. See *borescope light source.*

low-birefringence fiber. An *optical fiber* that has a *polarization beat length* of 50 m (*meter*) or greater. The beat length, L_B, is given by the relation:

$$L_B = 2\pi/(\beta_x - \beta_y)$$

where β_x and β_y are the *propagation constants* of the two *polarized electromagnetic waves* caused by the *birefringence* of the fiber. In a low-birefringence fiber the difference between the propagation constants is small. A low birefringence fiber has a negligible intrinsic *birefringence*. Thus, birefringence effects induced in the fiber by external forces can be calculated directly from the polarization state of the output *light*. Low-birefringence applications include *Faraday-effect* electric current sensors, *magnetic field* sensors, and optical fiber *isolators*. Two approaches have been developed and used to make very-low-birefringence fibers, namely (1) by preserving the circular symmetry of the fiber cross section to a very high degree and using carefully selected *dopants* to reduce transverse stress in the fiber, and (2) by rapidly spinning the *preform* while *pulling* the fiber during the *drawing process*. The spin pitch must be very much less than the fiber *beat length* that would be obtained if the preform were not spun. Also see *birefringence; high-birefringence fiber.*

magnetic. See *transverse magnetic (TM).*

magnetostriction. The phenomenon exhibited by some materials, such as ferromagnetic materials including iron, cobalt, and nickel, in which dimensional changes occur when the material is subjected to a *magnetic field,* usually becoming longer in the direction of the applied field. The effect can be used to launch a shock or sound wave each time the field is applied or changed, possibly giving rise to *phonons* that could influence energy levels in atoms of certain materials, such as semiconductors and *lasers.* Thus, magnetostriction can be used as a method of *modulation.* Along with *photon* or *electric field* excitation, the phonon energy could provide *lasing threshold* energy to cause electron energy level transitions, causing photon absorption or emission. If an *optical fiber* is *coated* with a bonded magnetostrictive material, the length of the fiber can be modulated by varying an applied magnetic field. Hence, a *coherent lightwave* exiting from the end can be made to shift phase. An *interferometer* can be used to detect the phase shift. Thus, the output of a *photodetector* can be modulated by the applied magnetic field that is causing the magnetostriction. The magnetostrictive effect can also be used to apply pressure to an optical fiber causing changes in its *refractive index,* which can also be used as a modulation method.

modulation. See *intensity modulation (IM); wavelength modulation (WM).*

monolithic integrated receiver. An *optical receiver* in which a *PIN photodiode* is combined with a monolithic *transimpedance amplifier* in a single housing. The packaging is performed in such a manner that the performance of the combination is better than that of the individual components packaged separately. Also see *PIN-FET integrated receiver.*

noise. See *polarization noise.*

off-axis fiber-optic rotary joint. A *rotary joint* for *coupling fiber optic cables* but an *optical fiber* is not used at the axis of the rotating *interfaces.* Also see *on-axis fiber optic rotary joint.*

offset. See *core/cladding offset.*

offset ratio. See *fiber/coating offset ratio.*

on-axis fiber-optic rotary joint. A *rotary joint* for *coupling fiber optic cables* in which an *optical fiber* is used at the axis of the rotating *interfaces.* Also see *off-axis fiber-optic rotary joint.*

optical fiber. See *self-focusing optical fiber.*

parameter. See *polarization-holding parameter.*

pattern. See *eye pattern.*

photodetector transimpedance amplifier. A *transimpedance amplifier* that takes the output *photocurrent* of a *photodetector* and produces an output voltage signal at a higher power level, that is, with a power gain.

polarization beat length. For a *plane-polarized lightwave propagating* in an *optical fiber,* the distance required for the *polarization* to go through one complete cycle of change due to *modal birefringence,* such as from *linear polarization* in one transverse direction, through *elliptical polarization* to *circular polarization,* then through linear polarization in the opposite direction from before, through elliptical polarization to circular polarization and back to linear polarization in the original direction. Thus, the *polarization plane* rotates through an angle of 360° during one beat length. The beat length for a *single-mode optical fiber* is given by the relation:

$$L_B = 2\pi/(\beta_x - \beta_y)$$

where β_x and β_y are the *propagation constants* of the two polarized waves caused by the *birefringence* of the fiber. Beat lengths of single-mode optical fibers are typically a few centimeters.

polarization-holding parameter. A parameter of a *polarization-preserving (PP) fiber* that indicates the ability of a *high-birefringence optical fiber* to maintain the *polarization* of a *lightwave launched* into the fiber. The parameter describes the average rate at which *optical power* is transferred from one excited *mode* to the other orthogonally polarized mode. The relative cross-coupled *optical power* increases with the length of the fiber according to the relation:

$$P_x/(P_x + P_y) = (1 - e^{-2hL})/2$$

where P_x and P_y are the optical powers in the two orthogonal modes of the high-birefringence fiber, h is the polarization-holding parameter, and L is the fiber length. A polarization-holding parameter value, or h-parameter value, of 1×10^{-5} m^{-1} corresponds to a 1% average cross-coupled optical power after 1 km of fiber. Synonymous with *h-parameter; h-value.*

polarization-maintaining (PM) fiber. See *polarization-preserving (PP) fiber.*

polarization noise. Fluctuations of the direction of the *polarization plane* of an *electromagnetic wave* emanating from the end of an *optical fiber.* The fluctuations are caused by the variations in the direction of *polarization* of the input *lightwave* and from added fluctuations caused by longitudinal and transverse variations in the *refractive-index profile* along the fiber. Polarization noise above certain threshold levels will interfere with the operation of *polarization-preserving (PP) optical fibers.* Synonymous with *birefringence noise.* Also see *birefringence.*

polarization-preserving (PP) fiber. An *optical fiber* with a flattened, elliptical, or rectangular *core* such that the direction of the *polarization plane* of a *plane-polarized lightwave* entering the fiber remains fixed with respect to the direction of the *optical axis* of the fiber as the lightwave *propagates* along the fiber. *Launch conditions* at the entrance of the fiber must be consistent with the direction of the transverse axis of the fiber cross section. Synonymous with *polarization-maintaining (PM) fiber.*

PP fiber. See *polarization-preserving (PP) fiber.*

ratio. See *fiber/coating offset ratio.*

receiver. See *monolithic integrated receiver.*

restricted launch. See *xx/yy restricted launch.*

rod. See *GRIN® rod.*

rotary joint. See *off-axis fiber-optic rotary joint; on-axis fiber-optic rotary joint.*

self-focusing optical fiber. An *optical fiber* capable of focusing *lightwaves* at one or more points in the fiber or at the output end face. The fiber is made by precisely controlling the geometry, *refractive-index profile,* and other parameters. Focusing for a given fiber will occur only at a given *wavelength.* Thus, the fiber is wavelength sensitive. Self-focusing optical fibers, rods, and lenses are used in *integrated optical circuits (IOC)* and other *fiber optic* and *optoelectronic* components, such as *fiber optic connectors, attenuators, switches, multiplexers,* and *sensors.* Also see *GRIN® rod; SELFOC® lens.*

SELFOC® lens. A cylindrically shaped lens, such as a short length of *optical fiber,* with a *graded-refractive-index profile parameter* that will cause all *light rays* entering within a given *acceptance cone* to be focused at one point for a given operating *wavelength*. The SELFOC® lens has the capability to *collimate* and focus a *light beam* emanating from a point *source* of light. The SELFOC® lens was jointly developed by Nippon Sheet Glass Company and Nippon Electric Company, Ltd. The profile parameter is approximately 2, making the refractive-index profile close to that of a parabola. Also see *GRIN® rod; self-focusing fiber.*

SEL® fiber. An *optical fiber* produced by Standard Electric Lorenz.

single-mode fiber. See *dispersion-unshifted single-mode fiber.*

source. See *borescope light source.*

switching time. The time interval between initiation of the switch actuation method, such as application of a force field, and the settling of the active output of the switch to within 10% of the nominal steady state *transmittance*. The switching time, T_S, is given by the relation:

$$T_S = T_O + T_B$$

where T_O is the *switch operating time* and T_B is the *switch bounce time.*

TE. *Transverse electric.*

time. See *switching time.*

time-division demultiplexing. A method of reversing the process of *time-division multiplexing*. Thus, messages, conversations, packets, blocks or other units of *data* that time shared one *channel* are each assigned a separate channel so that they can each be routed to a different destination, that is, a form of serial-to-parallel conversion. Also see *time-division multiplexing (TDM).*

TM. *Transverse magnetic.*

transimpedance. The ratio of the output voltage to the input current of a device, such as a *photodetector transimpedance amplifier*. See *optical transimpedance.*

transimpedance amplifier. An amplifier that accepts an input signal current and produces a corresponding output signal voltage at a higher power level, that is, with a power gain. See *photodetector transimpedance amplifier.*

transverse electric (TE). Pertaining to an *electromagnetic wave* in which the *electric field vector* is perpendicular to the direction of *propagation* and therefore, in an *optical fiber,* always perpendicular to the longitudinal axis of the fiber regardless of the direction of the *magnetic field vector*. Also see *transverse magnetic.*

transverse magnetic (TM).   Pertaining to an *electromagnetic wave* in which the *magnetic field vector* is perpendicular to the direction of *propagation* of the wave, and therefore, in an *optical fiber,* always perpendicular to the longitudinal axis of the fiber regardless of the direction of the *electric field vector.*

unshifted single-mode fiber.   See *dispersion-unshifted single-mode fiber.*

wavelength modulation (WM).   A method of changing the *wavelength* of an *electromagnetic wave* in accordance with the instantaneous value of an input signal. The wavelength may be varied by a *modulating* signal applied to the source of the waves, such as a *light source,* or by modulating the wavelength after the wave is *launched* from the source.

WM.   *Wavelength modulation.*

xx/yy restricted launch.   In the *launching* of a *light beam* from a *light source* into an *optical fiber,* a launch with an xx percent *launch spot* size and source *aperture* equal to yy percent of the fiber *numerical aperture.* Also see *launch spot; numerical aperture.*